Basics of X-ray Diffraction
and
its Applications

Basics of
X-RAY
Diffraction and its Applications

K. Ramakanth Hebbar

Formerly Professor
Department of Metallurgical Engineering
NITK, Surathkal

I.K. International Publishing House Pvt. Ltd.

NEW DELHI · BANGALORE

Published by

I.K. International Publishing House Pvt. Ltd.
S-25, Green Park Extension, Uphaar Cinema Market
New Delhi - 110 016 (India)
E-mail : ik_in@vsnl.net

ISBN 81-89866-07-9

10   9   8   7   6   5   4   3   2

Published by Krishan Makhijani for I.K. International Publishing House Pvt. Ltd., S-25, Green Park Extension, Uphaar Cinema Market, New Delhi - 110 016 and Printed by Rekha Printers Pvt. Ltd., Okhla Industrial Area, Phase II, New Delhi - 110 020.

Dedicated to
My son, Late Master Karthik
and
Dear students

FOREWORD

Physical metallurgy, which has its essence in structure-property correlations, remained an empirical science well into the early decades of this century, dependent as it was essentially on the optical microscope and on various mechanical testing methods. The first impetus on its transformation into a more exact science was given by the application of x-ray diffraction methods in the determination of crystal structures of phases, in the study of solid solubility and precipitation phenomena, in the determination of crystallographic textures, to mention only a few of the topics studied. Such work progressed rapidly in the 1930s and 1940s, helping to refine, and to some extent, revolutionize various concepts in physical metallurgy. In spite of the development of various other methods of investigation, notably electron microscopy, x-ray diffraction continues to be a very important tool in the hands of the physical metallurgist. Like in other fields, there has been a steady and continuing progress in the refinement of x-ray diffraction procedures in the solution of problems in physical metallurgy and in the wider field of materials science.

From the 1950s well into the 1970s, students of x-ray diffraction had the text books by Klug and Alexander "X-ray diffraction procedures", and by Cullity "Elements of x-ray diffraction", at their disposal, and the excellent chapters on the subject in Barret's "The structure of metals" were the metallurgist's (and the materials scientist's) delight. These books are now out of print, and apart from the absence of equivalents in the book market, the high cost of available foreign books precludes their possession by an average Indian student.

It is in this context that the present book on x-ray diffraction by Dr. Ramakanth Hebbar assumes great importance. Although written by a metallurgist primarily for metallurgists and materials scientists, it will certainly be useful to anyone wishing to use x-ray diffraction as an experimental tool.

Professor Hebbar has taught courses in x-ray diffraction to undergraduates and postgraduates in metallurgical and materials engineering for several years, and thus knows at first hand the difficulties the students face when they are first exposed to the theory and applications of x-ray diffraction procedures.

After a brief introductory chapter on the history of the discovery of x-rays by Roentgen and the subsequent formalization of diffraction laws by Von Laue, Bragg and others, two chapters cover the essential fundamentals of crystallography and the theory of x-ray generation and of the interaction of x-rays with matter. X-ray diffraction theory is covered in some detail in the fourth chapter, serving as an excellent background to the following chapters the application procedures such as the Laue, the rotating crystal and the powder techniques. Reciprocal lattice theory is introduced with the necessary vector algebra and the relationship between reciprocal lattice and diffraction, which the student has generally some difficulty in understanding in the beginning, has been brought out very elegantly. The chapters on film techniques are followed by one on the x-ray diffractometer and its applications.

Specific applications such as crystal structure determination, accurate determination of lattice parameters, single crystal studies, studies of crystallographic texture, stress measurement, order-disorder transformation, phase diagram determination are covered in sufficient detail in the next chapters. There is a final chapter on the use of x-ray diffraction in chemical analysis.

As the above account indicates, the list of topics covered is quite comprehensive. The treatment of topics in each chapter is sufficiently exhaustive for undergraduate and graduate courses in x-ray diffraction, not only for metallurgists and materials scientists/engineers, but also for other disciplines requiring the study of x-ray diffraction.

The book is very well written and the examples, solved and unsolved, at the end of various chapters will benefit the students greatly in understanding the concepts underlying them.

For years there has been no textbook of this type available for students in India. This book thus fills a gap and it does so admirably. One should not be surprised if the book finds a ready overseas market as well.

Professor Hebbar deserves congratulations for a difficult task well done. I wish the book all success.

Prof. T. Ramchandran,
Former Head,
Department of Metallurgical Engineering,
KREC (Presently NITK), Surathkal

PREFACE

Having taught the subject for a long time, I have been thinking of writing a book on x-ray diffraction and its applications for Indian students. It so happens that books on this subject available in the market are expensive for an average Indian or Asian student. This book, I hope, fills the gap.

The book is meant as a text for undergraduate and postgraduate engineering students in general, and metallurgy students in particular. The book can also be used as a reference in the practical techniques of diffraction for postgraduate students of chemistry, physics and materials science.

The initial eight chapters deal with the basics of crystallography, generation and diffraction of x-rays, techniques of diffraction and equipment used. The remaining chapters deal with the applications of x-ray diffraction in various fields of materials science and metallurgy. Most of the chapters contain solved examples at the end together with certain exercises for clear understanding of the theoretical principles involved.

I am grateful to my doctoral student Dr. Jagannath Nayak, Asst. Professor of Metallurgical & Materials Engineering Department, NITK, Surathkal for the timely help during the last stages of preparation of the manuscript. I am thankful to Late Mr. Dinesh Hosadurga for typing most of the manuscript neatly. I am also grateful to Prof. T. Ramchandran, former Head of the Department of Metallurgical Engineering at KREC (presently NITK), Surathkal, for carefully scrutinizing the manuscript and writing an excellent foreword to my book.

K.R. Hebbar

CONTENTS

1

C H A P T E R

INTRODUCTION

1.1 HISTORICAL BACKGROUND

Certain invisible rays were discovered by German Physicist Roentgen in 1895, which affected photographic plates in a manner similar to the visible light. But, unlike visible light, these rays were able to penetrate matter (human body, wood, steel, etc.). These invisible rays were named x-rays as their exact nature was unknown at that time. The x-rays were put to use almost immediately for the purpose of detecting internal flaws in opaque objects like castings and the human body through the technique now, well-known as radiography. In 1912, Max von Laue, working under A. Sommerfeld in Munich University, Germany, noted that crystals were composed of atoms periodically arranged in three dimensions approximately one angstrom apart. He already knew that the wavelengths of x-rays are also of the order of one angstrom as a result of various experiments being conducted at Munich University and elsewhere those days.

Max von Laue then made an important and historic guess that x-rays could be diffracted by crystals just like the diffraction of the visible light by ruled gratings where, the order of wavelength of light and that of the distance between gratings were nearly the same. His experiments in 1912, using this guess, proved beyond doubt that x-rays could be diffracted by crystals and laid the foundation for the vast fields of applications of x-ray diffraction. In the same year 1912, two British physicists W.H. Bragg and W. L. Bragg, father and son respectively, successfully analyzed the Laue experiment and came up with an important condition for diffraction, now known as Bragg's law of diffraction.

1.2 X-RAY DIFFRACTION AND CRYSTALLOGRAPHY

Soon after its advent, Bragg's law became an important investigative tool for research work in the field of x-ray diffraction. Determination of structures and lattice parameters of crystalline substances became commonplace. The powder method, devised by Debye and Scherrer (1916-17) and independently by Hull

(1917), widened the field of application of x-ray diffraction. This method is widely used for the determination of crystal structures and lattice parameters, qualitative and quantitative analysis of unknown substances, phase diagram determinations, order-disorder transformation analysis, residual stress measurement and the determination of the nature of polycrystalline aggregates like particle size, perfection and texture.

Bragg's law (i.e. $\lambda = 2\,d\,\mathrm{Sin}\,\theta$) is a relation between the wavelength (λ), the interplanar distance (d) and the Bragg angle (θ) to be satisfied during the diffraction of an x-ray beam by a crystalline material. By knowing the wavelength, λ, used for diffraction and by measuring the Bragg angle θ at which the diffraction takes place from a crystal plane, the interplanar distance d of the diffracting plane can be calculated. The lattice parameters of the crystal structure can then be determined using standard relations between them and the d values. The determination of the crystal structure involves a knowledge of the possible crystallographic planes which can diffract x-rays in each crystal structure and identification of these from the diffraction pattern of the unknown crystalline substance.

Orientation of single crystals can be determined using the Laue method by a study of the geometry of diffraction from the crystal. The Laue method is also capable of assessing the perfection of the crystal qualitatively.

The rotating crystal method also uses single crystals for diffraction work and needs mathematical analysis to solve structural complications. Though not routinely used, it is a powerful technique to study the atomic arrangements in crystals with two or more atomic species and to determine complex unknown crystal structures.

The pinhole technique of powder method is capable of establishing the particle (fine grain) size within certain ranges by continuity or otherwise of diffraction lines. The extent of perfection of the polycrystalline structure can be determined by studying the broadening of the diffraction lines. Preferred orientation (texture) of certain crystallographic planes in metallic products can be studied using the intensity variations in the diffraction ring (Debye ring) from a given diffracting crystallographic plane.

Whenever a crystalline material is stressed, it deforms elastically in the direction of the stress in accordance with Hooke's law. Due to this elastic deformation, the interplanar distance, d, of atomic planes changes. This causes a change in θ value at which diffraction occurs provided the wavelength, λ, of the x-ray beam remains a constant. The change in θ value can then be used to calculate the change in d values which can be used to determine the stress present in the crystalline substance.

Order-disorder transformation can be studied by the powder diffraction technique. The diffraction patterns of the ordered alloy would contain additional lines which do not exist in a disordered alloy. Thus, the presence of order and the kinetics of ordering in a solid solution could be studied on the basis of the presence and the intensities of such additional diffraction lines.

X-ray diffraction is a powerful tool for the study of phase diagrams, particularly in the solid state transformation regions. The diffraction patterns (of alloys) taken in the single phase regions show variations in the d values of specific planes with changes in composition at a given temperature, but in two-phase regions, the d values of a particular phase remain constant. Thus, the boundary between the single and the two-phase regions (points on the solvus curves) could be established by the point where the d value of the single phase becomes constant while going from this single phase region to the two-phase region at a given temperature. Solvus curves can also be established by the fact that as the composition of the alloy in the system is varied from a two-phase region to a single phase region the corresponding diffraction patterns would indicate the disappearance of the diffraction lines of one of the phases.

The powder diffraction pattern of a particular crystalline phase (may be an element or a compound or a solid solution) is a characteristic of that phase. This fact is utilized in the qualitative and quantitative analysis of unknown crystalline substances. For this purpose, the diffraction patterns of known phases are indexed alphabetically and d value-wise so that the patterns of unknown substances can be compared with the indexed patterns and identified. Further the intensities of the diffraction lines depend on the concentration of the particular phase. This fact is utilized in quantitative estimation of phases in substances.

The chapters 2 to 8 describe the x-ray diffraction fundamentals including the equipment necessary for the experimentation and the different techniques used for various studies. The applications of x-ray diffraction discussed in the paragraphs above are presented in the later chapters individually.

CRYSTALLOGRAPHY AND STEREOGRAPHIC PROJECTION

2.1   INTRODUCTION

A crystalline substance is one which has a regular periodic arrangement of either atoms or molecules or groups of them in three dimensions. Whenever, this kind of symmetric or periodic arrangement is not found in a substance, it is referred to as an amorphous substance. In general, liquids and gases are amorphous although there are exceptions, e.g., liquid crystals. Most of the solid substances seen around in common use are crystalline except most of the polymers and glasses. The nature and structures of crystalline matter were the focus of attention after the diffraction experiment of Von Laue and the subsequent formulation of the law of diffraction by Bragg in 1912 as mentioned in chapter 1.

2.2   SPACE LATTICE

As the essential characteristic of a crystal is the periodic nature, it is convenient to relate the atomic arrangement inside the crystal to a network of points in space called space lattice in which the surroundings of each point is identical to any other. A space lattice is shown in Fig. 2.1.

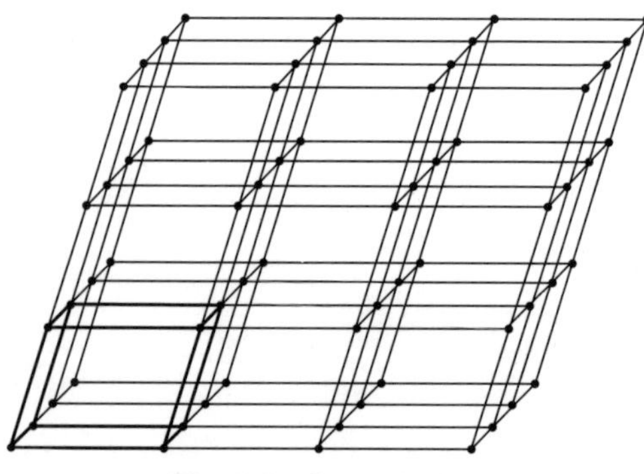

Fig. 2.1   Space lattice

Here, the intersections of the lines indicate the positions of atoms or molecules or groups of them, as the case may be, in a given crystalline solid. A space lattice consists of lines drawn in space along the directions in which the repetition of atoms or molecules occur. The repetition distances, along the three axial directions (called crystal axes), are termed lattice parameters a, b and c as shown in the figure. The crystal axes may or may not be at right angles to each other.

2.3 CRYSTAL SYSTEMS

The smallest building block of the space lattice of a crystalline substance is termed the unit cell (Fig. 2.2).

The unit cell dimensions and the interaxial angles are characteristic of a given crystalline substance. In all, seven different crystal systems are identified in crystalline materials. They are described in Table 2.1.

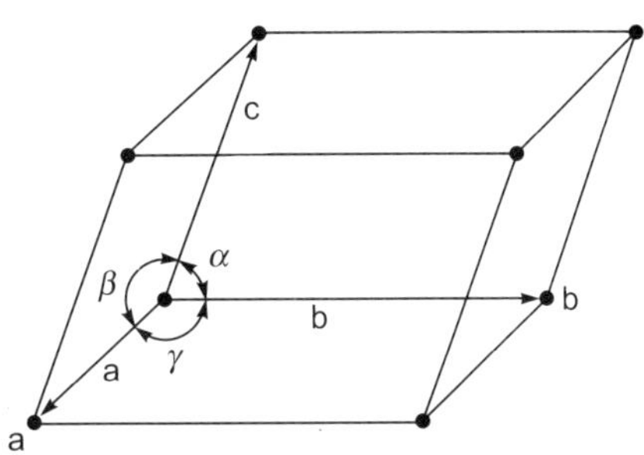

Fig. 2.2 Unit cell

Table 2.1 Crystal Systems and Bravais Lattices

System	*Axial lengths and angles*	*Bravais lattice*
Cubic	Three equal axes at right angles $a = b = c, \ \alpha = \beta = \gamma = 90°$	Simple Body-centered Face-centered
Tetragonal	Three axes at right angles, two equal $a = b \neq c, \ \alpha = \beta = \gamma = 90°$	Simple Body-centered
Orthorhombic	Three unequal axes at right angles $a \neq b \neq c, \ \alpha = \beta = \gamma = 90°$	Simple Body-centered Base-centered Face-centered
Rhombohedral	Three equal axes, equally inclined $a = b = c, \ \alpha = \beta = \gamma \neq 90°$	Simple
Hexagonal	Two equal coplanar axes at 120° third axis at right angles $a = b \neq c, \ \alpha = \beta = 90°, \ \gamma = 120°$	Simple
Monoclinic	Three unequal axes, one pair not at right anlges $a \neq b \neq c, \ \alpha = \gamma = 90° \neq \beta$	Simple Base-centered
Triclinic	Three unequal axes, unequally inclined and none at right anlges $a \neq b \neq c, \ \alpha \neq \beta \neq \gamma \neq 90°$	Simple

(Reference: B.D.Cullity, Elements of X-ray Diffraction, second edition, Addison Wesley Publishing Company Inc, California, 1978)

A unit cell is always drawn with lattice points at each corner, but in some cases, additional lattice points at the center of certain faces or at the center of the body of the cells may be present. Unit cells with lattice points only at the corners are called primitive cells and cells with additional lattice points are termed as non-primitive cells. Further, it is found that a total of fourteen types of lattices are possible and no more types of arrangements of points in space are possible without violating the rule of identical surroundings of lattice points. These fourteen types of arrangements were first recognized by Bravais and are termed Bravais Lattices illustrated in Fig.2.3 and included in Table 2.1.

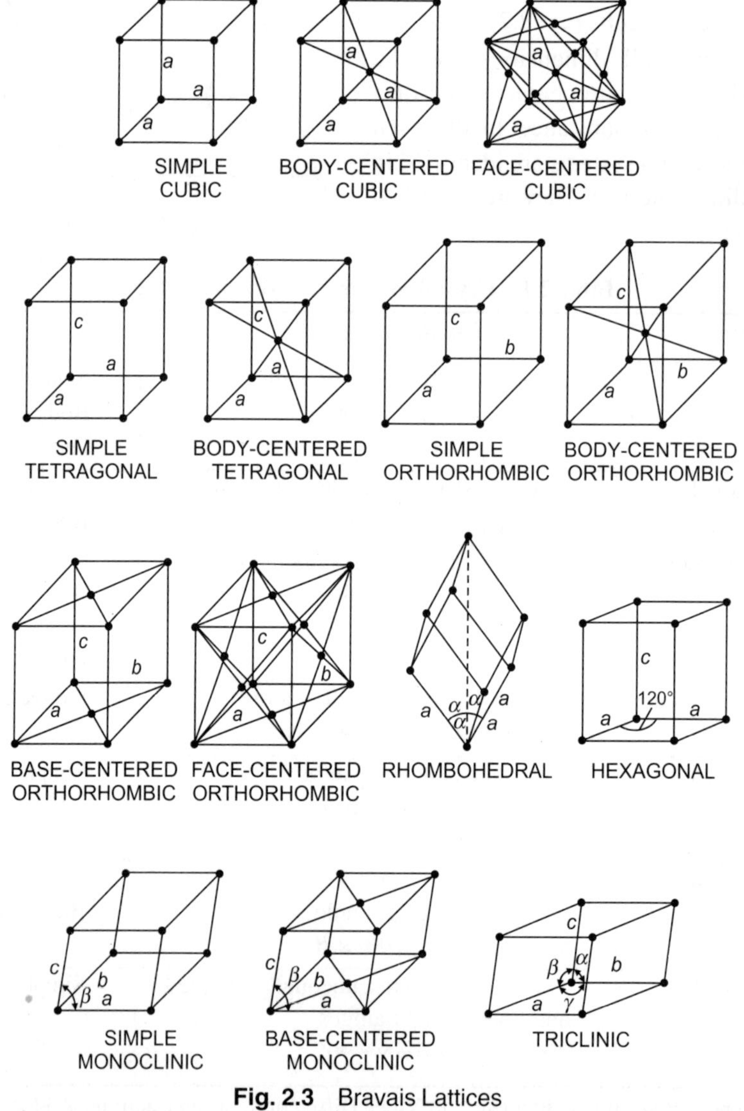

Fig. 2.3   Bravais Lattices
(Ref: B.D.Cullity, Elements of X-ray Diffraction, second edition, Addison Wesley Publishing Company Inc, California, 1978)

It should be noted that a corner lattice point is shared by eight unit cells. A primitive cell, therefore, has only one lattice point (i.e., 8 corners × 1/8). Further, a lattice point at the center of the face of a unit cell is shared by the adjacent unit cell and thus has a contribution of only one half to the unit cell considered. An interior lattice point within the cell belongs only to the cell considered. So, it can be summarized by saying that the number of lattice points per (atoms) unit cell is given by

$$N = N_i + \frac{N_f}{2} + \frac{N_c}{8},$$

where, N_i = number of internal lattice points (atoms),

N_f = number of face centered lattice points (atoms),

and N_c = number of corner lattice points (atoms).

2.4 MILLER AND MILLER-BRAVAIS INDICES

It is necessary to identify various planes and directions in a crystal structure for studies concerning the properties of crystalline matter. The Miller system of designating indices for crystal planes and directions is universally accepted for this purpose.

The first step in the determination of Miller indices of any direction is to draw a straight line or direction passing through the origin for convenience, parallel to the given direction. This newly drawn direction passing through the origin will have the same Miller indices as the given direction initially taken because of the simple fact that all parallel directions indicate one and the same direction in space. If the direction drawn through the origin passes through a point with coordinates a,b,c (unit cell dimensions in the three lattice directions x, y and z), then the Miller indices of the direction are designated as [1 1 1]. In case, the direction through the origin happens to pass through a point (2a, 3b, 5c), then the Miller indices of this direction are [2, 3, 5]. The Miller indices of a direction from the origin passing through a point (a/2, b, c/4) should be [1/2, 1, 1/4], but since Miller indices are integers, these fractional indices could be multiplied by the factor 4 so that all of them get converted to integers, i.e., [241] which are the Miller indices for this direction. It can be noted that a direction which passes through point (a,b,c) also passes through points (2a, 2b, 2c) as well as (3a, 3b, 3c), and so on. But the Miller indices of this direction is written as [111] and not as [222] or as [333]. Fig. 2.4 shows some crystallographic directions and their Miller indices in a unit cell of a cubic crystal. Generally, square brackets [ ] are used for indicating a single direction. A family of directions like [100], [010] and [001] can all be indicated together by enclosing one of these directions by carets, i.e., <100>.

To determine the Miller indices of a plane, the following procedure may be adopted:

1. Find the intercepts of the given plane on the three crystal axes in terms of the multiples of unit cell dimensions, a, b, c.

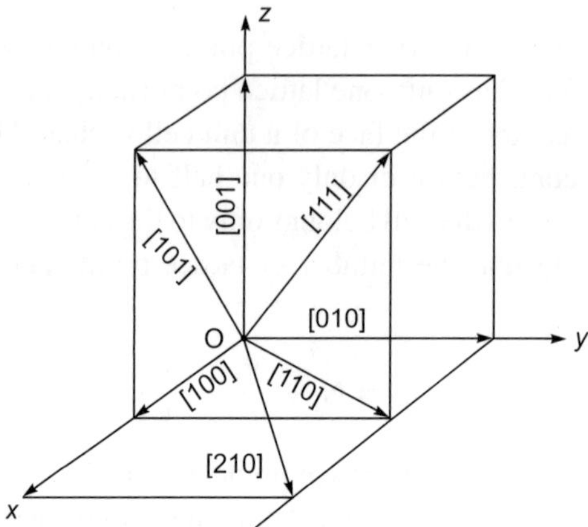

Fig. 2.4   Miller Indices of crystallographic directions

2. Take reciprocals of those intercepts.
3. Reduce the reciprocals to the smallest possible integers using a common multiplication factor.

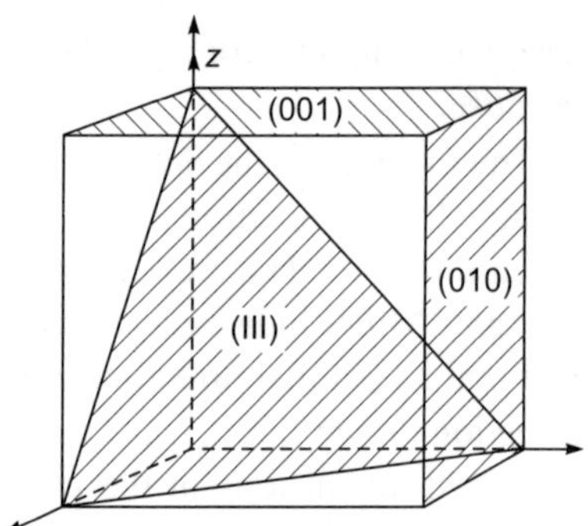

Fig. 2.5   Miller indices of planes

Fig.2.5 illustrates some planes and their Miller indices. A plane with intercepts 2a, 3b and 4c where a, b, c are unit cell dimensions along the x, y and z directions respectively, has intercepts 2, 3, and 4 in terms of unit cell dimensions. The reciprocals of these intercepts are 1/2, 1/3 and 1/4, respectively. On multiplying these reciprocals by a common factor 12, the Miller indices of the plane (643) are obtained. The Miller indices of a plane are enclosed conventionally by parentheses i.e.( ). A family of planes are denoted by enclosing the indices with braces { }.

So far, the discussion in the previous paragraphs was based on a three-axes system. There is an indices system based on four-axes instead of three which is very convenient for use in hexagonal crystals. This four-axes indices system is known as Miller-Bravais indices system. The Fig. 2.6 shows the four-axes system used for hexagonal crystals. The three axes a_1, a_2 and a_3 are all on the basal plane of the hexagonal lattice 120° apart from each other. The c axis is perpendicular to the basal plane. The procedure for the determination of Miller-Bravais indices of any plane is similar to the one used for Miller indices determination. For example, the prism plane ABHG intersects a_1 axis and a_3 axis at 1 and -1 unit distances, respectively and is parallel to a_2 and c axes. Taking the reciprocals of these intersects on the axes a_1, a_2, a_3 and c, in that order, the Miller-Bravais indices of the plane ABHG are $(10\bar{1}0)$. Taking another example, if the intercepts of the plane

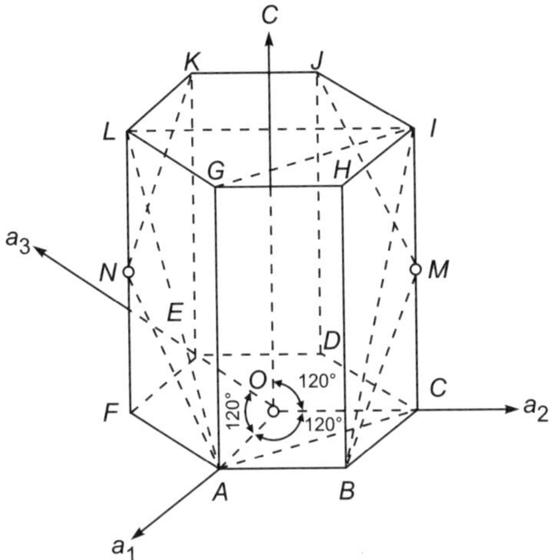

Fig. 2.6 Miller-Bravais indices in hexagonal systems

ACIG on the axes a_1, a_2, a_3, and c are 1, 1, –1/2, and ∞ respectively, the Miller-Bravais indices of the plane are $(10\bar{2}0)$. Similarly, planes ABIL and ABMJKN have indices $(10\bar{1}1)$ and $(10\bar{1}2)$ respectively. In general, if the four indices of a plane are (hkil), it can be proved by the geometry of the axial system that h+k = –i. This holds good for the indices of the directions also.

Fig. 2.7 shows two important directions in hexagonal system. While writing the Miller-Bravais indices for directions, care should be taken to see that the rule, h+k = –i, is satisfied. Because of this constraint, the Miller-Bravais indices of the a_1

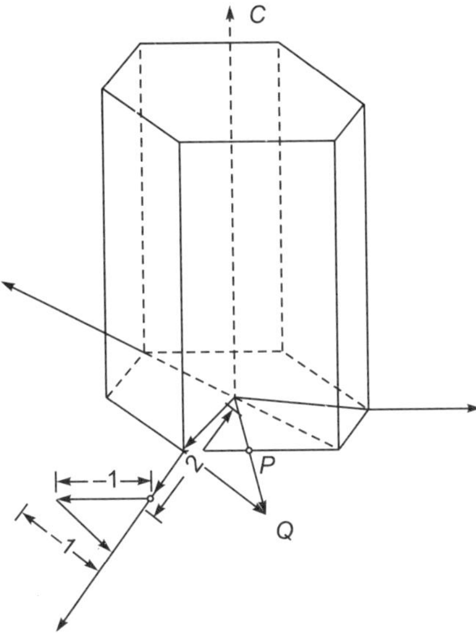

Fig. 2.7 Miller-Bravais indices for crystallographic directions

axis is written as $[2\bar{1}\bar{1}0]$ and not as [1000]. The Miller-Bravais indices of the direction OP, where P is the mid-point of side AB, is $[10\bar{1}0]$. Both these directions lie in the basal plane of the hexagonal crystal containing the axes a_1, a_2, and a_3. The c axis has the Miller-Bravais indices of [0001].

2.5   INTERPLANAR SPACINGS

The interplanar spacings (d_{hkl}) of any plane (hkl) depend on the lattice parameters a, b, c, α, β and γ of the crystal concerned and the Miller indices of the plane (hkl). The expressions for the interplanar spacings of different crystal systems are given in Table 2.2.

2.6   STEREOGRAPHIC PROJECTIONS

2.6.1   General

For studying the diffraction patterns with reference to crystal structure, it is necessary to have a method by which the angular relationships between planes and directions in a three-dimensional crystal are described on a two-dimensional planar projection. Isometric or perspective views are not convenient as the angles between various planes and directions cannot be measured directly on the projection. Stereographic projection is very convenient as the angles between various planes and directions can be measured directly on the projection with an accuracy of $\pm\, 0.5^\circ$ or better. These projections are useful in solving problems involving angular relationships and in the determination of orientations of single crystals and textures of polycrystalline samples from corresponding diffraction patterns.

2.6.2   Reference Sphere and Poles

A crystal can be represented by its unit cell. The orientation of any plane in the unit cell of the crystal can be described by the orientation of the normal to that plane. The interplanar angles can be obtained by the angles between the normals to the corresponding planes. A reference sphere of a convenient radius can be drawn with the unit cell at its center. The normals to the various crystallographic planes passing through the center of the reference sphere intersect the reference sphere at points called poles. Fig. 2.8 shows a unit cell of a crystal at the center of the reference sphere. The circle A, which is the intersection of the reference sphere and a crystallographic plane through the center of the sphere, is termed the trace of the plane concerned. A circle of maximum possible diameter (equal to the diameter of the reference sphere) drawn on a given reference sphere is called a great circle.

All other circles of lesser diameter drawn on the reference sphere are termed small circles. Thus, the trace of a plane is always a great circle. The pole P_1 is the

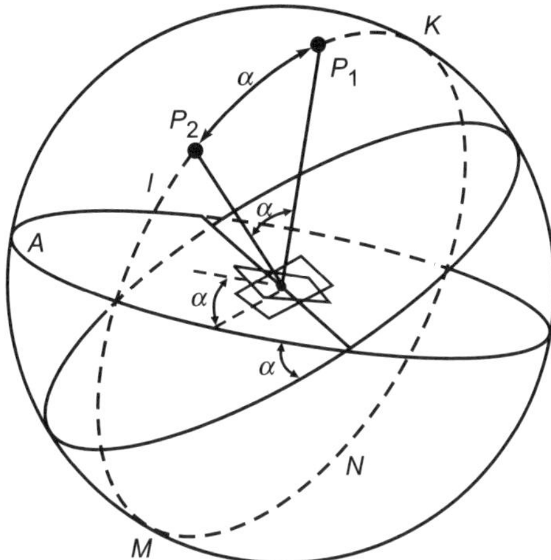

Fig. 2.8   Reference sphere with traces of planes A and B

intersection point of the reference sphere and the normal to the given plane. Fig. 2.8 shows another crystallographic plane intersecting the reference sphere in a great circle (trace of the plane) B. The angle α between the traces A and B at the point of intersection is nothing but the angle between the planes which can also be measured by the angle between the corresponding normals as shown in Fig. 2.8. Let the normal to the plane causing the trace B intersect the reference sphere at pole P_2. The angle α in degrees is also equivalent to the number of parts lying between P_1 and P_2 on the great circle, passing through them, if this great circle is divided into 360 parts.

2.6.3   Stereographic Projection

In practice, it is difficult to make use of a three-dimensional reference sphere for problems involving interplanar angles. It is rather convenient to have a two-dimensional projection of the reference sphere on which poles can be represented provided the poles can still maintain the original angular relationships and are measurable. Stereographic projection is one such convenient two-dimensional projection of the three-dimensional reference sphere containing poles on its surface. Fig. 2.9 shows how stereographic projection of a pole on a reference sphere can be obtained. Let the reference sphere be assumed to be transparent and the pole P, a dark spot on its surface. A light source is assumed to be placed on the surface of the reference sphere at A. The projection plane touches the reference sphere at the point B which is located diametrically opposite to the point A. Due to the light source the pole P casts a shadow pole P′ on the projection plane. The pole P′ is the stereographic projection of the pole P of the reference sphere. Similarly, all the poles on the reference sphere may be stereographically projected on to the projection plane. A plane perpendicular to the diameter AB may be imagined to cut the reference sphere in a circle CDEF which projects stereographically as the

Table 2.2   Interplanar spacings d of any set of (hkl) planes

Cubic:
$$\frac{1}{d^2} = \frac{h^2 + k^2 + l^2}{a^2}$$

Tetragonal:
$$\frac{1}{d^2} = \frac{h^2 + k^2}{a^2} + \frac{l^2}{c^2}$$

Hexagonal:
$$\frac{1}{d^2} = \frac{4}{3}\left(\frac{h^2 + hk + k^2}{a^2}\right) + \frac{l^2}{c^2}$$

Rhombohedral:
$$\frac{1}{d^2} = \frac{(h^2 + k^2 + l^2)\sin^2\alpha + 2(hk + kl + hl)(\cos^2\alpha - \cos\alpha)}{a^2(1 - 3\cos^2\alpha + 2\cos^3\alpha)}$$

Orthorhombic:
$$\frac{1}{d^2} = \frac{h^2}{a^2} + \frac{k^2}{b^2} + \frac{l^2}{c^2}$$

Monoclinic:
$$\frac{1}{d^2} = \frac{1}{\sin^2\beta}\left(\frac{h^2}{a^2} + \frac{k^2\sin^2\beta}{b^2} + \frac{l^2}{c^2} - \frac{2hl\cos\beta}{ac}\right)$$

Triclinic:
$$\frac{1}{d^2} = \frac{1}{V^2}(S_{11}h^2 + S_{22}k^2 + S_{33}l^2 + 2S_{12}hk + 2S_{23}kl + 2S_{13}hl)$$

In the equation for triclinic crystals

V = volume of unit cell (see below),

$S_{11} = b^2c^2\sin^2\alpha,$

$S_{22} = a^2c^2\sin^2\beta,$

$S_{33} = a^2b^2\sin^2\gamma,$

$S_{12} = abc^2(\cos\alpha\cos\beta - \cos\gamma),$

$S_{23} = a^2bc(\cos\beta\cos\gamma - \cos\alpha),$

$S_{13} = ab^2c(\cos\gamma\cos\alpha - \cos\beta).$

Unit cell volumes of various crystal systems are

Cubic: $V = a^3$

Tetragonal: $V = a^2c$

Hexagonal:
$$V = \frac{\sqrt{3}\,a^2c}{2} = 0.866a^2c$$

Rhombohedral: $V = a^3\sqrt{1 - 3\cos^2\alpha + 2\cos^3\alpha}$

Orthorhombic: $V = abc$

Monoclinic: $V = abc\sin\beta$

Triclinic: $V = abc\sqrt{1 - \cos^2\alpha - \cos^2\beta - \cos^2\gamma + 2\cos\alpha\cos\beta\cos\gamma}$

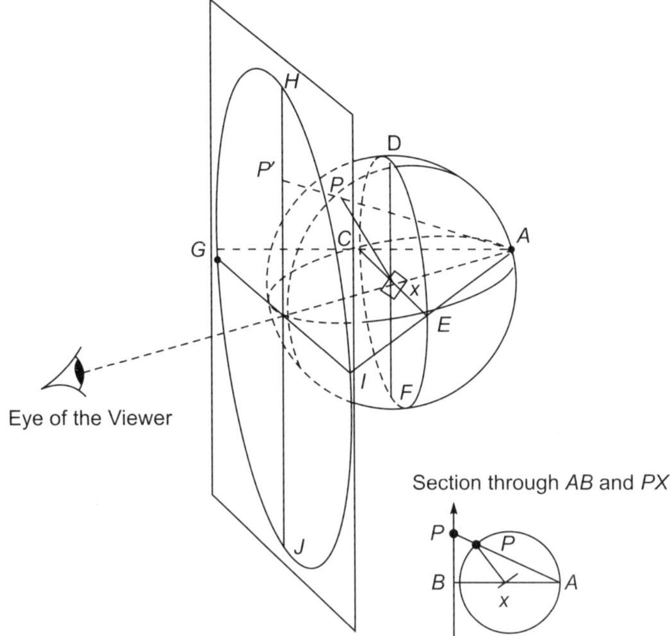

Fig. 2.9 Stereographic projection method

circle GHI J which is termed as the basic circle. It can now be observed that only the poles on the hemispherical surface to the left of circle CDEF project stereographically inside the circle GHI J. The poles lying on the right hemisphere project outside the circle GHI J. It could be further observed that it needs an infinite sized projection plane for projecting all the poles lying on the right hemisphere. However, it is not necessary to project the poles on the right hemisphere as the projection of those on the left hemisphere would suffice. This is because diametrically opposite poles are the intersections on the reference sphere of a given plane normal at either end and it is sufficient to plot the position of one of them. Since, Miller indices written with a bracket refer to a crystallographic plane or direction, the poles, by convention, are referred by their Miller indices written without a bracket.

2.6.4 Wulff net and Polar net

In section 2.6.2, it was shown that the interplanar angles could be determined by their interpolar angles by drawing great circles through the poles of interest. It is convenient to draw longitudes and latitudes on a reference sphere so as to facilitate the measurement of angles between poles. The stereographic projection of such reference sphere, with longitudes and latitudes ruled on it, is necessary to study the angular relationships between poles obtained by stereographic projection. A reference sphere with latitudes and longitudes ruled on it at 10 degree interval is shown in Fig. 2.10.

Reference spheres can be drawn with longitudes and latitudes at smaller intervals of 2 or 1 degree for convenience of measurement. The stereographic projection of such a ruled reference sphere, made with the North-South axis parallel to the plane of projection, is termed the Wulff net. Such a Wulff net with 2° graduations is shown in Fig. 2.11. A polar net can be obtained by the stereographic projection of the ruled reference sphere with the North-South axis perpendicular to the plane of projection. A polar net with 2° graduation is shown in Fig. 2.12.

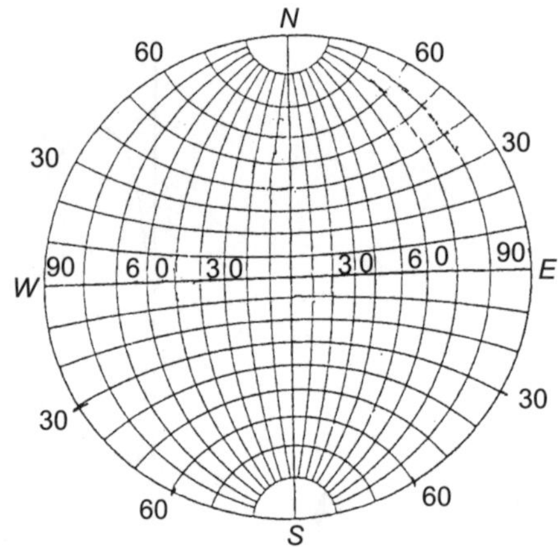

Fig. 2.10   Stereographic projection of a reference sphere ruled (10° interval)

Both the Wulff net and the polar net are available in various sizes and graduations depending on the measurement precision required.

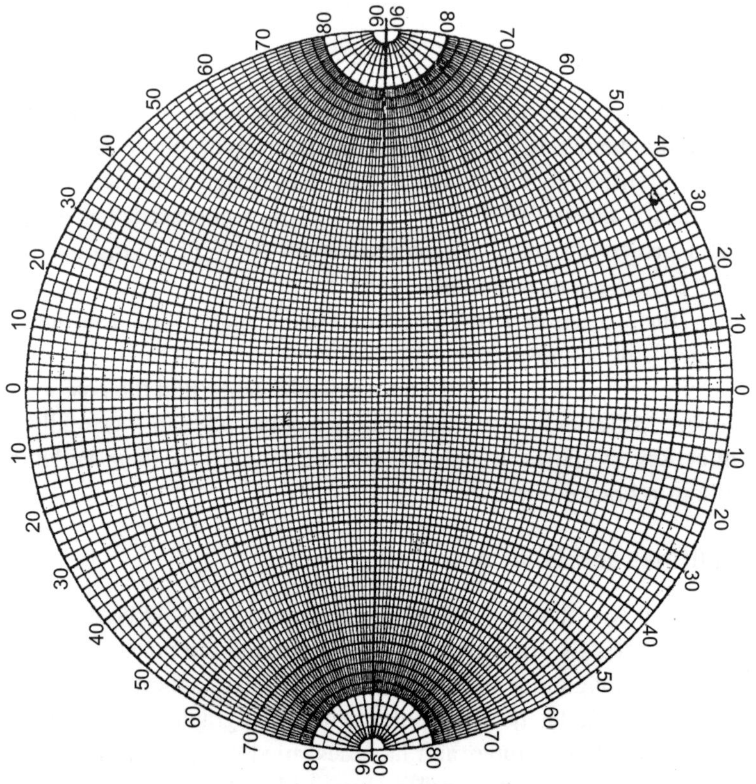

Fig. 2.11   Wulff net with 2° graduations

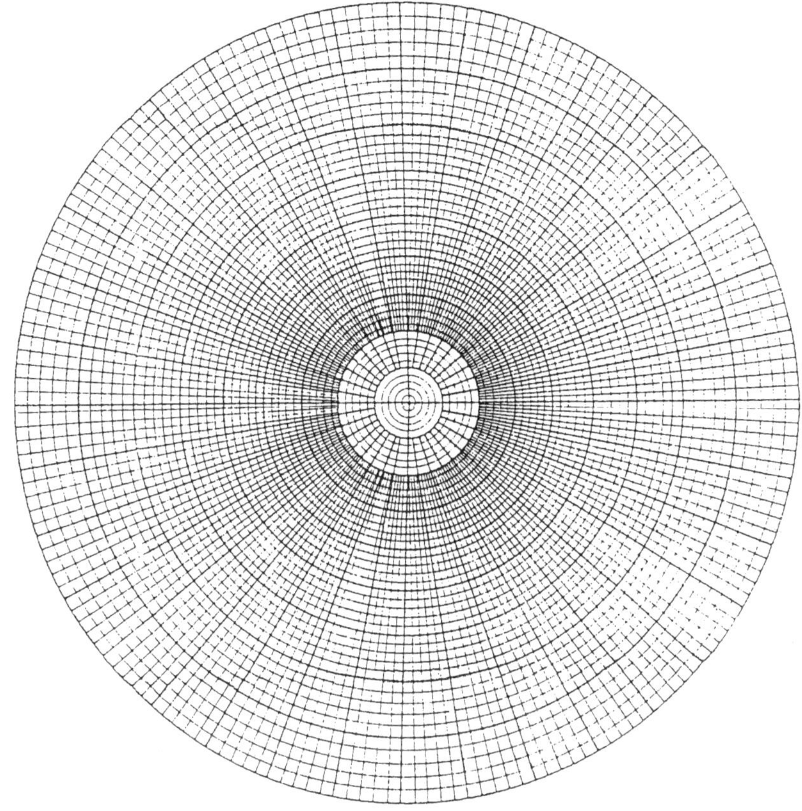

Fig. 2.12   Polar stereographic net with 2° graduations

2.6.5   Applications

In order to study interplanar angles, it is first necessary to obtain the stereographic projections of the poles of interest on a tracing sheet. The size of the basic circle used for the projection should be the same as that of the Wulff net to be used. The tracing containing the poles of interest may be placed over the Wulff net such that the basic circles and the North-South (NS) axis of both coincide. Fig. 2.13 shows such a superimposition. The poles A to H are shown on the stereographic projection (generally, on a tracing sheet for convenience) which is superimposed over the Wulff net with 10 degree graduations. From the underlying Wulff net angles between A & B, C & D, E & F and G & H can be measured directly as shown. While measuring the angle between any two poles, care must be taken to see that the measurement is done only when both the concerned poles are lying on a single great circle. It should be noted that all longitudes are great circles including the NS axis. Except the equator EW which is a great circle, all latitudes are small circles.

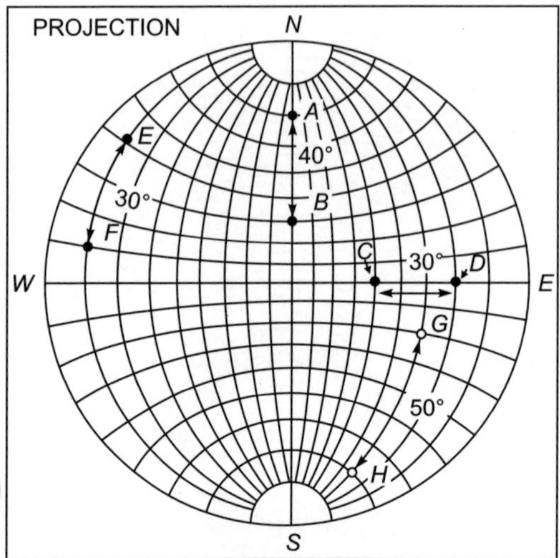

Fig. 2.13 Superimposition of tracing containing the projection of poles over Wulff net

In case the poles are not lying on a great circle, the angle between them can be measured by rotating the projection about the center over the Wulff net such that the poles fall on a great circle of the underlying Wulff net as shown in Fig. 2.14(a) and (b). The rotation of the projection can be conveniently done by fixing a pin at the center of the projection and the underlying Wulff net.

A pole on a stereographic projection is a representation of the normal to the concerned plane. The plane itself can be represented as a trace on the projection as discussed earlier in section 2.6.2. It is possible to draw the trace of a plane on a stereographic projection with the help of the pole of that plane or vice versa. Fig. 2.15 shows the method which consists of rotating the projection over the center of the Wulff net such that the pole concerned falls on the equator of the underlying

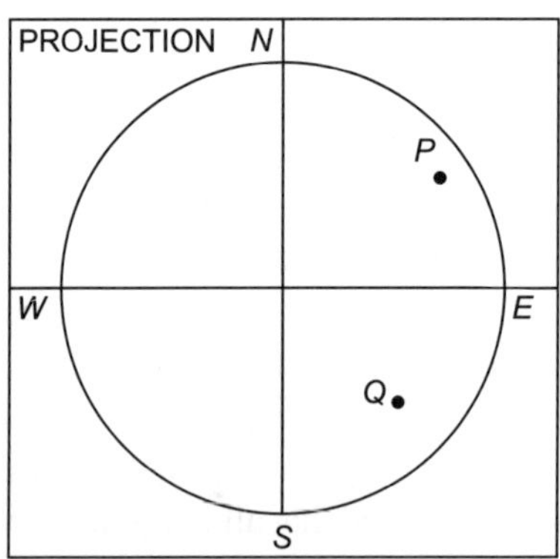

Fig. 2.14 (a) Stereographic projection of poles P & Q

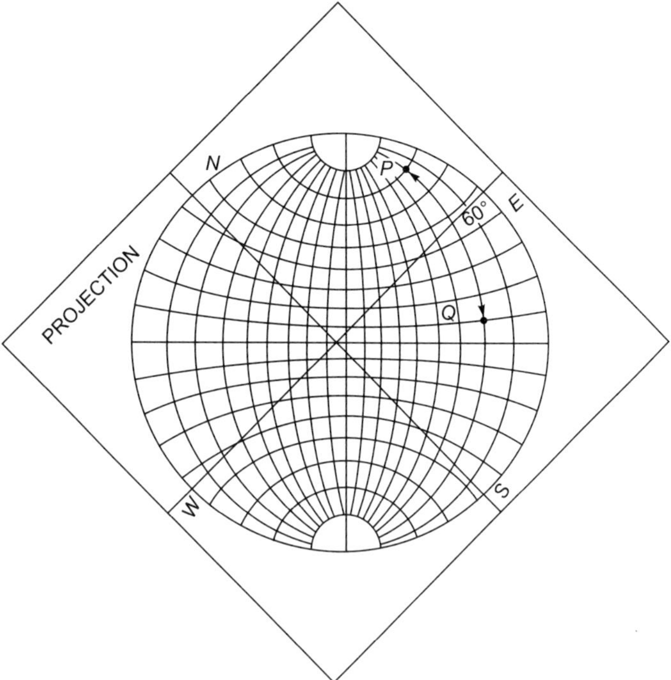

Fig. 2.14   (b) Rotation of projection for measuring the angle PQ

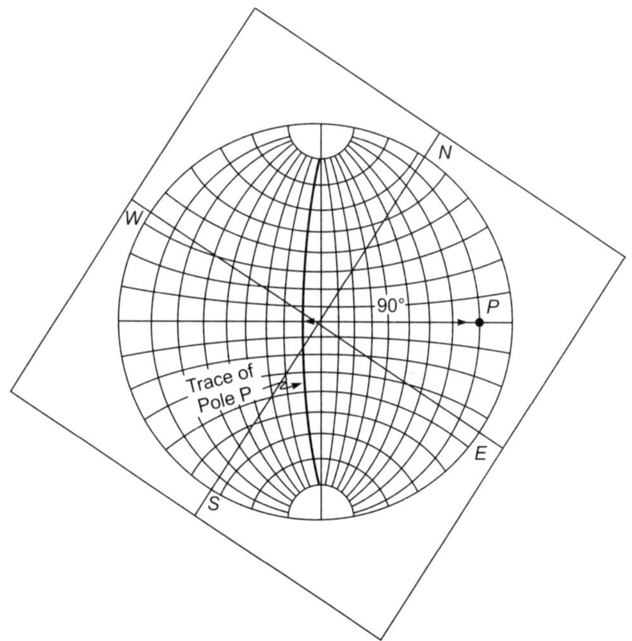

Fig. 2.15   Trace of pole P of Fig. 2.14 (a)

Wulff net and tracing the longitude which is 90 degrees away from the pole. This longitude is the trace of the pole concerned. This method is based on the fact that every point on the trace of a plane is at 90 degrees from the pole of that plane.

In some of the problems involving crystallography it may be required to rotate the crystal about certain axis. During such a rotation the poles of the planes would take up new positions in the stereographic projection of that crystal. The rotation about an axis normal to the plane of projection can be done by rotating the poles concerned about the center of the Wulff net. Fig. 2.16 shows that the position P is changed to position P′ after a rotation of 60° in the clockwise direction about the axis normal to the plane of projection. The rotation of a pole about the NS axis can be done by moving the pole through the required angle along the latitude in which it lies. Fig. 2.17 shows this kind of rotation of poles P and Q about NS axis 60° clockwise looking from N to S. It should be noted that the pole Q reaches the

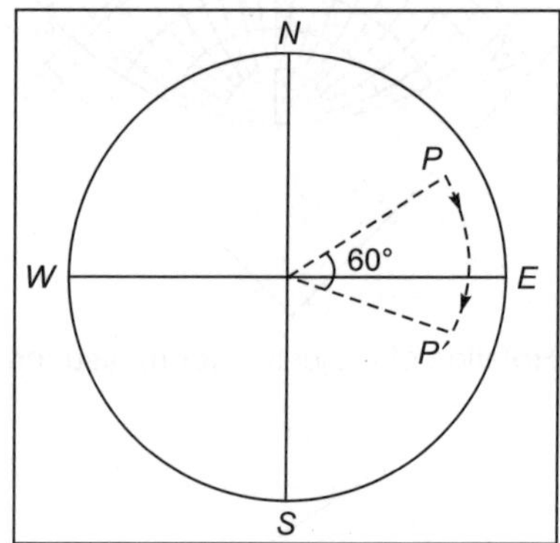

Fig. 2.16 Position of a pole about the projection plane normal

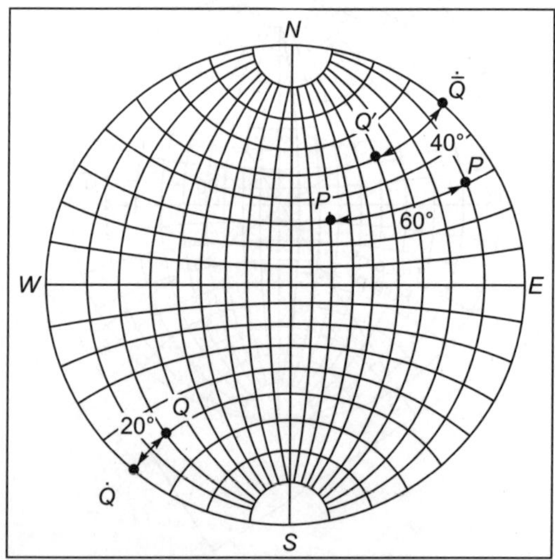

Fig. 2.17 Rotation of a pole about the NS axis

position Q with a rotation of 20°. The rest of the 40° rotation is accomplished by rotating the diametrically opposite pole Q, so that the new position after rotation is Q′. This is because of the fact that the diametrically opposite poles are the intersections of a given plane normal with the reference sphere at two diametrically opposite locations. The rotation of a pole about an inclined axis can be accomplished by initially rotating the given axis into coincidence with one of the axis discussed above (i.e., either the axis perpendicular to the plane of projection or the NS axis) and then performing the rotation of the poles and finally rotating the axis back to the original position. Fig. 2.18 indicates the procedure to be followed in such cases. Here pole A is to be rotated 60° counter-clockwise about an inclined axis B. Initially the axis B is made to fall on the equator by rotating the projection about the center. Then the entire projection can be rotated about the NS axis through the required angle (40° in this case), so that the axis B falls on the center of projection and at the same time the pole A moves over to position A'. This procedure has brought the axis B into coincidence with the axis perpendicular to the plane of projection. Now the intended 60° rotation about the new position of the axis B could be performed which brings the pole A′ to position A″. The axis has to be taken back to its original position B during which procedure A″ moves over to A‴ on its own latitude.

The relative positions of the important planes in a given crystal can be studied conveniently with a standard projection. The standard projection can be obtained by stereographically projecting the poles of importance (or of low indices) on a selected crystallographic plane. Such a standard projection needs a knowledge of the interplanar angles of the concerned crystal system. The important interplanar angles in cubic system are given in Table 2.3.

The interplanar angles in different crystal systems can be calculated using their respective lattice parameters and the Miller indices of the planes concerned. The necessary equations for this purpose are listed in Table 2.4. The zonal relations can be used for the convenience of the construction of the standard projections. The planes belonging to a zone are crystallographic planes parallel to a given axis called the zone axis. All the normals (and the poles) of the planes of a zone will be located at 90° to their zone axis. Thus, the poles of the planes belonging to a zone will all lie on a great circle (called the zone circle) located at 90° from the pole corresponding to the zone axis. If (hkl) is a plane belonging to a zone whose zone axis pole is uvw, it can be shown by the reciprocal lattice treatment (see section 6.3) that $hu + kv + lw = 0$. This equation simplifies the plotting of poles on a standard projection.

A standard 001 projection is shown in Fig. 2.19 for a cubic crystal in which low indices poles of the type 100, 110 and 111 are plotted. Here, the plane of projection is (001) and the pole of 001 plane is naturally at the center in projection. The standard convention is to have the 010 and 100 poles at the east and the south positions of the projection, respectively in order to fix the orientation

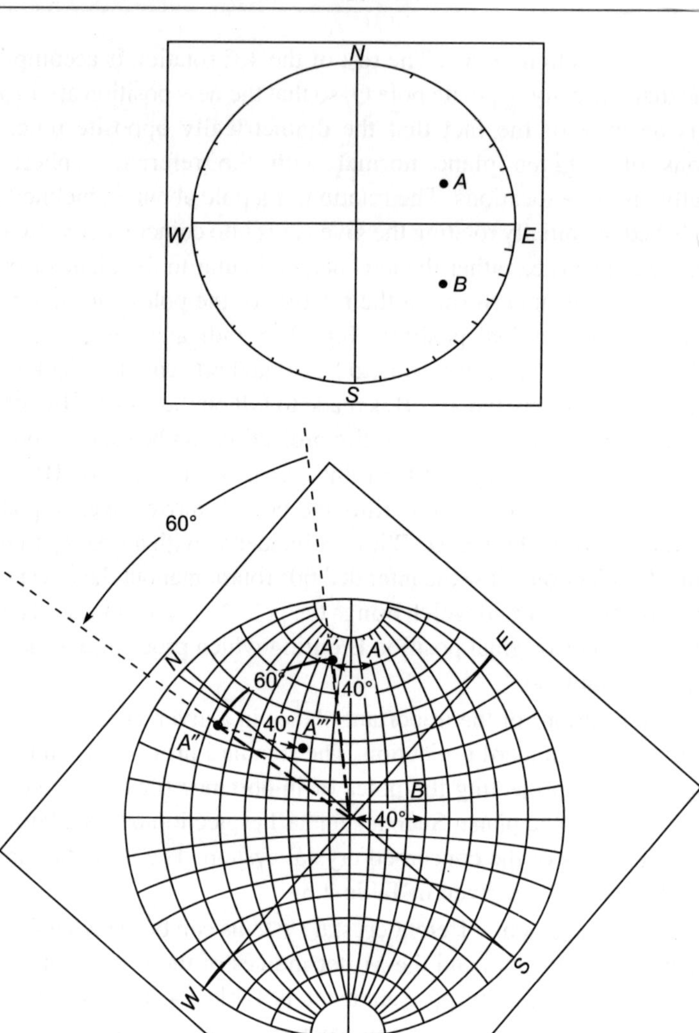

Fig. 2.18   Rotation of a pole about an inclined axis B

of the crystal with respect to the projection. The west and the north positions of the projection are marked as $0\bar{1}0$ and $\bar{1}00$ respectively to indicate the diametrically opposite poles of 010 and 100. The plane (110) makes an angle of 45° with both (100) and (010) and further it is perpendicular to the plane (001). So, the pole 110 can be located at the midpoint of the part of the basic circle between 100 and 010 poles satisfying all the angular relationships mentioned. Similarly, all poles of the family of planes {110} can be plotted. The important zone circles can be plotted

Table 2.3 Interplanar angles (in degrees) in cubic crystals between plane of the form $\{h_1\ k_1\ l_1\}$ and $\{h_2\ k_2\ l_2\}$ (Reference: B.D. Cullity, Elements of X-ray Diffraction)

$\{h_2\ k_2\ l_2\}$	$\{h_1\ k_1\ l_1\}$						
	100	*110*	*111*	*210*	*211*	*221*	*310*
100	0 90						
110	45 90	0 60 90					
111	54.7	35.3 90	0 70.5 109.5				
210	26.6 63.4 90	18.4 50.8 71.6	39.2 75.0	0 36.9 53.1			
211	35.3 65.9	30 54.7 73.2 90	19.5 61.9 90	24.1 43.1 56.8	0 33.6 48.2		
	48.2 70.5	19.5 45 76.4 90	15.8 54.7 78.9	26.6 41.8 53.4	17.7 35.3 47.1	0 27.3 39.0	
310	18.4 71.6 90	26.6 47.9 63.4 77.1	43.1 68.6	8.1 58.1 45	25.4 49.8 58.9	32.5 42.5 58.2	0 25.9 36.9
311	25.2 72.5	31.5 64.8 90	29.5 58.5 80.0	19.3 47.6 66.1	10.0 42.4 60.5	25.2 45.3 59.8	17.6 40.3 55.1
320	33.7 56.3 90	11.3 54.0 66.9	61.3 71.3	7.1 29.8 41.9	25.2 37.6 55.6	22.4 42.3 49.7	15.3 37.9 52.1
321	36.7 57.7 74.5	19.1 40.9 55.5	22.2 51.9 72.0 90	17.0 33.2 53.3	10.9 29.2 40.2	11.5 27.0 36.7	21.6 32.3 40.5
331	46.5	13.1	22.0				
510	11.4						
511	15.6						
711	11.3						

Largely from R. M. Bozorth, *Phys. Rev.* 26, 390 (1925); rounded off to the nearest 0.1°.

Table 2.4   Equations for determination of interplanar angle θ between the plane ($h_1 k_1 l_1$) of spacing d_1 and the plane ($h_2 k_2 l_2$) of spacing d_2 (Reference: B.D. Cullity, Elements of X-ray Diffraction)

Cubic:

$$\cos \phi = \frac{h_1 h_2 + k_1 k_2 + l_1 l_2}{\sqrt{(h_1^2 + k_1^2 + l_1^2)(h_2^2 + k_2^2 + l_2^2)}}$$

Tetragonal:

$$\cos \phi = \frac{\dfrac{h_1 h_2 + k_1 k_2}{a^2} + \dfrac{l_1 l_2}{c^2}}{\sqrt{\left(\dfrac{h_1^2 + k_1^2}{a^2} + \dfrac{l_1^2}{c^2}\right)\left(\dfrac{h_2^2 + k_2^2}{a^2} + \dfrac{l_2^2}{c^2}\right)}}$$

Hexagonal:

$$\cos \phi = \frac{h_1 h_2 + k_1 k_2 + \dfrac{1}{2}(h_1 k_2 + h_2 k_1) + \dfrac{3a^2}{4c^2} l_1 l_2}{\sqrt{\left(h_1^2 + k_1^2 + h_1 k_1 + \dfrac{3a^2}{4c^2} l_1^2\right)\left(h_2^2 + k_2^2 + h_2 k_2 + \dfrac{3a^2}{4c^2} l_2^2\right)}}$$

Rhombohedral:

$$\cos \phi = \frac{a^4 d_1 d_2}{V^2}\left[\sin^2 \alpha\,(h_1 h_2 + k_1 k_2 + l_1 l_2) + \right.$$

$$\left.(\cos^2 \alpha - \cos \alpha)\,(k_1 l_2 + k_2 l_1 + l_1 h_2 + l_2 h_1 + h_1 k_2 + h_2 k_1)\right)$$

Orthorhombic:

$$\cos \phi = \frac{\dfrac{h_1 h_2}{a^2} + \dfrac{k_1 k_2}{b^2} + \dfrac{l_1 l_2}{c^2}}{\sqrt{\left(\dfrac{h_1^2}{a^2} + \dfrac{k_1^2}{b^2} + \dfrac{l_1^2}{c^2}\right)\left(\dfrac{h_2^2}{a^2} + \dfrac{k_2^2}{b^2} + \dfrac{l_2^2}{c^2}\right)}}$$

Monoclinic:

$$\cos \phi = \frac{d_1 d_2}{\sin^2 \beta}\left[\frac{h_1 h_2}{a^2} + \frac{k_1 k_2 \sin^2 \beta}{b^2} + \frac{l_1 l_2}{c^2} - \frac{(l_1 h_2 + l_2 h_1)\cos \beta}{ac}\right]$$

Triclinic:

$$\cos \phi = \frac{d_1 d_2}{V^2}\left[S_{11}h_1 h_2 + S_{22}k_1 k_2 + S_{33}l_1 l_2 + S_{23}(k_1 l_2 + k_2 l_1)\right.$$

$$\left. + S_{13}(l_1 h_2 + l_2 h_1) + S_{12}(h_1 k_2 + h_2 k_1)\right]$$

next. For example, (010), (101) and (010) are belonging to a zone whose axis (uvw) is 101 which can be checked by the relation hu + kv + lw = 0. The straight line passing through poles 010, 011, 001, 0$\bar{1}$1 and 0$\bar{1}$0 is also a zone circle whose axis is 100. The plane (111) belongs to the zone of planes (011) and (100) whose axis is (0$\bar{1}\bar{1}$) which can also be verified by the relation hu + kv + lw = 0. With the same relation it can be checked that the plane (111) also belongs to the zone of planes (010) and (101). The pole 111 should therefore be present on the zone circle of poles 011 and 100 and also on the zone circle of poles 010 and 101. This argument enables the plotting of the pole 111 at the intersection of these two zone circles. Similarly, the other planes of the {111} family can be located.

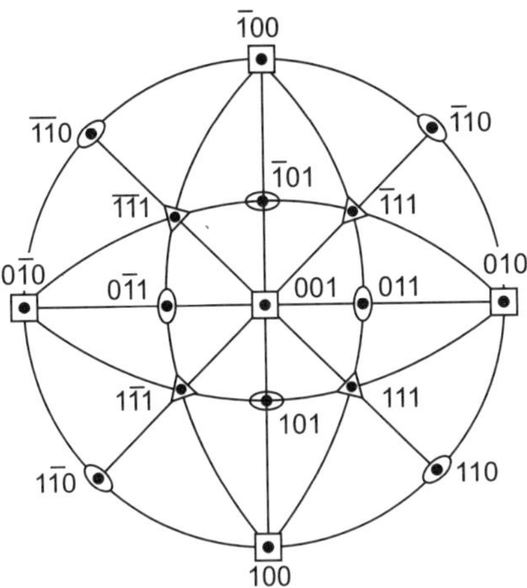

Fig. 2.19 Standard 001 projection
(Ref.: B.D. Cullity, Elements of X-ray Diffraction)

The 011 standard projection (i.e. projection on (011) plane) can be obtained by rotating all the poles of the 001 standard projection by an angle of 45° about the NS axis such that the pole 011 comes to the center of projection. The 011 standard projection is shown in Fig. 2.20. Standard projection on 111 plane also can be obtained by proper manipulations of either 001 or 011 standard projections. It should be noted that, irrespective of the lattice parameter, all materials belonging to the class of cubic structure will have identical standard projections.

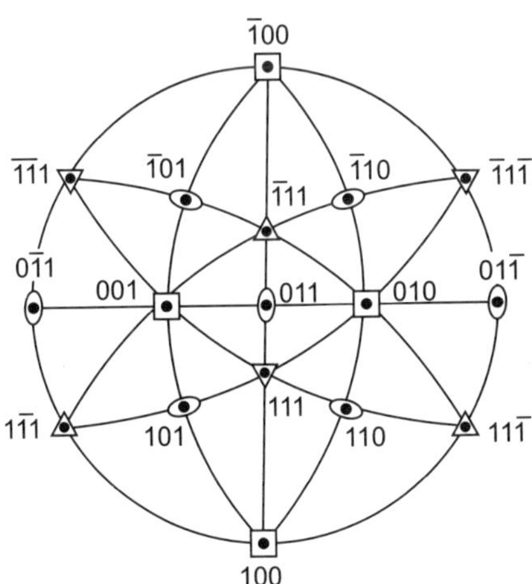

Fig. 2.20 Standard 011 projection
(Ref.: B.D. Cullity, Elements of X-ray Diffraction)

Standard projections can also be obtained for crystals other than cubic structure provided a knowledge of axial ratios and angular parameters of the structure are known. But, it is an involved exercise and useful only for a given material since each material will have its own standard projection even amongst the materials of the same crystal structure because of the differences in axial ratios and angular parameters. Standard 0001 projection of zinc (hexagonal, c/a = 1.86) is shown in Fig. 2.21. The 0001 projection for titanium and magnesium will look different, though these two also have hexagonal structures, because their axial ratios (c/a) are different from that of zinc.

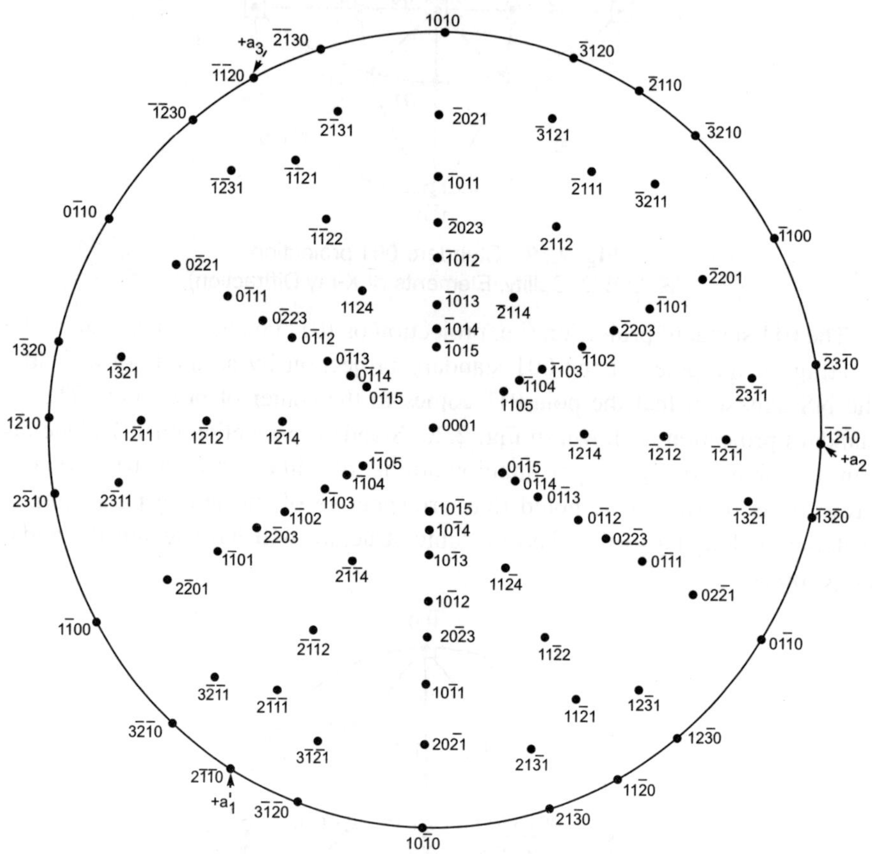

Fig. 2.21   Standard 0001 projection of zinc
(Ref.: B.D. Cullity, Elements of X-ray Diffraction)

The reorientation of the crystal lattice due to twinning can be studied with the help of stereographic projection. Fig. 2.22 indicates the 100 poles of the cubic crystal, in a standard 001 projection, represented by open symbols. Assuming the crystal to be face centered cubic, one of its twin planes is (111) which is represented by both the pole and the trace. The new positions of the planes in the twinned region are a mirror reflection of the old positions in the untwinned state with the 111 plane acting as a mirror. Thus, the new positions of the 100 cube poles in the twinned region could be obtained by rotating the projection on a Wulff net until the pole of the twin plane lies on the equator and then moving the cube

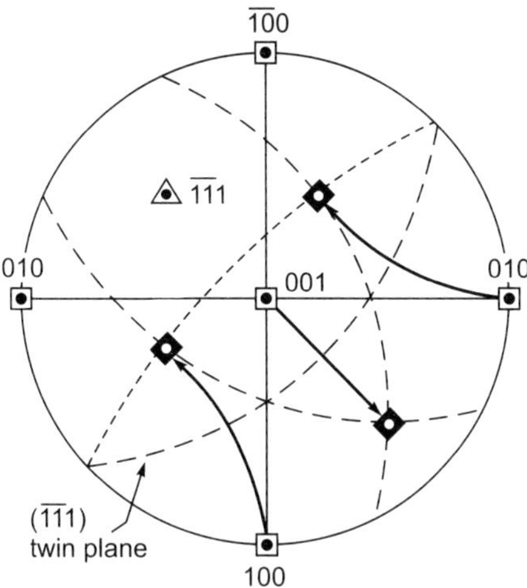

Fig. 2.22   Old and new positions of cube poles on twinning

poles along the latitudes (in which they lie) of the Wulff net to positions reflected on the trace of (111) twin plane as shown in Fig. 2.22. Here, the term reflection refers to subtending equal angles on either side of the trace of the twin plane (111).

It is possible to find the Miller indices of an unknown pole lying on a standard projection of a given crystal. Suppose, the Miller indices of a pole A lying in the standard projection of a cubic crystal, as shown in Fig. 2.23 (a), is to be determined. If a standard (001) projection like the one shown in Fig. 2.24 is available, it is a simple matter to identify the pole by superimposing the given projection over the standard projection. If it is not available, an alternate method, discussed below, can be used. Let (hkl) be the plane represented by the pole A. Let the direction [hkl] make angles α, σ, and θ with the crystallographic axes x, y and z as shown in Fig. 2.23(b). Since, the angle between the poles is equal to the angle between the planes, the angles α, σ, and θ can be measured on the projection as

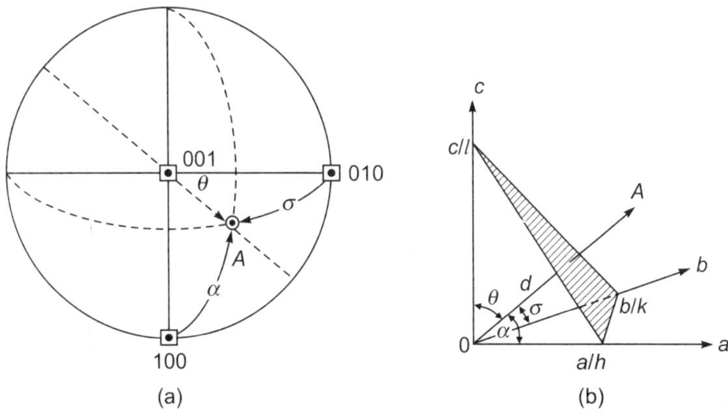

Fig. 2.23   Determination of Miller indices of a pole
(Ref.: B.D. Cullity, Elements of x-ray diffraction)

shown in Fig. 2.23(a). Let a, b and c be the unit cell parameters of the crystal in general (need not be cubic). Let the perpendicular distance between the origin and the plane (hkl) nearest to the origin be 'd'. Then, from Fig. 2.23(b),

$$\cos \alpha = \frac{d}{a/h}, \quad \cos \sigma = \frac{d}{b/k}, \quad \cos \theta = \frac{d}{c/l}$$

or   h:k:l = a cos α : b cos σ : c cos θ.

Since   a = b = c for a cubic crystal,

          h:k:l = cos α : cos σ : cos θ.

In addition to the uses mentioned above, the stereographic projection is helpful in identifying the orientation of single crystals by Laue method and in recognizing the type of texture in metallic products by powder method. These and other applications will be discussed in later chapters wherever relevant.

SOLVED EXAMPLES

Example 2.1   Determine the number of atoms per unit cell in (i) a simple cubic (SC) and (ii) a face centered cubic (FCC) crystal structures.

   (i) A simple cubic unit cell has only eight lattice points at each of the eight corners of the cubic cell. So, using the relation $N = N_i + N_f/2 + N_c/8$, = 0 + 0 + 8/8 = 1.

       $\therefore$ The number of atoms per unit cell of simple cubic = 1.

   (ii) A FCC unit cell has 6 face centered lattice points at the centers of the six faces of the cubic cell plus eight-corner lattice points

       $\therefore$ N = 0 + 6/2 + 8/8 = 0 + 3 + 1 = 4.

Therefore, FCC system has 4 lattice points per unit cell.

It should be noted here that, if there had been atoms at the lattice points in each of the cases indicated above, then the results could be interpreted as the number of atoms per the unit cell concerned.

Example 2.2   If the origin of the coordinates of a three axes system is assumed at the corner of a cubic unit cell, determine the fractional indices of face centered lattice points and the body centered lattice point.

By moving through a distance of half a unit (in terms of unit cell dimension) along the x axis and further half a unit along the y axis, a face centered lattice point position is reached. So, this face centered position has the fractional indices of (1/2, 1/2, 0). Similarly, the other two face centered positions are (1/2, 0, 1/2) and (0, 1/2, 1/2). The lattice points lying at the opposite faces of these three mentioned, have indices (1/2, 1/2, 1), (1/2, 1, 1/2) and (1, 1/2, 1/2), respectively. But, these three are not generally taken into consideration while mentioning the fractional indices for face centered positions as only three-face centered lattice points are counted per unit cell and the former three-fractional indices are convenient and representative.

By moving through a distance of half a unit in all the three axes, a body centered position is reached. So, the fractional indices of body centered lattice point is (1/2, 1/2, 1/2).

Example 2.3

(i) In a cubic crystal, a crystallographic plane cuts the x, y and z axes at 5, 4 and 3 unit distances, respectively. Determine the Miller indices of this plane.

The intercepts of the plane with the axes are 5, 4, 3.

The reciprocals of these intercepts are 1/5, 1/4, 1/3.

Expressing these as integers by multiplying all these fractions by 60, the Miller indices of the plane are (12, 15, 20).

(ii) Determine the Miller indices of directions OA, OB, OC and OD shown in the figure below where O is the origin and A, B, C, and D are the midpoints of the top edges of a cubic unit cell.

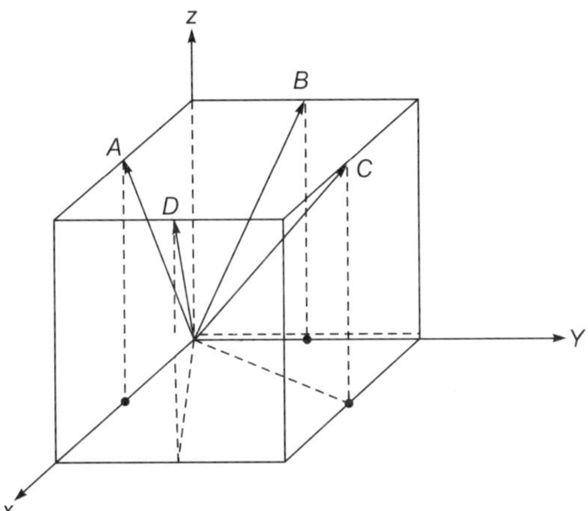

The direction OA can be thought of as a vector obtained by moving half a unit distance along x-axis and one unit distance along z-axis. Thus, the distance indices could be (1/2, 0, 1) These fractional indices can be converted to Miller indices by using a multiplying factor of 2. So, the Miller indices of OA are [102]. By similar procedure, Miller indices of OB can be determined as [012]. It can be further determined by similar procedure that the Miller indices of OC and OD are [122] and [212] respectively.

EXERCISES

2.1 Determine the number of lattice points per unit cell in (i) a body centered tetragonal (BCT) and (ii) a base centered orthorhombic crystal structures.

[Ans: (i) 2 (ii) 2]

2.2 In a cubic crystal, determine the Miller indices of (i) a plane containing the two diagonally opposite sides, (ii) a plane which cuts the two diagonally opposite sides into two halves and passes through two diagonally opposite corners and (iii) a plane which cuts the x, y and z axes 1/2, 1/3 and 1/5 unit distances respectively.   [Ans: (110), (112), (235)]

2.3 Determine the Miller indices of the body diagonal of a rectangular cell made by stacking three cubic unit cells along the y-axis.   [Ans: (131)]

2.4 Draw a hexagonal unit cell and represent (i) the directions $[10\bar{1}0]$, $[2\bar{1}\,\bar{1}0]$ and $[10\bar{1}2]$, (ii) the planes (0001), $(2\bar{1}\,\bar{1}2)$, (1011) and $(10\bar{1}2)$.

2.5 (i) Determine the interplanar spacings of the following planes of a tetragonal unit cell whose lattice parameters a and c are 4 and 5 angstrom units respectively: (100), (101), (110), (111), (301) and (103).

    (ii) If the unit cell is of cubic type with a lattice parameter of 4 angstroms, what would be the interplanar spacings of the planes noted above? Compare the results of (i) and (ii).

2.6 Construct (i) a Wulff net and (ii) a Polar net each 10 cm in diameter and graduated at 30 intervals using compass, scale and divider only. Show dotted construction lines.

2.7 (i) Locate the poles A (34° N 50° E), B (0° N 60° E), C (40° S 30° E), D (56° S 46° W) and E (0° S 40° W).

    (ii) Find the angle between each of the following pairs of poles:
A&B, B&C, C&D, D&E, A&E and B&E.

2.8 Plot the pole A (30° N 40° S) on a stereographic projection. Rotate the pole A 100° about the following axes and determine the coordinates of the pole position after each rotation. Show the path traced during rotation by dotted lines.

    (i) NS axis, anticlockwise looking from N to S.

    (ii) The axis normal to the plane of projection, clockwise.

    (iii) Inclined axis B (20° S 20° W), clockwise.

2.9 Draw standard 001, 011 and 111 projections of a cubic crystal showing all the poles of the type 100, 110 and 111. Draw important zone circles between these poles.

2.10 On a standard 001 projection of a cubic crystal, plot the following poles and determine their Miller indices:

    (i) (30° E 40° S)

    (ii) (26.6° E 53.3° S)

    (iii) (53.5° E 74.8° S)

2.11 Draw 001 projection of a cubic crystal. Locate the position of (112) plane. Find the locations of the cube poles of a (112) BCC reflection twin and determine their coordinates.

2.12 Draw a standard 0001 projection of a hexagonal crystal with an ideal c/a ratio showing important poles.

GENERATION AND NATURE OF X-RAYS

3.1 X-RAY TUBES

X-rays are generated when fast moving electrons collide with matter. The x-ray generating tube is an evacuated container having provisions to produce electrons, to accelerate these electrons to high speeds and a target to decelerate or stop these electrons. In practice, all x-ray tubes must have

    (i) a source of electrons,

    (ii) a high voltage source to accelerate the electrons, and

    (iii) a metal target.

X-ray tubes may be classified into two types: gas tubes and filament tubes. Gas tubes are those x-ray tubes in which electrons are produced by ionizing a small quantity of gas present in the tube. Filament tubes are those x-ray tubes in which a hot filament cathode produces electrons by thermionic emission.

3.1.1 Gas Tubes

This consists of a concave-faced aluminium cathode for focusing electrons onto a water cooled copper anode faced with a suitable target metal. It operates at a pressure of 1×10^{-2} torr, maintained by controlling a needle valve and a continuously operating rotary vacuum pump. Cooling fins are provided to the cathode, so as to dissipate the heat produced by the bombardment of the ions. Fig. 3.1 shows the schematic sketch of a gas tube.

During the operation of this tube, the ionized nitrogen or oxygen molecules, few of which are always present in the air, get attracted to the aluminium cathode, which is at a high negative potential (>5 kV). These gaseous ions gain high velocities due to the attraction by the cathode and strike the cathode releasing electrons from its surface. These released electrons in turn are attracted by the anode of the tube and accelerate towards it. On their way towards the anode these electrons encounter other air molecules and ionize them. This process produces

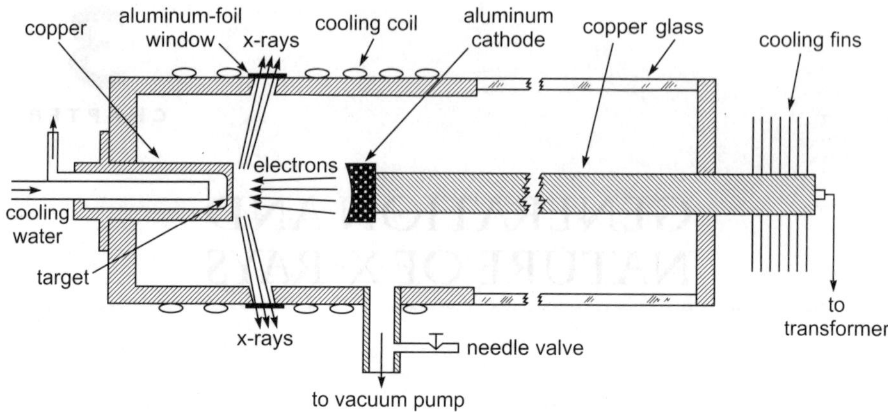

Fig. 3.1   Schematic sketch of a x-ray gas tube (Ref.: B.D. Cullity, Elements of x-ray Diffraction)

more ions and electrons. The number of ions and electrons produced quickly reach a limiting value at which time the x-rays produced by the phenomenon of electrons hitting the target also reach a maximum intensity for a given voltage differential of the tube.

Gas tubes are difficult to operate and control, mainly because of the fact that a critical gas pressure is required inside the tube for satisfactory operation of the tube. If the pressure inside the tube is more than the critical, the amount of ionization is so high that the maintenance of a high voltage inside the tube becomes difficult, due to a high current flow as a result of high ionization. If the pressure is too low, the number of ionized air molecules available for starting the entire process of x-ray production may be insufficient and the process may not occur at all for all practical purposes. In these tubes there is no method by which the tube current and the voltage may be independently controlled. The values of the tube current and the voltage are linked to the pressure inside the tube. Because of these difficulties, gas tubes have become obsolete. But these are the original x-ray tubes man ever invented.

These tubes also have their own advantages. These tubes are all demountable, meaning, they can easily be dismantled and the anode target material can be replaced when necessary. They require a simple mechanical vacuum pump and need no filament transformer which is a must in filament tubes. Thus, gas tubes are the cheapest kind available for x-ray generation. Another important merit of these tubes is that they generate the purest x-radiation since, the target in these tubes never gets contaminated.

3.1.2   Filament Tubes

The filament tube (Fig. 3.2), invented by Coolidge in the year 1913, is a highly evacuated glass envelope containing a cathode of tungsten filament and an anode of water cooled copper block faced with the desired target metal. The negative

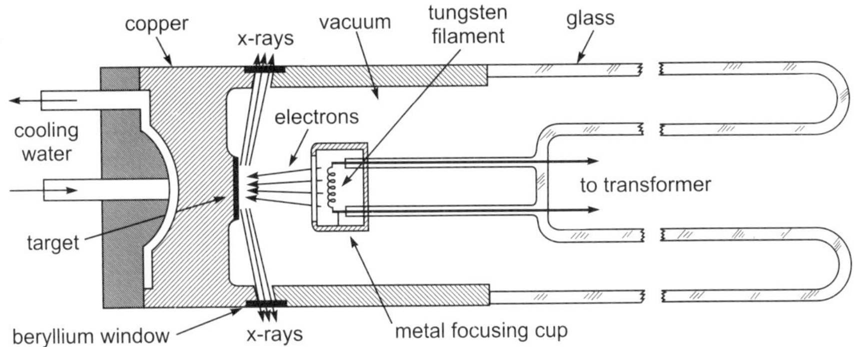

Fig. 3.2   Schematic sketch of a x-ray filament tube

lead of a high voltage transformer is connected to the filament and the other lead to the ground in order to maintain a high voltage between the cathode and the anode. The filament is heated by the current (approx. 3 amps) supplied by a filament transformer. The hot tungsten filament emits electrons by thermionic emission. The tungsten filament cathode is surrounded by a small metal cup, maintained at the same negative potential as that of the filament, which tends to focus the electrons generated by the filament due to a repulsive action on them. The electrons thus get focused on to a small region on the target metal called the focal spot. X-rays are emitted from this focal spot because of the deceleration of the electrons by the target atoms. X-rays escape from the tube through windows (generally two, three or four) usually made of beryllium or aluminium or mica which have low coefficient of absorption and provide the necessary vacuum sealing.

The Fig. 3.3 shows a simplified circuit diagram of a filament x-ray generator. Two transformers are present in the circuit, one step-up, high voltage type for supplying the tube voltage and another, step-down, low voltage type for supplying the filament current of the order of 3 amperes. The milliammeter in the figure is for measuring the tube current, which is of the order of 10 to 25 milliamperes. The filament rheostat controls the voltage supplied to the filament which in turn decides the filament current. The number of electrons emitted by the filament increases as the filament temperature increases which is a function of the filament current. The tube current depends on the number of electrons emitted by the filament since, these electrons are basically responsible for current flow in the tube. Thus, the filament rheostat can be used to control the tube current. The tube voltage can be independently controlled by an autotransformer, which controls the voltage applied to the primary of the high voltage transformer. The voltmeter shown in the figure measures the input voltage to the transformer, but can be calibrated to show the output voltage applied to the tube by the transformer. Thus, the tube voltage and the tube current can be independently controlled in filament tubes in contrast to the dependence of the tube voltage and current on each other,

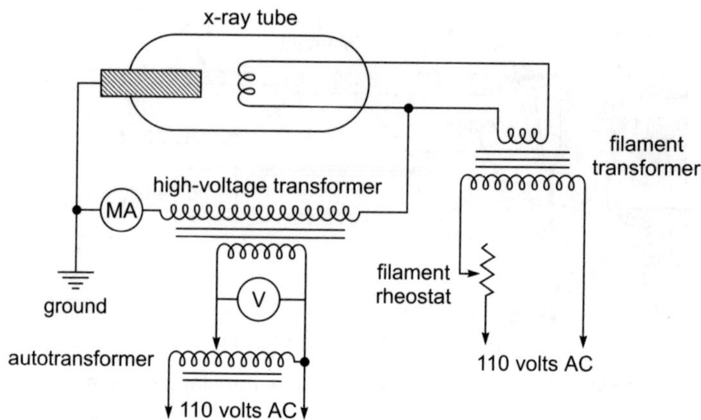

Fig. 3.3 Simplified circuit diagram of a filament tube x-ray generator

and on the pressure of air inside the tube, in the case of gas tubes. This is the basic reason behind the increased use of filament tubes in comparison to gas tubes in x-ray work.

Though it looks as if the filament tube works only with a d.c. (direct current) source, in practice it works also with an a.c. (alternating current) source. With an a.c. source the filament tube works only during one half of the a.c. cycle, when the filament becomes a cathode. During the other half of the cycle, when the metal target becomes the cathode, there would be no source of electrons for the production of x-rays. In effect, the tube acts as a rectifier during that particular half cycle of a.c. in which the target becomes cathode. Filament tubes are available either as sealed-off types or demountable, the former being more common. The sealed-off tubes are vacuum sealed by the manufacturer and no replacement of target or filament is possible. So, one has to keep a stock of tubes with as many varieties of target materials as are needed. In contrast, demountable tubes can be dismantled and damaged filaments or target materials can be replaced as and when needed. But, these tubes need a high vacuum system (a diffusion pump backed by a rotary pump) for evacuating the tube during x-ray generation.

In filament tubes, the L characteristic radiation of tungsten (which occasionally evaporates from the filament and gets deposited on the target metal) also accompanies the x-radiation emitted by the target material. This is a drawback in the filament tube compared to the purest kind of radiation emitted from the gas tubes. But this is not a serious drawback as it can be easily corrected by using a suitable x-ray filter to filter out the unwanted radiation.

The focal spot of an x-ray tube can be made as small as possible in order to concentrate the electron energy in a small area to get a high intensity x-ray source. For this purpose, the metal cup surrounding the filament can be maintained at a slightly lower potential compared to the filament, so that the electrons emitted by the filament get focused by repulsion from the cup, to a smaller focal spot on the target. But this may cause the focal spot on the target to get heated to very high temperatures thus necessitating an efficient cooling system for the target in order to prevent its fusion. This problem can be circumvented by rotating the metal target anode continuously during the operation, so that the heat is distributed uniformly over a large area and thus, reducing the burden on the cooling system.

3.2 NATURE AND ORIGIN OF X-RAYS

3.2.1 General

X-rays are invisible electromagnetic radiations of wavelengths ranging from about 0.001 nm to 10 nm. They are of shorter wavelengths compared to the visible light, which is also an electromagnetic radiation of wavelengths ranging from 10^2 to 10^3 nm. Any electromagnetic radiation has an electric field E and a magnetic field H, both of which are perpendicular to each other and to

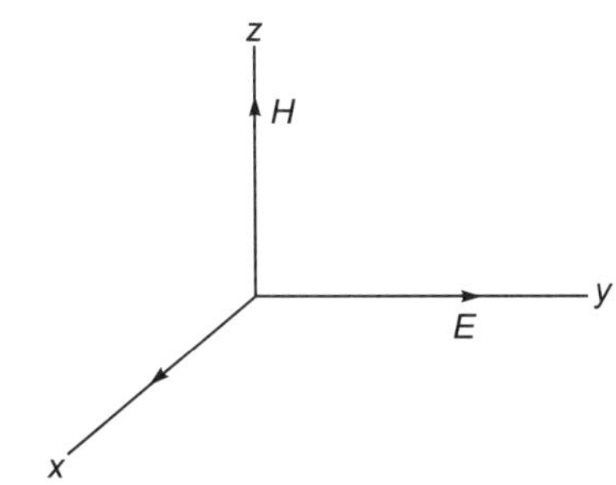

Fig. 3.4 Electric and magnetic fields of an electromagnetic radiation

the direction of propagation of the radiation (Fig. 3.4). Both the electric and the magnetic field vectors vary with time 't' and distance along the propagation of the radiation 'x' in periodical fashion. If it is assumed that these variations are sinusoidal as shown in Fig. 3.5, then the magnitude of E at a time 't' and distance 'x' can be expressed as

$$E = A. \sin 2\pi\left(\frac{x}{\lambda} - vt\right) \tag{3.1}$$

where A is the amplitude, v is the frequency and λ is the wavelength of the radiation. The wavelength and the frequency of the radiation are related as $\lambda = c/v$, where 'c' is the velocity of light $= 3 \times 10^8$ m/s.

The rate of flow of the electromagnetic energy through an unit area perpendicular to the direction of propagation is called the intensity 'I' of the

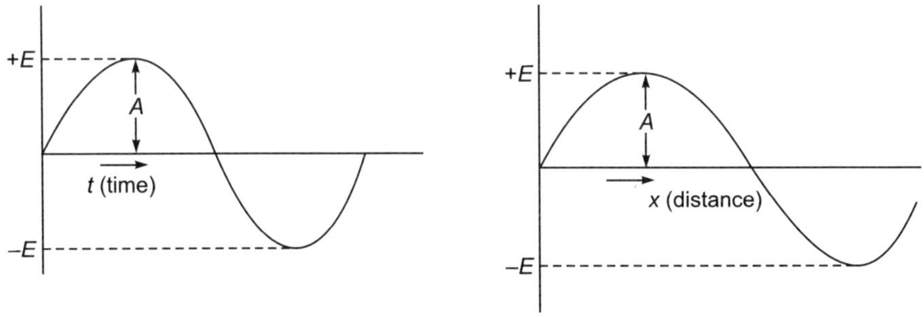

Fig. 3.5 Sinusoidal variations of the electric field of x-radiation

radiation. The average intensity is proportional to the square of the amplitude (A) of the electromagnetic wave. The intensity is measured in Joules/m^2/sec in absolute units. Practically this kind of absolute units are neither measured nor used directly. Instead, x-ray intensity measurements are done on the basis of the degree of darkening of a photographic film or the amount ionization caused in an ionization counter. The foregoing discussion assumes a wave nature for x-radiations. Electromagnetic radiations can also be thought of as a stream of particles called photons as per quantum theory. Each photon, according to this theory, has an energy 'hv' associated with it, where 'h' is Planck's constant (6.63 × 10^{-34} Joule.sec) and 'v' is the frequency of the photon. Thus, generally a dual 'wave-particle' nature is assumed for x-radiations.

3.2.2   The Continuous Spectrum

X-rays issuing out of a x-ray tube have the characteristic shown in Fig. 3.6 which is basically drawn for a tube with molybdenum target. Each curve in the figure represents the wavelength versus intensity of the x-rays at the tube voltages indicated. All the curves seem to start abruptly at a specific wavelength and the intensity increases quickly to a high value as the wavelength increases and after reaching a maximum intensity, it decreases with wavelength in a gradual manner and approaches zero. These radiations are aptly called the continuous or heterochromatic or white radiations as they contain radiations of a number of wavelengths continuously (like visible light) starting from a critical wavelength known as the short-wavelength-limit (λ_{SWL}).

In addition to this, the curve at 25 kV shows two intensity peaks at certain specific wavelengths marked in the figure as Kβ and Kα. These almost monochromatic (single wavelength) radiations are termed as characteristic radiations since these occur at definite wavelengths and are characteristic of the target material used (molybdenum for the case considered in Fig. 3.6).

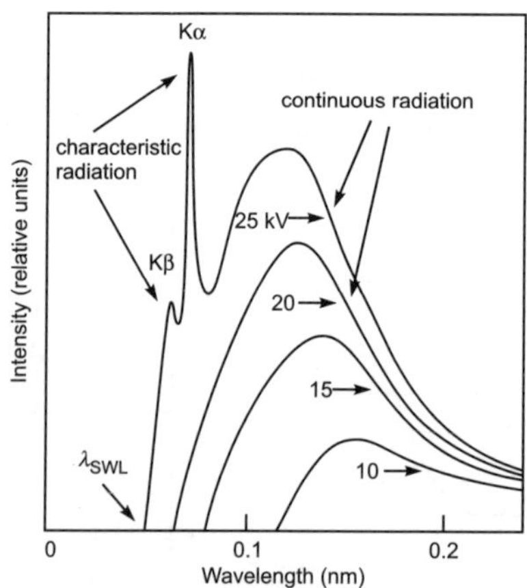

Fig. 3.6 Continuous spectrum emanating from an x-ray tube (schematic)

In an x-ray tube, generally of filament type used presently, as a high voltage of the order of a few kilovolts is applied between the cathode and the anode, the electrons are rapidly attracted towards the anode and attain high velocities. The energy applied to the tube gets converted into kinetic energy of electrons and the relation between the two can be expressed as

$$eV = 1/2\ mv^2 \tag{3.2}$$

where 'e' is the charge on the electron (1.6×10^{-19} coulomb), V is the voltage applied between the cathode and the anode, m is the mass of the electron (9.11×10^{-31} kg) and 'v' is the velocity of the electron just before hitting the target material. The continuous spectrum occurs due to the sudden stoppage (in case of a head-on collision with a target atom) or deceleration of these high velocity (of the order of the velocity of light) electrons by the target material. Most of the kinetic energy of these electrons gets converted into heat and a small amount of less than 1% is converted into x-rays. If all the kinetic energy of an electron (i.e., $1/2\ mv^2$ expressed by the eq. 3.2) gets converted into an x-ray photon, then this photon will have the maximum energy E_{max}, which can be expressed as $h\nu_{max}$ where h is the Planck's constant and ν_{max} is the frequency of the x-ray photon generated. But, from the equation 3.2 the maximum energy supplied to the tube is eV. Thus, it can be observed that

$$E_{max} = h\nu_{max} = hc/\lambda_{min} = eV \tag{3.3}$$

where λ_{min} is the wavelength of the x-ray photon of frequency ν_{max}. This situation can occur only under a very low probability condition of an electron making a head-on collision with a target atom, so as to convert all its energy into a x-ray photon. The maximum frequency ν_{max} under a given tube voltage can be calculated by using the equation 3.3.

By putting the values of the constants in the equation 3.3 and simplifying, the wavelength of the minimum energy, x-ray photon under a given voltage V (practical units) can be written as:

$$\lambda_{min}\ (\text{in nm}) = \frac{1239.8}{V\ \text{in volts}} \tag{3.4}$$

The equation 3.4 gives the minimum wavelength of x-ray photon that could be generated at a given voltage V applied to the x-ray tube. This minimum wavelength is termed as the short wavelength limit (λ_{swl}). This wavelength is generated only when a head-on collision of the high velocity electron with the target atom. In all other kinds of collisions, the electron loses its energy by parts due to various types of glancing collisions with target atoms. These small amounts of energy lost by electrons is converted into either heat or x-rays of wavelengths longer than λ_{swl} (since, these x-rays have energies and consequently frequencies lesser than those produced by a head-on collision). Thus, x-rays produced in a x-ray tube will contain all wavelengths starting from a short wavelength limit. There seems to be no limit on the long wavelength end as it is theoretically possible to have x-rays with energies approaching zero. The totality of all these wavelengths starting from a short wavelength limit can be termed as continuous, or white, or heterochromatic spectrum. From Fig. 3.6 it can also be seen that as the tube voltage increases, the curves of intensity versus wavelength shift to the left and upwards. As the voltage increases, the energy supplied to the tube and consequently the kinetic energy of the electrons increases which causes the maximum energy of the x-ray photon to increase resulting in the decrease of the minimum wavelength λ_{swl}. In addition, both the number of photons produced per second and the energy per photon increase resulting in the shifting of the curve upward. The total x-ray energy emitted per second (or the total intensity) I_{cs} is proportional to the area under the curve and is known to be proportional to the tube current i, the atomic number Z and the applied voltage (V) to a power m. Or,

$$I_{cs} = A \, i \, Z \, V^{m} \tag{3.5}$$

where A is a proportionality constant and m has a value of about 2. Thus, if it is desired to increase the intensity of continuous spectrum a heavy metal target (high Z) coupled with a high voltage may be used.

3.2.3 The Characteristic Spectrum

The intensity versus wavelength curve for a tube voltage of 25 kV shows two peaks denoting very high intensity x-rays of specific wavelengths. These peaks do not exist at voltages lower than 20 kV, which is a clear indication that a new phenomenon occurs when the tube voltage goes beyond a critical value (which is 20.01 kV for molybdenum). These peaks occur at definite wavelengths superimposed on the continuous spectrum and are characteristic of the target material used (in this case, molybdenum), which is why they are referred as the characteristic spectrum. The minimum voltage necessary to excite these radiations is termed as the critical voltage.

The origin of the characteristic spectrum is connected with the interaction of the fast moving electrons with the orbital electrons of the target metal. Fig. 3.7 shows schematically a target metal atom with its electron shells. Each shell electron of the atom is bound to the nucleus of the atom with a certain specific energy characteristic of the atom of the target metal. If a fast moving electron in the x-ray tube has sufficient kinetic energy to knockout, say a K-shell electron of the target atom from its orbit, it would do so and then the target atom would become unstable with a K-shell vacancy thus attaining a high energy state called the K-state. The atom can go to a low energy state (say, a L-state or a M-state) by transfer of one of its L or M shell electrons into the K-shell vacancy. The difference in the energies of the K-state and the new state (L or M as the case may

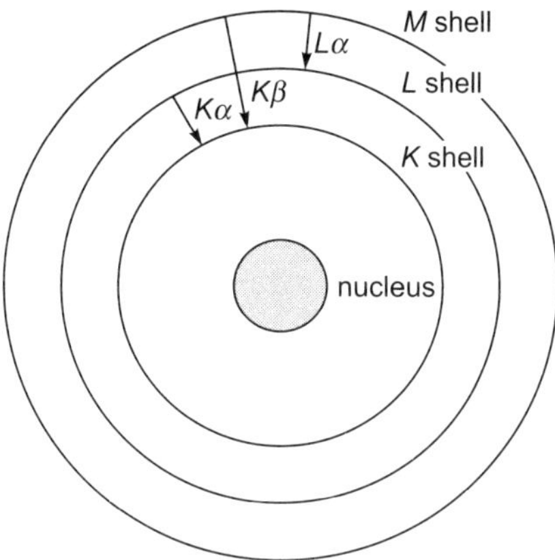

Fig. 3.7 An atom of a target metal (Ref.: B.D. Cullity, Elements of x-ray Diffraction)

be) results in x-radiation of a definite energy called the *K-characteristic radiation.* This characteristic radiation has a definite energy (equal to the energy level difference discussed) and therefore, a definite wavelength characteristic of the target element concerned. If the K-characteristic radiation is emitted by the target, as a result of a transfer of an L-shell electron to the K-shell vacancy then it is termed Kα radiation. But if the transfer is between M-shell to K-shell it is termed Kβ radiation. Characteristic radiations can be similarly generated due to a L-shell vacancy in the target atom created by the electron bombardment. These radiations are referred to the L-characteristic radiations. The totality of all such radiations Kα, Kβ, Lα, Lβ, Mα, Mβ and so on, is termed as the characteristic spectrum of the target metal concerned. The intensity of any characteristic line, like Kα, is proportional to the tube current, i, and a power n (approx. equal to 1.5) of the voltage applied, V, above the critical excitation voltage V_K. Thus,

$$I_{K\alpha} = B \, i \, (V - V_K)^n \tag{3.6}$$

where B is a proportionality constant. The characteristic x-radiations are intense and almost monochromatic (width at half the maximum intensity, I_{max}, is approx. equal to .0001 nm), which is why they are conveniently used for x-ray diffraction experiments. H. G. Mosley found that the square root of the characteristic line frequency is proportional to the atomic number Z of the target metal. i.e.,

$$\sqrt{v} = C \, (Z - p) \tag{3.7}$$

where C is the proportionality constant and p is another constant. This relation is known as Mosley's Law. The equation 3.7 shows that a plot of square root of v versus atomic number Z for any particular K or L characteristic radiation results in a straight line with a slope C as shown in Fig. 3.8.

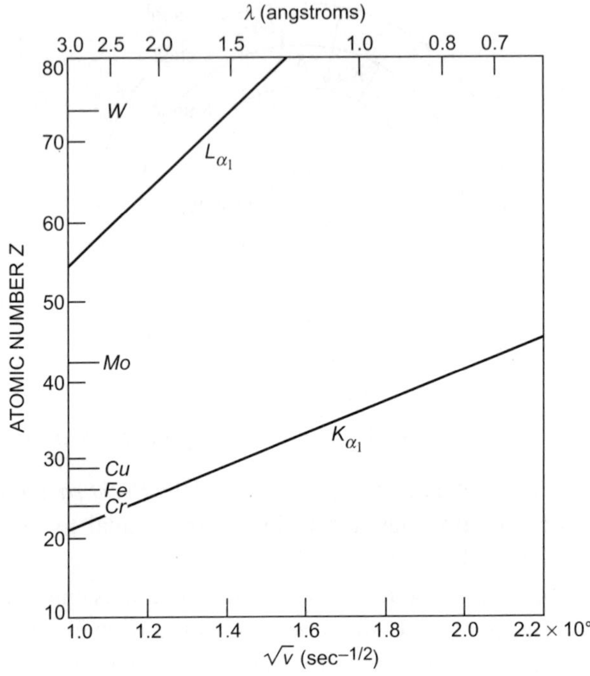

Fig. 3.8   Illustration of Mosley's law (Ref.: B.D. Cullity, Elements of x-ray Diffraction)

3.3   INTERACTION OF X-RAYS WITH MATTER

3.3.1   Absorption and Fluorescence

Whenever x-rays pass through a substance, a part of their energy is absorbed. The extent of absorption depends on the distance travelled through the substance matter. It has been found that the fractional decrease in intensity (I) of the x-ray beam is proportional to the distance 'x' travelled through the matter. Or, writing in differential form,

$$-dI/I = \mu dx \tag{3.8}$$

where the proportionality constant μ is called the linear absorption coefficient of the substance concerned. This equation when integrated between limits I_o and I_x, where I_o is the original intensity of the x-ray beam and I_x is the intensity after passing through a distance 'x' through the matter, gives,

$$\int_{I_o}^{I_x} dI/I = -\mu \int_0^x dx$$

Integrating both the sides,

i.e.,
$$[\ln I]_{I_o}^{I_x} = -\mu x$$

or,
$$\ln (I_x/I_o) = -\mu x$$

or,
$$I_x = I_o \, e^{-\mu x} \qquad (3.9)$$

The linear absorption coefficient 'μ' varies linearly with the density 'ρ' of the matter through which x-ray passes. But, μ/ρ is a constant for a given material whatever be the state of that material (solid, liquid or gaseous). Thus, it is the value of mass absorption coefficient, μ/ρ, which is generally tabulated. So, to make use of the tabulated values of μ/ρ, the equation 3.9 can be modified as

$$I_x = I_o \, e^{-(\mu/\rho).\rho.x} \qquad (3.10)$$

The mass absorption coefficient, μ/ρ, of a substance containing more than one element can be calculated from the weighted average of the mass absorption coefficients of its constituent elements. For example, if w_1, w_2, w_3 ......etc. are the weight fractions of the elements 1, 2, 3, ...etc. in a given substance and $(\mu/\rho)_1$, $(\mu/\rho)_2$, $(\mu/\rho)_3$ ......etc. are their mass absorption coefficients, then the mass absorption coefficient, $(\mu/\rho)_\alpha$, of the aggregate substance is given by

$$(\mu/\rho)_\alpha = (\mu/\rho)_1.w_1 + (\mu/\rho)_2.w_2 + (\mu/\rho)_3.w_3 + ... \qquad 3.11$$

The interaction of x-rays with matter can be best illustrated with a study of the variation of the mass absorption coefficient of a substance with the wavelength of radiation. Fig. 3.9 is a schematic illustration of such a typical variation for a material. The figure shows two branches of the variation separated by a vertical line. This vertical line occurs at a definite wavelength of incident x-radiation called the absorption edge which is characteristic of the absorbing matter. On either side of the absorption edge, the relationship between the mass absorption coefficient (μ/ρ) and the wavelength 'λ' of the incident radiation can be expressed as

$$(\mu/\rho) = D \, \lambda^3 \, Z^3 \qquad (3.12)$$

where D is a proportionality constant and Z is the atomic number of the absorbing element. The equation indicates that longer the wavelength of radiation, greater is the amount of absorption. Or, in other words, shorter the wavelength, more is the

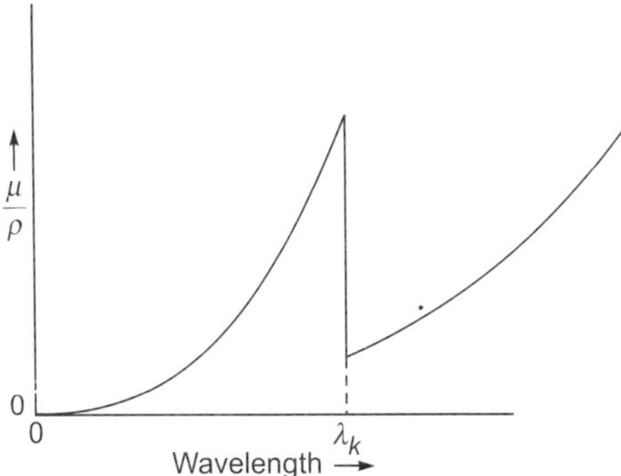

Fig. 3.9 X-ray absorption characteristics of an absorber (schematic)

penetrating power of the radiation. Thus, long wavelength radiations are easily absorbed by matter and are termed as 'soft' radiations and short wavelength radiations as 'hard' radiations. Further, it can also be seen from the above equation that the absorbing power depends on the atomic number of the absorbing element. This fact is made use of in safety devices for protecting personnel working with the x-ray equipment from the harmful effects of x-radiations. Lead (Z = 82) sheets are generally used for this reason as barriers over x-ray equipment in order to absorb stray radiations.

The absorption phenomenon, in general, can be classified into two categories, viz., scattering and true absorption. The scattering phenomenon occurs due to the scattering of x-rays by the atoms of the absorber much in a similar fashion as visible light is scattered by dust particles in air. X-rays falling on an atom are scattered in all directions and the intensity of x-rays in the original direction of travel is reduced by the amount scattered. On the other hand, true absorption occurs when the incident x-ray beam has sufficient energy to knockout a shell (K or L or M or others as the case may be) electron from the atom of the absorber. Fig. 3.9 indicates that as the wavelength of the incident x-radiation decreases the absorption decreases as per the equation 3.12. Further, as the wavelength decreases, the frequency, and consequently the energy of the x-ray photon, both keep increasing since energy is given by $E = h.v = h.c/\lambda$.

As the wavelength of the incident x-radiation is decreased from an initial high value, the energy of the x-ray photon increases and reaches a value E_K which is just needed to knockout a K-shell electron from the absorber atom. The incident energy at this juncture is fully utilized in doing this work of knocking out a K-shell electron. This is responsible for the high absorption occurring at a definite wavelength λ_K which is referred as the K-absorption edge of the absorber and the x-ray photon of this wavelength has the energy E_K. That is to say,

$$E_K = h.v_K = h.c/\lambda_K \qquad (3.13)$$

where v_K and λ_K are the frequency and the wavelength respectively of the k-absorption edge. If the wavelength of the incident x-radiation is further decreased below the value λ_K, the energy and hence the penetrating power increases over and above that required to just knockout the K-shell electron and more energy gets transmitted.

It should be noted that as soon as the K-shell vacancy (in the absorber atoms) occurs by the true absorption phenomenon explained above, the filling up of this vacancy by a L-shell or M-shell electron takes place and a $K\alpha$ or a $K\beta$ characteristic radiation is emitted by the absorber as discussed in section 3.2.3. These characteristic radiations produced as a result of the incident x-radiations, rather than electrons, are termed fluorescent radiations. Thus, the term characteristic radiation is used for such radiations caused by incident electrons and the fluorescent radiations are characteristic radiations caused by incident x-radiations.

The values of absorption edges (wavelengths) can be determined for a given absorber material by simple experiments and these are of help in constructing the energy level diagram of the atom of the absorber. A schematic energy level diagram is presented in Fig. 3.10(a). Here the highest energy level is the energy level corresponding to the K-shell vacancy state or generally referred to as the K-state. This K-state energy level should be equal to the energy needed to knock out the K-shell electron of the atom of the absorber. This energy E_K must be equal to $h.\nu_K$ or $h.c/\lambda_K$ as discussed earlier according to equation 3.13. If λ_K is determined for an absorber, it is possible to calculate E_K. Similarly, the other energy states of the atom, viz., E_L, E_M, E_N, etc., could be determined. Fig. 3.10(a) indicates schematically the process of production of characteristic radiations. The atom of an element is said to be in K-state after its K-excitation. If one of the L-shell electrons fills up the K-shell, then the atom changes its energy state to L and in the process emits a $K\alpha$ radiation, which derives its energy from the difference of energies of K and L states. Similar arguments hold for L, M and other characteristic radiations. Fig. 3.10(a) is oversimplified in the sense that the energy states L, M, N and others are not single energy states but contain a number of sub-states. For example, the L state consists of 3 sub-states L_I, L_{II} and L_{III}. Further, $E_{K\alpha1}$ (energy of $K\alpha1$ radiation) is equal to the energy level difference between the K state and the L_{III} state.

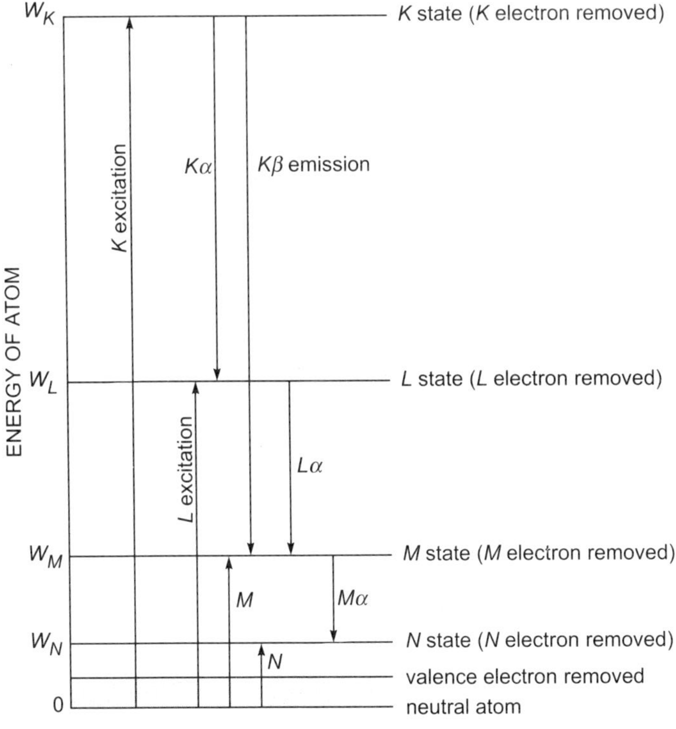

Fig. 3.10   (a) Energy level diagram of an absorber atom

i.e.,
$$E_{K\alpha 1} = E_K - E_{LIII}$$

or,
$$h\,c/\lambda_{K\alpha 1} = h\,c/\lambda_K - h\,c/\lambda_{LIII}$$

$$1/\lambda_{K\alpha 1} = 1/\lambda_K - 1/\lambda_{LIII} \tag{3.14}$$

So, equation 3.14 and the like can be utilized to calculate the wavelength of characteristic radiations from the corresponding absorption edges. If the absorption curve of nickel is plotted for longer wavelengths than what is shown in Fig. 3.9, other sharp discontinuities will be found corresponding to L and M (and others) absorption edges as shown in Fig. 3.10(b) (Three L levels and five M levels).Thus, true x-ray absorption by matter and fluorescent radiation are related to each other in the sense that the former is the cause and latter the effect.

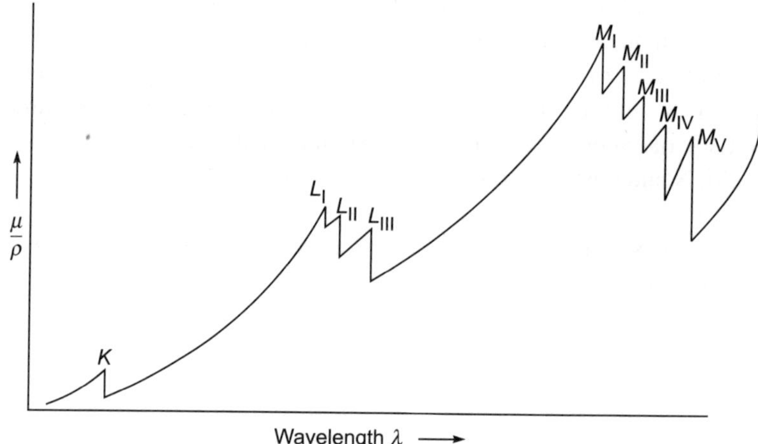

Fig. 3.10   (b) X-ray absorption characteristics (schematic)

3.3.2   Filtering of X-rays

X-ray diffraction experiments which make use of polycrystalline samples generally require reasonably monochromatic x-radiations. This kind of x-radiations can be obtained by filtering the white radiations generated in an x-ray tube by a suitable filter material. A strong monochromatic radiation can be obtained by operating the x-ray tube at a voltage far above the critical K-excitation voltage V_K. The critical excitation voltage V_K is the voltage necessary for supplying sufficient energy to the fast moving elements in order to enable them to knockout the K shell electrons of the target atom. This can be calculated on the basis of λ_K. The energy necessary to knockout the K shell electron is given by $h.c/\lambda_K$ and the electrons must possess this minimum energy for K excitation with the help of the voltage applied, i.e.,

$$h.c/\lambda_K = eV_K \tag{3.15}$$

and
$$V_K = 1239.8/\lambda_K \tag{3.16}$$

where V_K is in practical units of voltage and λ_K is in nm. And by equation (3.6) it can be seen that the intensity of K characteristic line is a function of voltage above the K critical voltage V. Most of the x-ray diffraction works need x-radiations of short wavelengths of the order of 0.05 to 0.3 nm. K characteristic radiations of metals are generally suitable.

Standard Tables 3.1 A and B are available which give the wavelengths of characteristic radiations and K absorption edge values of elements. The K characteristic radiations generally are composed of strong (in terms of intensity) $K\alpha_1$ and $K\alpha_2$ radiations, close enough to merge into one another in terms of wavelengths, and $K\beta$ radiation occurring at a slightly shorter wavelength than those of $K\alpha$ radiations, as shown schematically in Fig. 3.11.

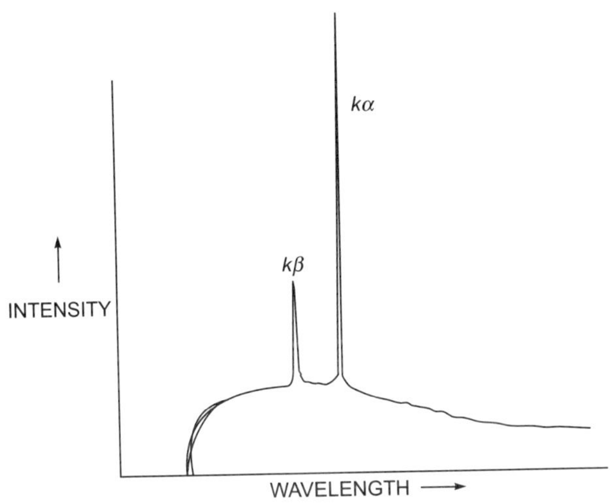

Fig. 3.11 Schematic sketch of variation of intensity with wavelength

Filtering is done to reduce the intensity of $K\beta$ radiation, so that nearly monochromatic radiation $K\alpha$ is obtained. If a filter material can be found which has its K absorption edge lying exactly in between the wavelengths of $K\alpha$ and $K\beta$ radiation of a given target element, it can reduce most of the $K\beta$ component and transmit most of the $K\alpha$ radiation falling on it. Fig. 3.12 illustrates schematically the intensity versus the wavelength of x-radiation before filtering and after filtering.

It can be seen from the figure that the $K\beta$ radiation is absorbed to a large extent by passing through the filter because the filter material has high absorption coefficient at the $K\beta$ wavelength range. But a large fraction of original $K\alpha$ intensity passes through the filter since the absorption coefficient of the filter is low in the $K\alpha$ wavelength range. Such a situation occurs whenever a filter material has its atomic number 1 or 2 less than the target material. For example, for a

Table 3.1a Wavelengths (in angstroms) of some characteristic emission lines and absorption edges

Element	Z	Kα (weighted average)*	Kα₂ strong	Kα₁ very strong	Kβ₁ weak	K edge	Lα₁ very strong	L_III edge
Na	11		11.909	11.909	11.617			
Mg	12		9.8889	9.8889	9.558	9.5117		
Al	13		8.33916	8.33669	7.981	7.9511		
Si	14		7.12773	7.12528	6.7681	6.7446		
P	15		6.1549	6.1549	5.8038	5.7866		
S	16		5.37471	5.37196	5.03169	5.0182		
Cl	17		4.73050	4.72760	4.4031	4.3969		
A	18		4.19456	4.19162	–	3.8707		
K	19		3.74462	3.74122	3.4538	3.43645		
Ca	20		3.36159	3.35825	3.0896	3.07016		
Sc	21		3.03452	3.03114	2.7795	2.7573		
Ti	22		2.75207	2.74841	2.51381	2.49730		
V	23		2.50729	2.50348	2.28434	2.26902		
Cr	24	2.29092	2.29351	2.28962	2.08480	2.07012		
Mn	25		2.10568	2.10175	1.91015	1.89636		
Fe	26	1.93728	1.93991	1.93597	1.75653	1.74334		
Co	27	1.79021	1.79278	1.78892	1.62075	1.60811		
Ni	28		1.66169	1.65784	1.50010	1.48802		
Cu	29	1.54178	1.54433	1.54051	1.39217	1.38043	13.357	13.2887
Zn	30		1.43894	1.43511	1.29522	1.28329	12.282	12.1309
Ga	31		1.34394	1.34003	1.20784	1.19567	11.313	
Ge	32		1.25797	1.25401	1.12889	1.11652	10.456	
As	33		1.17981	1.17581	1.05726	1.04497	9.671	9.3671
Se	34		1.10875	1.10471	0.99212	0.97977	8.990	8.6456
Br	35		1.04376	1.03969	0.93273	0.91994	8.375	
Kr	36		0.9841	0.9801	0.87845	0.86546		
Rb	37		0.92963	0.92551	0.82863	0.81549	7.3181	6.8633
Sr	38		0.87938	0.875214	0.78288	0.76969	6.8625	6.3868
Y	39		0.83300	0.82879	0.74068	0.72762	6.4485	5.9618
Zr	40		0.79010	0.78588	0.701695	0.68877	6.0702	5.5829
Nb	41		0.75040	0.74615	0.66572	0.65291	5.7240	5.2226
Mo	42	0.71069	0.713543	0.70926	0.632253	0.61977	5.40625	4.9125
Tc	43		0.676	0.673	0.602			
Ru	44		0.64736	0.64304	0.57246	0.56047	4.84552	4.3689
Rh	45		0.617610	0.613245	0.54559	0.53378	4.59727	4.1296
Pd	46		0.589801	0.585415	0.52052	0.50915	4.36760	3.9081
Ag	47		0.563775	0.559363	0.49701	0.48582	4.15412	3.6983

Contd.

Table 3.1a   Contd.

Ele-ment	Z	Kα (weighted average)*	Kα₂ strong	Kα₁ very strong	Kβ₁ weak	K edge	Lα₁ very strong	L_III edge
Cd	48		0.53941	0.53498	0.475078	0.46409	3.95628	3.5038
In	49		0.51652	0.51209	0.454514	0.44387	3.77191	3.3244
Sn	50		0.49502	0.49056	0.435216	0.42468	3.59987	3.1559
Sb	51		0.47479	0.470322	0.417060	0.40663	3.43915	2.9999
Te	52		0.455751	0.451263	0.399972	0.38972	3.28909	2.8554
I	53		0.437805	0.433293	0.383884	0.37379	3.14849	2.7194
Xe	54		0.42043	0.41596	0.36846	0.35849	–	2.5924
Cs	55		0.404812	0.400268	0.354347	0.34473	2.8920	2.4739
Ba	56		0.389646	0.385089	0.340789	0.33137	2.7752	2.3628
La	57		0.375279	0.370709	0.327959	0.31842	2.6651	2.2583
Ce	58		0.361665	0.357075	0.315792	0.30647	2.5612	2.1639
pr	59		0.348728	0.344122	0.304238	0.29516	2.4627	2.0770
Nd	60		0.356487	0.331822	0.293274	0.28451	2.3701	1.9947
Il	61		0.3249	0.3207	0.28209	–	2.2827	
Sm	62		0.31365	0.30895	0.27305	0.26462	2.1994	1.8445
Eu	63		0.30326	0.29850	0.26360	0.25551	2.1206	1.7753
Gd	64		0.29320	0.28840	0.25445	0.24680	2.0460	1.7094
Tb	65		0.28343	0.27876	0.24601	0.23840	1.9755	1.6486
Dy	66		0.27430	0.26957	0.23758	0.23046	1.90875	1.579
Ho	67		0.26552	0.26083	–	0.22290	1.8447	1.5353
Er	68		0.25716	0.25248	0.22260	0.21565	1.78428	1.48218
Tm	69		0.24911	0.24436	0.21530	0.2089	1.7263	1.4328
Yb	70		0.24147	0.23676	0.20876	0.20223	1.6719	1.38608
Lu	71		0.23405	0.22928	0.20212	0.19583	1.61943	1.34135
Hf	72		0.22699	0.22218	0.19554	0.18981	1.56955	1.29712
Ta	73		0.220290	0.215484	0.190076	0.18393	1.52187	1.25511
W	74		0.213813	0.208992	0.184363	0.17837	1.47635	1.21546
Re	75		0.207598	0.202778	0.178870	0.17311	1.43286	1.17700
Os	76		0.201626	0.196783	0.173607	0.16780	1.39113	1.14043
Ir	77		0.195889	0.191033	0.168533	0.16286	1.35130	1.10565
Pt	78		0.190372	0.185504	0.163664	0.15816	1.31298	1.07239
Au	79		0.185064	0.180185	0.158971	0.15344	1.27639	1.03994
Hg	80		–	–	–	0.14923	1.24114	1.00898
Tl	81		0.175028	0.170131	0.150133	0.14470	1.20735	0.97930
Pb	82		0.170285	0.165364	0.145980	0.14077	1.17504	0.95029
Bi	83		0.165704	0.160777	0.141941	0.13706	1.14385	0.92336
Th	90		0.137820	0.132806	0.117389	0.11293	0.95598	0.76062
U	92		0.130962	0.125940	0.111386	0.1068	0.91053	0.72216

* In averaging, $K\alpha_1$ is given twice the weight of $K\alpha_2$.

Characteristic *L* Lines of Tungsten

Line	Relative Intensity	Wavelength (A)
$L_{\alpha 1}$	Very strong	1.47635
$L_{\alpha 2}$	Weak	1.48742
$L_{\beta 1}$	Strong	1.28176
$L_{\beta 2}$	Medium	1.24458
$L_{\beta 3}$	Weak	1.26285
$L_{\gamma 1}$	Weak	1.09852

(Reference: B.D. Cullity, Elements of X-ray Diffraction, second edition, Addison Wesley Publishing Company Inc, California, 1978).

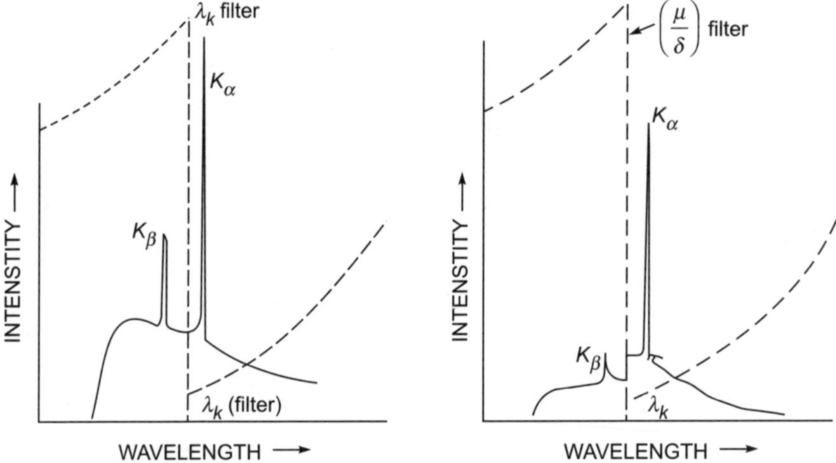

Fig. 3.12   Intensity versus wavelength before and after β-filtering (schematic)

copper target (atomic no. = 29) nickel filter (atomic no. 28) can be used conveniently. Table 3.2 shows the commonly used target materials and filters. It can also be noted from the table that thickness of the filter is another variable. The thicker the filter, the higher the ratio of intensity of Kα to that of Kβ radiations in the transmitted beam. But it must be remembered, however, that as the thickness the filter increases, the transmitted intensity decreases for all wavelengths in accordance with equation (3.9). Thus, a compromise must be arrived at in selecting the thickness of the filter material such that, on the one hand, the intensity ratio $I_{K\alpha}/I_{K\beta}$ is sufficiently high to suppress Kβ efficiently, and, on the other, a sufficiently high Kα intensity is transmitted for use in diffraction work. Filter material can be in the form of thin foils of a given metallic element. If the metal is not a stable one at ambient temperature and pressure, an oxide of the metal may be used corresponding to the required mass of the metal per unit area. However, the technique of β filtering is only able to suppress the Kβ characteristic radiation to a large extent, but does not eliminate the Kβ radiation. In addition, a

Table 3.1b   Mass absorption coefficients (cm²/gm) and densities

Absorber	Density (gm/cm³)	Mo Kα 0.711 Å	Mo Kβ 0.632 Å	Cu Kα 1.542 Å	Cu Kβ 1.392 Å	Co Kα 1.790 Å	Co Kβ 1.621 Å	Cr Kα 2.291 Å	Cr Kβ 2.085 Å
1 H	0.08375×10^{-3}	0.3727	0.3699	0.3912	0.3882	0.3966	0.3928	0.4116	0.4046
2 He	0.1664×10^{-3}	0.2019	0.1972	0.2835	0.2623	0.3288	0.2966	0.4648	0.4001
3 Li	0.533	0.1968	0.1866	0.4770	0.3939	0.6590	0.5283	1.243	0.9639
4 Be	1.85	0.2451	0.2216	1.007	0.7742	1.522	1.152	3.183	2.388
5 B	2.47	0.3451	0.2928	2.142	1.590	3.357	2.485	7.232	5.385
6 C	2.27 (graphite)	0.5348	0.4285	4.219	3.093	6.683	4.916	14.46	10.76
7 N	1.165×10^{-3}	0.7898	0.6054	7.142	5.215	11.33	8.330	24.42	18.23
8 O	1.332×10^{-3}	1.147	0.8545	11.03	8.062	17.44	12.85	37.19	27.88
9 F	1.696×10^{-3}	1.584	1.154	15.95	11.66	25.12	18.57	53.14	39.99
10 Ne	0.8387×10^{-3}	2.209	1.597	22.13	16.24	34.69	25.72	72.71	54.91
11 Na	0.966	2.939	2.098	30.30	22.23	47.34	35.18	98.48	74.66
12 Mg	1.74	3.979	2.825	40.88	30.08	63.54	47.38	130.8	99.62
13 Al	2.70	5.043	3.585	50.23	37.14	77.54	58.08	158.0	120.7
14 Si	2.33	6.533	4.624	65.32	48.37	100.4	75.44	202.7	155.5
15 P	1.82 (yellow)	7.870	5.569	77.28	57.44	118.0	89.05	235.5	181.6
16 S	2.09	9.625	6.835	92.53	68.90	141.2	106.6	281.9	217.2
17 Cl	3.214×10^{-3}	11.64	8.261	109.2	81.79	164.7	125.3	321.5	250.2
18 A	1.663×10^{-3}	12.62	8.949	119.5	89.34	180.9	137.3	355.5	275.8
19 K	0.862	16.20	11.51	148.4	111.7	222.0	169.9	426.8	334.2
20 Ca	1.53	19.00	13.56	171.4	129.0	257.4	196.4	499.6	389.3

Contd.

Table 3.1b *Contd.*

21 Sc	2.99	21.04	15.00	186.0	140.8	275.5	212.2	520.9	410.7
22 Ti	4.51	23.25	16.65	202.4	153.2	300.5	231.0	571.4	449.0
23 V	6.09	25.24	18.07	222.6	168.0	332.7	254.7	75.06	501.0
24 Cr	7.19	29.25	20.99	252.3	191.1	375.0	288.1	85.71	65.79
25 Mn	7.47	31.86	22.89	272.5	206.7	405.1	311.2	96.08	73.75
26 Fe	7.87	37.74	27.21	304.4	233.6	56.25	345.5	113.1	86.77
27 Co	8.8	41.02	29.51	338.6	258.7	62.86	47.71	124.6	96.06
28 Ni	8.91	47.24	34.18	48.83	282.8	73.75	56.05	145.7	112.5
29 Cu	8.93	49.34	35.77	51.54	38.74	78.11	59.22	155.2	119.5
30 Zn	7.13	55.46	40.26	59.51	45.30	88.71	68.00	171.7	133.5
31 Ga	5.91	56.90	41.69	62.13	46.65	94.15	71.39	186.9	144.0
32 Ge	5.32	60.47	44.26	67.92	51.44	102.0	77.79	199.9	154.5
33 As	5.78	65.97	48.57	75.65	57.01	114.0	86.76	224.0	173.3
34 Se	4.81	68.82	51.20	82.89	62.32	125.1	95.11	246.1	190.4
35 Br	3.12 (liquid)	74.68	55.56	90.29	68.07	135.8	103.5	266.2	206.2
36 Kr	3.488×10^{-3}	79.10	58.64	97.02	73.22	145.7	111.2	284.6	220.7
37 Rb	1.53	83.00	62.07	106.3	80.16	159.6	121.8	311.7	241.8
38 Sr	2.58	88.04	65.59	115.3	86.77	173.5	132.2	339.3	263.4
39 Y	4.48	97.56	72.57	127.1	96.19	190.2	145.4	368.9	286.9
40 Zr	6.51	16.10	75.20	136.8	103.3	204.9	156.6	398.6	309.7
41 Nb	8.58	16.96	81.22	148.8	112.3	222.9	170.4	431.9	336.4
42 Mo	10.22	18.44	13.29	158.3	119.7	236.6	181.0	457.4	356.5
43 Tc	11.50	19.78	14.30	167.7	126.9	250.8	191.9	485.5	378.0

Contd.

Table 3.1b *Contd.*

44 Ru	12.36	21.33	15.40	180.8	137.0	269.4	206.6	517.9	404.4
45 Rh	12.42	23.05	16.65	194.1	147.1	289.0	221.8	555.2	433.7
46 Pd	12.00	24.42	17.63	205.0	155.6	304.3	234.0	580.9	455.1
47 Ag	10.50	26.38	19.10	218.1	165.8	323.5	248.9	617.4	483.5
48 Cd	8.65	27.73	20.13	229.3	174.0	341.8	262.1	658.8	513.5
49 In	7.29	29.13	21.18	242.1	183.3	362.7	277.1	705.8	548.0
50 Sn	7.29	31.18	22.62	253.3	193.1	374.1	288.7	708.8	556.6
51 Sb	6.69	33.01	23.91	266.5	203.6	391.3	303.1	733.4	578.8
52 Te	6.25	33.92	24.67	273.4	208.4	404.4	311.7	768.9	602.7
53 I	4.95	36.33	26.53	291.7	221.9	434.0	333.2	835.2	650.8
54 Xe	5.495×10^{-3}	38.31	27.86	309.8	235.9	459.0	353.4	755.4	685.2
55 Cs	1.91 ($-10°$C)	40.44	29.51	325.4	247.5	483.8	371.5	802.7	725.1
56 Ba	3.59	42.37	31.00	336.1	256.0	499.0	383.6	587.3	644.6
57 La	6.17	45.34	33.10	353.5	270.8	519.0	401.9	222.9	661.5
58 Ce	6.77	48.56	35.54	378.8	289.6	559.1	431.5	240.4	509.4
59 Pr	6.78	50.78	37.09	402.2	306.8	596.2	458.8	260.5	205.3
60 Nd	7.00	53.28	38.88	417.9	319.8	531.7	475.9	271.3	213.4
61 Pm		55.52	40.52	441.1	336.4	401.4	503.2	284.7	223.8
62 Sm	7.54	57.96	42.40	453.5	346.6	411.8	446.3	295.0	231.5
63 Eu	5.25	61.18	44.74	417.9	369.0	165.2	476.9	312.7	244.9
64 Gd	7.87	62.79	45.95	426.7	377.2	169.5	346.7	318.9	250.3
65 Tb	8.27	66.77	48.88	321.9	399.9	178.7	367.1	338.9	265.2
66 Dy	8.53	68.89	50.38	336.6	360.2	184.9	142.9	351.7	275.0

Contd.

Table 3.1b   *Contd.*

67 Ho	8.80	72.14	52.76	128.4	272.4	189.8	146.3	363.3	283.4
68 Er	9.04	75.61	55.07	134.3	291.7	198.4	153.0	379.7	296.1
69 Tm	9.33	78.98	57.94	140.2	288.5	207.4	159.8	397.0	309.6
70 Yb	6.97	80.23	59.22	144.7	110.9	214.0	164.9	409.6	319.4
71 Lu	9.84	84.18	62.04	152.0	116.5	224.6	173.2	429.5	335.1
72 Hf	13.28	86.33	64.15	157.7	121.0	232.9	179.6	445.0	347.2
73 Ta	16.67	89.51	66.07	161.5	123.9	238.3	183.9	454.7	355.0
74 W	19.25	95.76	70.57	170.5	131.5	249.7	193.7	470.4	369.1
75 Re	21.02	98.74	72.47	178.3	137.3	261.8	202.7	495.5	388.0
76 Os	22.58	100.2	74.13	183.8	141.3	270.3	209.0	512.4	401.1
77 Ir	22.55	103.4	77.20	192.2	147.6	283.4	218.8	539.6	421.6
78 Pt	21.44	108.6	80.23	198.2	151.2	295.2	226.4	571.6	443.9
79 Au	19.28	111.3	82.33	207.8	160.6	303.3	235.7	568.0	446.7
80 Hg	13.55	114.7	85.30	216.2	166.6	317.0	245.7	597.9	468.9
81 Tl	11.87	119.4	88.25	222.2	171.1	326.3	252.7	616.9	483.3
82 Pb	11.34	122.8	90.55	232.1	178.6	340.8	263.8	644.5	504.9
83 Bi	9.80	125.9	93.50	242.9	187.5	355.3	275.8	667.2	524.2
86 Rn	4.40	117.2	100.7	263.7	203.0	387.1	299.7	731.4	573.2
	(liq., − 62°C)								
90 Th	11.72	99.46	73.34	306.8	236.6	449.0	348.4	844.1	663.0
92 U	19.05	96.67	72.63	305.7	236.2	446.3	346.9	774.0	657.1
94 Pu	19.81	48.84	78.99	352.9	271.2	519.6	401.6	803.2	771.6

(Reference: B. D. Cullity, Elements of X-ray Diffraction, second edition, Addison Wesley Publishing Company Inc, California, 1978)

Table 3.2 Filters for Suppression of $K\beta$ Radiation

Target	Filter	Incident beam* $\dfrac{I(K\alpha)}{I(K\beta)}$	Filter thickness for $\dfrac{I(K\alpha)}{I(K\beta)} = \dfrac{500}{1}$ in trans. beam		$\dfrac{I(K\alpha)\ \text{trans.}}{I(K\beta)\ \text{incident}}$
			mg/cm^2	$in.$	
Mo	Zr	5.4	77	0.0046	0.29
Cu	Ni	7.5	18	0.0008	0.42
Co	Fe	9.4	14	0.0007	0.46
Fe	Mn	9.0	12	0.0007	0.48
Cr	V	8.5	10	0.0006	0.49

(Reference: B.D. Cullity, Elements of X-ray Diffraction, second edition, Addison Wesley Publishing Company Inc, California 1978)

certain intensity of continuous or white radiation also accompanies the filtered beam. So, the filtered beam is not strictly a monochromatic $K\alpha$ radiation. Strictly monochromatic radiation can be obtained by using crystal monochromators. These crystal monochromators diffract a chosen single wavelength at a definite Bragg angle in accordance with Bragg's law. But the intensity of radiation obtained by crystal monochromators is very low since diffraction is a very inefficient ($< 1\%$ efficiency) process.

3.4 DETECTION OF X-RAYS

Intensity of x-rays is to be detected and in certain cases quantified in various applications. X-ray beams can be detected by fluorescent screens, photographic films and ionization devices.

3.4.1 Fluorescent Screens

Fluorescent screens generally contain zinc sulphide, which fluoresces with visible yellow light whenever x-rays fall on it. These screens are used to detect the direction of the primary beam in order to locate and adjust the specimen for diffraction work.

3.4.2 Photographic Films

X-ray beams affect photographic films in much the same manner as visible light affects them. But, the extent of darkening of the film, which is basically related to the amount of absorption of the radiation, is dependent on the thickness of emulsions on the film. Since, x-rays are highly penetrating radiations, the extent of absorption of x-rays is very much less compared to that of visible light. Thus, x-ray films generally have thicker emulsion coatings on both sides of the film in order to increase the amount of absorption of x-rays and consequently the darkening of the film. But even with these thick emulsions, the exposure times necessary to get sufficient darkening of the film by x-rays are two to three orders of magnitude higher than those by visible light. Exposure is defined as the product of intensity of x-ray beam and time of interaction of x-rays with film. Blackening of the photographic film or the photographic density 'd' is measured by the amount of visible light transmitted through the film and is defined as

$$d = \log (I_o/I_t)$$

where I_o = intensity of incident light beam

and I_t = intensity of transmitted light beam

The density with zero exposure is the fog in the emulsion, which varies with the type and age of emulsion and the technique of development. The most useful range of densities lies between fog level and densities of the order of 1.0 (corresponding to 10% transmission by the above definition) and most x-ray films show a linear relationship between density and exposure throughout this range. Thus, photographic density is directly proportional to the exposure which is in turn proportional to the x-ray intensity. Density can be measured with the help of a microphotometer. A simple microphotometer consists of a light source and an arrangement by which a beam of light passes through the x-ray film and strikes a photocell or a thermopile connected to a recording galvanometer. Current through the galvanometer is proportional to the intensity of light striking the photocell or thermopile which is inversely proportional to the x-ray photographic density. Thus, with proper calibration, a relationship between the intensity of the x-ray beam and the distance along the x-ray photographic film can be obtained.

When great detail and precision are not required, it is possible to decrease the exposure time by the use of intensifying screens placed in contact with the film. Most intensifying screens make use of certain substances which fluoresce with visible light whenever x-rays fall on them. The photographic films, being more sensitive to visible light than to x-rays, require less time for attaining a given density whenever these intensifying screens are used. But, the sharpness of a diffraction line or spot is lost since visible light produced by the intensifying screen not only affects the area of the x-ray film on which x-ray beam falls, but also the neighbouring region close to the affected area.

3.4.3 Intensity Counters

These counters permit intensities to be measured within a fraction of a percent compared to the accuracy of 5 to 10 percent obtainable by photographic methods. Further, these methods permit easy computerization of the data and storage.

Proportional and Geiger counters are modifications of ionization chambers whereas scintillation counter is a device making use of fluorescent crystals. These are the three commonly used devices in electrical methods of detecting x-ray intensities.

Fig. 3.13 shows a basic circuit which can be used as a proportional or Geiger counter, with suitable operational modifications. It consists of a cylindrical chamber, a potential supply, a resistor, a capacitor and a detector circuit. The cylindrical chamber is made of hollow cylindrical shell of metallic material (the cathode) filled with a gas, closed at one end by an insulator and the other end by a material of high transparency to x-rays (like mica or beryllium). A fine metal wire (the anode) runs along the axis of the cylinder. This device can work as an ionization chamber when a potential dif-

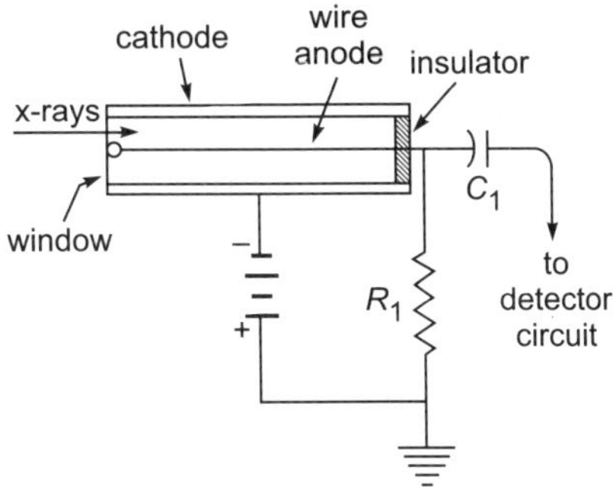

Fig. 3.13   Gas counter (proportional or Geiger) and basic circuit connections (Ref.: B.D. Cullity, Elements of x-ray Diffraction)

ference of around 200 V is applied between the anode and the cathode of the chamber. A large portion of the x-rays, which enter the chamber through the x-ray transparent window at one end, ionize the gas in the chamber. Under the influence of the electrical field the electrons move towards anode and ions towards cathode. A small current of the order 10^{-12} amps develops through the resistance R. This current is a measure of x-ray intensity. This kind of ionization chamber is not used in modern equipment, because of the low sensitivity.

The same device can work as a proportional counter, if the voltage is raised to 600 to 900 volts. The electrons produced by primary ionization are accelerated towards the anode wire at a much higher rate now compared to that in ionization chamber because of higher voltage present here. These electrons in turn cause further ionization by knocking out electrons from the gas atoms in their path. The process goes on repeatedly till the number of gas atoms ionized by the absorption of a single x-ray quantum reaches a value of about 10^3 to 10^5 times that obtained in an ionization chamber. This causes a detectable pulse of current in the circuit, which adds a charge to capacitor C before leaking through the resistor R. The whole process, starting with the absorption of an x-ray quantum to the detection of the charge added to the capacitor by a rate meter, takes an extremely short time of the order of microseconds. In ionization chambers, an x-ray quantum ionizes a certain number of gas atoms and the current produced in the circuit is proportional to this number. In proportional counters, each of the gas atoms ionized, would in turn, ionize about 10^3 to 10^5 atoms. Thus, the amplification produced is about 10^3 to 10^5 times, in a proportional counter, if the reference of no amplification (i.e., amplification = 1) is assumed in ionization chambers.

Fig. 3.14 shows the variation of the gas amplification factor with voltage applied. The important characteristics of the proportional counters are the following: (i) the pulses are proportional in amplitude to the energy of the absorbed

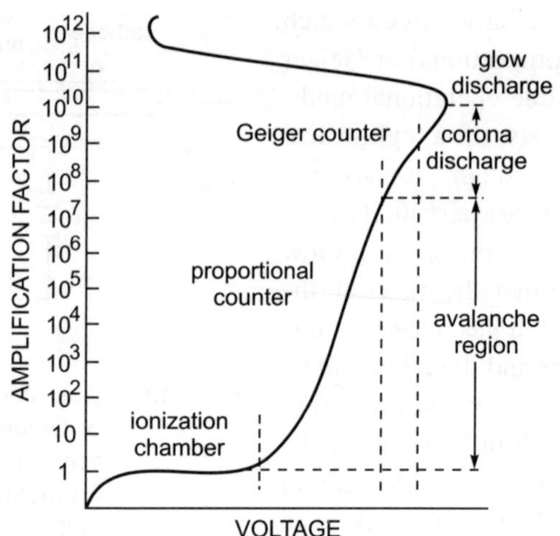

Fig. 3.14　Effect of voltage on gas amplification (Ref.: B.D. Cullity, Elements of x-ray Diffraction)

quantum (ii) the dead time is very short, of the order of 0.2 microseconds. The dead time is the time during which the counter is insensitive following a pulse in which time it is unable to count the intensity. Since, the dead time in a proportional counter is very short, the counter response is proportional to incident intensities upto very high counting rates which are often limited by the electrical circuits rather than by the counter itself (iii) the requirements regarding the constancy of applied voltage are more rigorous than with other counters (iv) the efficiency of the counter varies with the wavelength of the x-rays in a manner that depends on the gas in the counter. Thus, the proportional counter is a very fast counter of x-ray intensities with counting rates of pulses as high as 10^6 per second. This is possible because each avalanche of electrons (by amplification) is confined to an extremely narrow region of the wire about 0.1 mm and does not spread longitudinally to the whole length of the tube as occurs in the case of Geiger counters.

In the circuit of Fig. 3.13 if the voltage is raised to about 1000–1500 V, the counter will work as a Geiger counter. The operating voltage is to be decided on the basis of a simple procedure noted below. The counter is exposed to a beam of x-rays of constant intensity and the circuit is connected to measure the rate of pulses. The voltage in the circuit is gradually increased from a low voltage of about 100 V and the counting rate is measured. The variation of counting rate with voltage is as shown in Fig. 3.15. The counting rate increases with voltage until a threshold voltage is reached above which counting rate is practically almost independent of voltage. A Geiger counter is operated normally in this plateau region at an overvoltage of about 100 volts, i.e., 100 volts higher than the threshold voltage. In Geiger counters the gas amplification factor is high, of the order of 10^8 to 10^9. So less amplification will suffice in the external circuit for the purpose of measurement by counting-rate meters. The efficiency of the Geiger counter,

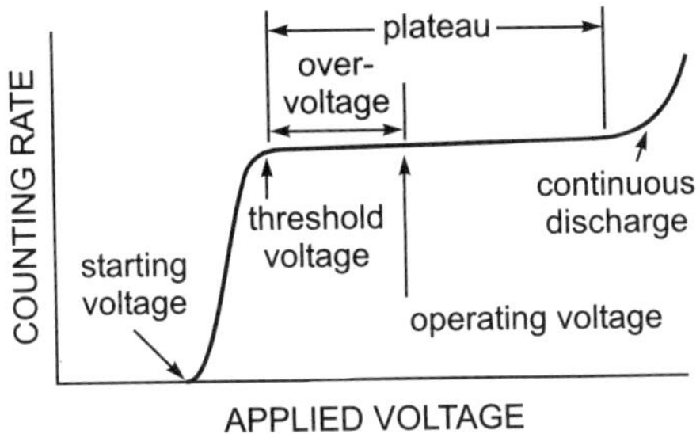

Fig. 3.15   Effect of voltage on counting rate for constant x-ray intensity

similar to that of proportional counter, varies with the wavelength of X-rays and the gas with which the counter is filled. The main disadvantage of the Geiger counter is the loss in counts that occurs because of the long dead time during which the counter is insensitive following a pulse. The dead time depends on the voltages applied, amplification which follows, gas content, etc. The usual range of dead time is between 50 and 300 microseconds and losses in counting may exceed 8% even with a counting rate as low as 500 per second.

The scintillation counter has its own advantages for detecting x-rays in the wavelength range of 0.02 to 0.2 nm. It consists of a fluorescent crystal mounted at the end of a photomultiplier tube. The crystal is usually a plate of sodium iodide that contains around 1% thallium in solid solution and is sealed to exclude moisture and light but has a window to admit x-rays. This crystal emits blue light when an x-ray beam falls on it. This light enters the photomultiplier tube and ejects electrons which are multiplied in number by 10^5 or more before passing on to amplifiers. The counting efficiency of these counters is approximately 100% throughout the range of wavelengths used in diffraction whereas, it is around 60-80% in proportional and Geiger counters. The decay time of the fluorescence from x-ray quantum absorption event is only about 10^{-7} seconds, so that by proper circuitry, it is possible to count at very high rates (~ 10,000/sec) without losses (< 1% loss). But a major drawback of this counter is the electronic noise generated in most photomultiplier tubes that restricts their use to x-ray wavelengths shorter than about 0.2 nm.

For measurement of x-ray intensity, the counters are to be attached to either a scaler or a rate meter. A scaler is an electronic device which counts the pulses produced by the counter over a period of time. The average counting rate is then obtained by dividing the number of counts by the time of counting. If the rate pulse production is low a fast mechanical counter can do the work of counting. For high counting rates, the scaler fulfills the function of scaling down the pulses by a known factor, i.e., a power of two (binary scaler) or a power of ten (decade scale). This scaled down pulse rate can be then handled by the mechanical counter.

The counting-rate meter is a device which indicates the average counting rate directly—without requiring separate measurements of number of counts and the time. A proper circuit is used here which smooths out the randomly spaced pulses from the counter into a steady current, the magnitude of which is proportional to the average pulse rate in the counter.

3.5 SAFETY

Personnel connected with x-ray diffraction work are constantly in danger of exposing themselves to harmful soft (long wavelength) x-radiation which can kill human tissues. Of course, this property of x-radiations is used advantageously in the therapeutic use of these in killing cancer cells. But for a healthy human being, exposure to x-ray beams, is harmful. The x-rays used in diffraction work are particularly harmful since they are relatively long wavelength and are easily absorbed by the human body. X-rays can cause burns, radiation sickness and even genetic mutations. Slight exposures to x-rays are not cumulative, but above a 'tolerance' dose, can be cumulative and produce permanent injury. Personnel connected with diffraction work must locate the primary beams with a fluorescent screen and avoid contact with the direct path of this beam. Further, the entire diffraction set-up must be well shielded by lead screens so that stray soft radiations are absorbed. It is better to check for stray radiations, with a portable radiation detector, periodically. Periodic check on blood count levels of personnel is advisable since the first noticeable effect of the harmful dosage is the lowering of the white blood cell count.

SOLVED EXAMPLES

Example 1   Calculate the velocity and kinetic energy of the electrons which strike the target of an x-ray tube operated at 30 kV. Determine the maximum energy per quantum of x-radiation emitted and the short wavelength limit (λ_{swl}) of the spectrum.

The energy of electrons at 30,000 Volts (see section 3.2.2)

$$= \text{charge on electron} \times \text{voltage}$$
$$= 1.6 \times 10^{-19} \times 30,000$$
$$= 4.8 \times 10^{-15} \text{ Joules.}$$

This is also the maximum energy per quantum of x-radiation.

This energy of 4.8×10^{-15} Joules is the kinetic energy with which the electrons strike the target. Thus,

Kinetic energy $= \frac{1}{2} \times m\, v^2$ where m is the mass of the electron and v is the velocity of the electron. Substituting the values,

$$4.8 \times 10^{-15} = \frac{1}{2} \times 9.11 \times 10^{-31} \times v^2, \text{ solving for v,}$$
$$v = 1.0265 \times 10^8 \text{ cm/sec.}$$

λ_{swl} in nanometre units $= 1239.8/\text{voltage applied in practical units}$
$$= 1239.8/30000 = 0.04133 \text{ nm.}$$

Example 2   Determine the transmission factor of 1 mm lead screen for Mo Kα, Cr Kα radiations.

Lead screens are generally used for protecting personnel working in x-ray laboratories. It is thus necessary to know the amount of radiation that can get transmitted through such screens.

The transmission factor = Intensity transmitted (I_x)/Intensity incident (I_o).

But, it is known that I transmitted = I incident $\times$ e$^{-(\mu/\rho).\rho.x}$, (Equation 3.10) where μ is the linear absorption coefficient of the screen, ρ is the density of the screen and x is the thickness of the screen (0.1 cm here).

It can be seen from Table 3.1B that the mass absorption coefficient values (μ/ρ) of lead for Mo Kα and Cr Kα are 141 and 585 cm^2/gm respectively and the density of lead is 11.34 gm/cm^3.

Using these values for Mo Kα,

$$I \text{ transmitted } / I \text{ incident } = e^{-(\mu/\rho).\rho.x}$$
$$= e^{-(141).(11.34).(0.1)} = 3.62 \times 10^{-70}.$$

For Cr Kα, the transmission factor, similarly,
$$= e^{-(585).(11.34).(0.1)} \simeq 0.$$

Example 3   A nickel filter is needed to increase the intensity ratio of Cu Kα to Cu Kβ from 7.5 in the incident beam to 500 in the filtered beam. Determine the thickness of the filter required and the transmission factor of such a filter for Cu Kα. (Assume the density of nickel as 8.9 gm/cm^3 and the mass absorption coefficients of nickel for Cu Kα and Cu Kβ as 49.3 and 286 cm^2/gm respectively).

Rewriting the equation 3.10 for Cu Kα radiation,
$$I_x \text{ (Cu K}\alpha) = I_o \text{ (Cu K}\alpha) \times e^{-(\mu/\rho)\text{CuK}\alpha.\rho.x} \tag{3.17}$$

Rewriting similar equation for Cu Kβ radiation,
$$I_x \text{ (Cu K}\beta) = I_o \text{ (Cu K}\beta) \times e^{-(\mu/\rho)\text{CuK}\beta.\rho.x} \tag{3.18}$$

Dividing equation 3.17 by 3.18,

$$\frac{I_x \text{ (Cu K}\alpha)}{I_x \text{ (Cu K}\beta)} = \frac{I_o \text{ (Cu K}\alpha) \times e^{-(\mu/\rho)\text{CuK}\alpha.\rho.x}}{I_o \text{ (Cu K}\beta) \times e^{-(\mu/\rho)\text{CuK}\beta.\rho.x}}$$

Putting the values of incident and transmitted intensity ratios of Cu Kα to Cu Kβ and simplifying,
$$500 = 7.5 \times e^{\rho.x.[(\mu/\rho)\text{CuK}\beta - (\mu/\rho)\text{CuK}\alpha]}$$

Substituting for the density of nickel and the mass absorption coefficient values,
$$500/7.5 = e^{8.9(286-49.3).x} = e^{2106.63x}$$

or, $$66.67 = e^{2106.63x}$$

Therefore, $$\ln 66.67 = 2106.63x \text{ or,}$$

$$x = 0.001994 \text{ cm} = 19.94 \text{ micrometre.}$$

The transmission factor for Cu Kα can be calculated as done in the previous example 3.2, i.e.,

I transmitted / I incident $= e^{-(\mu/\rho).\rho.x}$

$$= e^{-8.9(49.3)0.001994}$$

$$= 0.4169.$$

EXERCISES

3.1 Determine the frequency and energy of Cu $K\alpha$ (0.15418 nm) and Fe $K\alpha$ (0.19373 nm). [Ans: 1.95×10^{18} per second, 1.29×10^{-8} erg, 1.55×10^{18} per second, 1.025×10^{-15} Joules.]

3.2 An x-ray tube is operated at 40 kV in a diffraction laboratory. Determine

   (i) the kinetic energy of the electrons in the x-ray tube,

   (ii) the velocity attained by the electrons and

   (iii) the short wavelength limit of the spectrum emitted.

      [Ans: (i) 6.4×10^{-15} Joules, (ii) 1.185×10^{10} cm/sec (iii) 0.031 nm.]

3.3 Show that Mosley's law holds for $K\alpha$ lines of at least four elements.

3.4 Draw a graph of thickness (varying from zero to 1mm) of lead filter versus the transmission factor for Cu $K\alpha$ radiation.

3.5 For a Bismuth absorber, determine whether the law $(\mu/\rho) = D \lambda^3 Z^3$ holds good using at least three chosen radiations.

3.5 Determine the mass absorption coefficient of air for Cu $K\alpha$ radiation assuming that air is a mixture of 80% nitrogen and 20% oxygen by weight. What is the transmission factor of air with a path length of 57.3 mm (standard Debye Scherrer camera) for Cu $K\alpha$ radiation? (Assume (μ/ρ) values for nitrogen and oxygen for Cu $K\alpha$ as 8.51 and 12.7 cm^2/gm respectively and that the density values as 1.1649×10 and 1.331810 gm/cm^3 also respectively).

      [Ans: 9.35 cm^2/gm, 93.8%]

3.6 An absorber (of density 7.43) of 15 mm thickness reduces the intensity of Fe $K\alpha$ radiation to 49.2% of its original value. Identify the absorber and determine the reduction of intensity in percentage if the absorber thickness is increased to 50 mm.

      [Ans: Mn, 9.5%]

3.7 Usually ferric oxide (Fe_2O_3) powder layer on a cardboard backing is used as a filter rather than iron foil for Co $K\alpha$ radiations. If such a filter contains 5 mg Fe_2O_3 per cm^2, determine the transmission factor for Co $K\alpha$ line. Calculate the intensity ratio of Co $K\alpha$ to Co $K\beta$ in the filtered beam if the incident intensity ratio is 9.4. (Assume the density of Fe_2O_3 as 5.24 gm/cm^3. Assume also the (μ/ρ) values for Co $K\alpha$ radiation as 59.5 and 20.2 cm^2/gm for iron and oxygen respectively and the same for Co $K\beta$ radiation as 371 and 15 cm^2/gm respectively).

      [Ans: 78.78%, 27.75]

4

CHAPTER

X-RAY DIFFRACTION

4.1 GENERAL

The phenomenon of diffraction originates from the phenomenon of scattering. Whenever, an x-ray beam falls on an atom in a material, it scatters the beam in all directions, almost as a dust particle scatters the visible light. Since, crystalline materials have atoms arranged in a regular periodic fashion in three dimensions, the scattered x-rays can undergo constructive interference in certain specific directions and thus produce strong x-ray beams in those directions which are referred to as diffracted beams. The intensity of a diffracted beam is obviously very small (probably a fraction of a percent) with reference to the intensity of the incident beam since most of it is scattered in directions other than the direction of diffraction. So, the efficiency of the diffraction phenomenon (i.e., intensity of diffracted beam/intensity of incident beam) is less than one percent. The extent and importance of the application of x-ray diffraction are so great that this highly inefficient phenomenon is being put to use in a variety of ways to study the nature and properties of materials.

4.2 DIFFRACTION DIRECTION

When a monochromatic x-ray beam is incident on an atom (Fig. 4.1), the electrons present in the atom would start vibrating with a frequency corresponding to the frequency of the incident x-ray photon. Any charged particle (in this case,

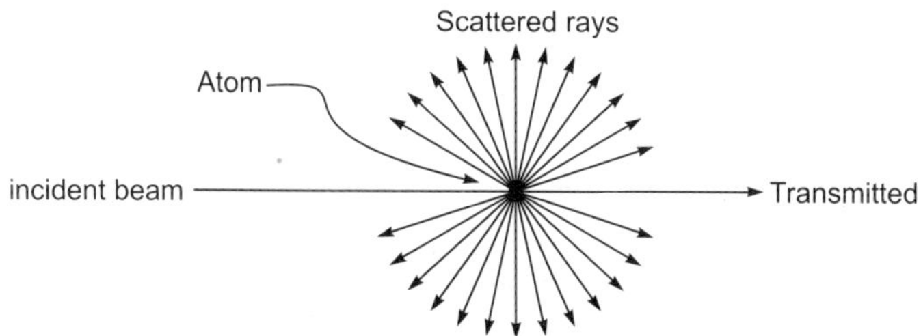

Fig. 4.1   Scattering of x-rays by an atom (schematic)

electrons of an atom) undergoing an oscillation in this fashion would become the source of a new set of electromagnetic waves and these waves are called scattered rays. Thus, the totality of the waves scattered by all the electrons in an atom is referred to as atomic scattering. For all practical purposes, the atomic scattering can be thought of as originating from a single point where the atom is present.

The situation of scattering from a single plane of atoms is shown in Fig. 4.2. Here, the incident monochromatic beam of x-rays strikes the atomic plane such that the angle between the plane and the beam is α and the Fig. 4.2 shows a section of the atomic plane considered. The plane of section contains the incident beam and is perpendicular to the atomic plane. The angle α is called the angle of incidence. The atoms in the path of the incident beam scatter X-rays in all directions. The scattered rays which make an angle α with respect to the atomic

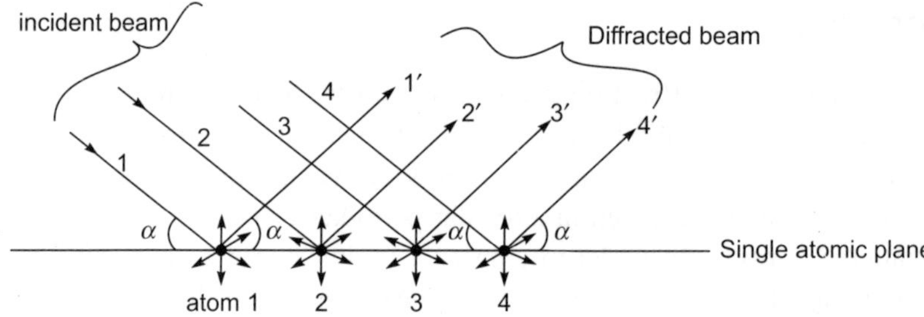

Fig. 4.2 Diffraction from a single atomic plane (schematic)

plane can be easily seen to be in phase with each other, and they constructively reinforce each other to produce a diffracted beam (i.e., 1', 2' and 3'). The scattered rays in any other direction (except for the direction of the travel of the incident beam) will destructively interfere with each other to different extents. Fig. 4.3 shows this state of affairs.

The rays 1 and 2 are from incident x-ray beam incident on the atomic plane at an angle α. The scattered rays 1' and 2' are at angle β, which is larger than the angle α, with respect to the atomic plane considered. It can be easily seen that the path difference between the scattered beams 1' and 2' by atoms A and B is given by the following:

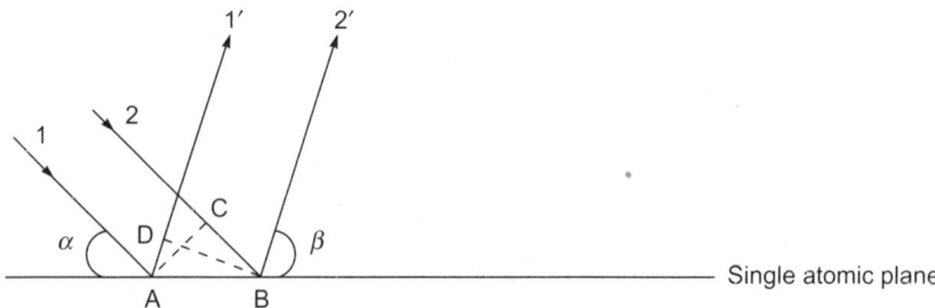

Fig. 4.3 Scattering at angle other than angle of incidence

$$\text{path difference} = BC - AD$$
$$= AB \cos \alpha - AB \cos \beta$$

This path difference is generally a fraction of the wavelength and always less than the wavelength of the x-rays used for the purposes of diffraction. Thus, destructive interference takes place between scattered rays 1' and 2'. So, scattered rays from an atomic plane can constructively reinforce each other only when the angle of incidence is equal to the angle of reflection (diffraction is also sometimes referred to as reflection because of similarity between these two), in which case the path difference is equal to zero, being [AB cos α – AB cos α].

Consider the interaction of a monochromatic x-ray beam with a crystalline material in which atoms are periodically present in a three-dimensional array. Fig. 4.4 shows such a situation of diffraction from a large number of atomic planes. From the previous discussions, it is clear that a single atomic plane will be able to

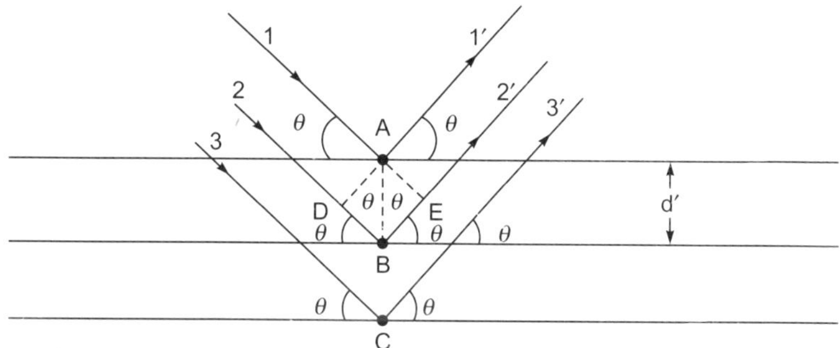

Fig. 4.4 X-ray diffraction from a crystalline material

produce a diffracted beam (constructive interference of scattered rays from all atoms in the plane) only when the angle of incidence is equal to the angle of scattering or diffraction. A set of a large number of atomic planes (three-dimensional crystalline matter), let us assume, also diffract x-rays under this condition. Consider the angle of incidence of the x-ray beam as θ. The incident x-rays 1, 2 and 3 strike the atoms A, B and C from atomic planes 1, 2, 3 respectively. The rays 1', 2' and 3' are the scattered rays, scattered at an angle θ, from the atomic planes 1, 2 and 3 respectively. If the scattered rays 1' and 2' have to interfere constructively, their path difference should be either zero or an integral multiple of the wavelength λ of x-ray beam. The path difference can be calculated as shown below:

If AD and AE are perpendiculars drawn from A to the incident ray 2 and scattered ray 2' respectively then the path difference between rays 1' and 2' = DB + BE.

From the Fig. 4.4,

$$\text{the path difference} = DB + BE$$

$$= AB \sin \theta + AB \sin \theta$$
$$= d' \sin \theta + d' \sin \theta$$
$$= 2d' \sin \theta$$

[AB is the interplanar distance d']

These two scattered rays 1' and 2' will be completely in phase or interfere constructively when the path difference $2d' \sin \theta$ is an integral multiple of λ.

$$n \lambda = 2d' \sin \theta \tag{4.1}$$

This relation was first formulated by W. L. Bragg and hence named as Bragg's Law.

It should be noted that a single atomic plane scatters (or diffracts) x-rays strongly in a direction making an angle with the plane which is equal to the angle of incidence. But, a number planes of atoms together diffract the x-ray beam at certain definite angles called Bragg angles. For given values of λ and d', we have only a set of Bragg angles θ_1, θ_2, θ_3, etc., which satisfy Bragg's law, for which n values are 1, 2, 3... etc., respectively.

i.e., $\quad\quad\quad\quad\quad\quad\quad\quad 1 \lambda = 2d' \sin \theta_1,$

$$2 \lambda = 2d' \sin \theta_2 \text{ and } 3\lambda = 2d' \sin \theta_3 \text{ and so on.}$$

It may also be noted that the path difference increases as the angle of incidence increases. The maximum possible value of the angle of incidence has a limit of 90°. So for a given λ and d' there can be only a fixed number of Bragg angles increasing from the lowest value θ, to the highest value θ which should be below 90°. The diffraction from the atomic planes such that the path difference between rays scattered by the top plane and the plane immediately below it is equal to one time the wavelength λ is called the first order reflection. In Fig. 4.4, for a first order reflection, the path difference between rays 1' and 2' is equal to λ, rays 1' and 3' is equal to 2λ and so on. The rays scattered by atoms from all planes of the crystal will constructively interfere with each other to produce a strong diffracted beam in the direction shown in the figure 4.4. Though the diffracted beam is strong in intensity compared to the scattered rays in other directions, it is very weak compared to the incident beam intensity, since the atoms scatter only a small fraction of the energy incident on them. This is the reason for the very low efficiency of the diffraction phenomenon.

From Bragg's law, it can be seen that

$$n\lambda/2d' = \sin \theta$$

The value of $\sin \theta$ cannot exceed unity. So, it can be seen that

$$n\lambda/2d' < 1$$

or $\quad\quad\quad\quad\quad\quad\quad\quad\quad n\lambda < 2d' \tag{4.2}$

For small values n, large values of λ can be used so as to satisfy equation (4.2). The smallest value of n is 1. So, the largest value of λ for diffraction should be less than 2d' from equation (4.2). Most of the crystal planes have their interplanar distances of the order of 0.4 nm or less. So, λ values cannot exceed 2d'. i.e., 0.8 nm and should not be too small, since small Bragg angles cannot be conveniently measured.

Bragg's law may also be written in the following form, i.e.,

$$\lambda = 2(d'/n) \; Sin \; \theta$$

or $\qquad\qquad\qquad\lambda = 2d \; Sin \; \theta \qquad\qquad\qquad (4.3)$

where $d = d'/n$.

The equation (4.3) is a convenient form of Bragg's law for usage. The meaning of this form can be illustrated as shown below. Fig. 4.5(a) shows the second order diffraction from (100) planes of a crystal, at a Bragg angle θ. Or, the path difference between rays 1' and 2' is equal to 2λ.

$$\therefore \qquad\qquad 2\lambda = 2 \; d_{100} \; Sin \; \theta \qquad\qquad\qquad (4.4)$$

Fig. 4.5(b) shows these rays and in addition a ray 3 which gets scattered by an imaginary plane which is present midway between the two (100) planes such that the interplanar distance between these new sets of imaginary planes is $d_{100}/2$. These sets of planes can be conveniently referred to as d_{200} planes. Further, the path difference between rays 1' and 3' is equal to 1λ by geometry. Thus, the rays 1' and 3' will be in phase with each other and constructively interfere and produce a diffracted beam. In other words,

$$1\lambda = 2d_{200} \; Sin \; \theta \qquad\qquad\qquad (4.5)$$

where d_{200} is the interplanar distance between the sets of imaginary planes (200). Equation (4.5) seems to be referring to a first order diffraction from imaginary (200) planes. But, this is in fact a second order diffraction from (100) planes. Thus, in general, a 'n' th order diffraction from (hkl) planes is equivalent to the

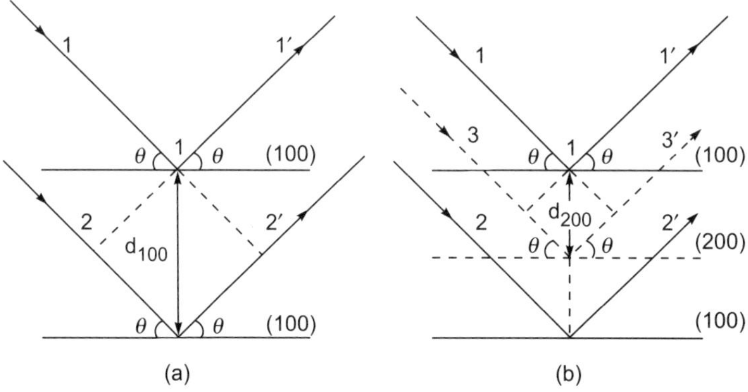

Fig. 4.5   (a) and (b)

first order diffraction from an imaginary sets of planes (nh nk nl) whose interplanar distance

$$d_{(nh\ nk\ nl)} = d_{(h\ k\ l)}/n.$$

It is convenient to treat all diffractions as first order diffractions.

It is to be noted that (i) the incident beam, the diffracted beam and the normal to the diffracting plane are all coplanar and (ii) the angle between the diffracted beam and the transmitted beam is 2θ and referred to as diffraction angle which is usually the quantity measured in diffraction experiments.

From the geometry of diffraction of x-rays, it looks as if it is similar to the reflection of visible light by mirrors. The angle of incidence in both cases is equal to the angle of reflection or diffraction. In fact, this is the reason that the term x-ray 'reflection' is used very commonly for the phenomenon of x-ray diffraction. Though the term reflection of x-rays is used, the phenomenon meant is diffraction of x-rays. X-rays can be totally reflected by a solid surface just like reflection of visible light by a mirror, but only at very small angles of incidence of the order one degree or less. It is useful to study the similarities and differences between visible light reflection and x-ray diffraction (reflection). The similarity seems to be only one, i.e., both occur when the angle of incidence is equal to the angle of reflection. The major differences between visible light reflection and x-ray diffraction are the following. (i) The reflection of visible light by a mirror is almost 100% efficient, but the intensity of the diffracted x-ray beam is extremely small compared to that of the incident beam. (ii) The reflection of visible light occurs from a thin surface layer of the mirror, but the diffracted beam is a sum total of scattered beams from large number of atoms of the crystal in the path of incident x-ray beam. (iii) The visible light reflection takes place at all angles of incidence. But diffraction phenomena of x-rays takes place only at definite angles called as Bragg angles which satisfy Bragg's law.

4.3 APPLICATION OF BRAGG'S LAW

Basically Bragg's law can be used to either analyze the wavelength of x-radiation or study the crystal structure. The analysis of wavelengths of x-radiation can be done by diffracting the radiations with the help of a set of planes of known interplanar spacing. Thus, if diffraction from a set of planes of crystal of known 'd' value occurs at an angle θ, then the wavelength λ of the radiation which causes diffraction can be determined, i.e, $\lambda = 2d \sin \theta$. The instrument used for this purpose is called the x-ray spectrometer.

The study of crystal structure can be done by making use of a beam of known wavelength x-rays which gets diffracted by the atomic planes of a crystal at the Bragg angle. Since, the wavelength, λ, is known and the Bragg angle, θ, can be measured, the d value can be determined utilizing Bragg's law. This is in fact, the basis of all fields of application of x-ray diffraction. In practice, there are three ways in which diffraction experiments may be conducted in order to study crystal structures. In Laue method, continuous (white) x-radiations are utilized for diffraction using a stationary single crystal. In rotating crystal method, a rotating single crystal diffracts a monochromatic x-ray beam. In the powder method a polycrystalline sample diffracts a monochromatic x-ray beam. These three methods differ from each other as shown below:

Methods	λ	θ
Laue method	Heterochromatic, variable	Fixed
Rotating crystal method	Monochromatic, fixed	Partly variable
Powder method	Monochromatic, fixed	Variable

4.3.1   The Laue Method

This was the original method of x-ray diffraction, used by Von Laue in 1912. This method utilizes heterochromatic or white radiation from an x-ray tube and a single crystal. The single crystal is fixed so that all the reflecting planes make fixed angles with respect to the incident heterochromatic x-rays. A particular set of atomic planes diffract x-rays of a particular wavelength which satisfy the Bragg's law. There are two variations of the technique. The transmission Laue technique consists of placing the crystal between the x-ray source and the photographic film. In the back reflection Laue technique, the incident x-ray beam passes through a hole in the photographic film and reaches the single crystal specimen. Fig. 4.6 shows both these techniques. In the transmission technique, the photographic film records the x-ray diffraction spots which occur on the transmission beam side of the specimen whereas in the back reflection technique, the diffraction spots on the incident beam side of the specimen are recorded as shown. In both cases, the diffraction spots are seen to lie on certain imaginary ellipses or hyperbolas.

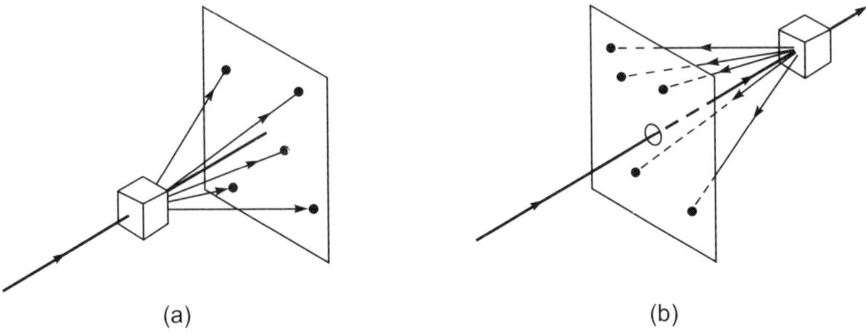

Fig. 4.6   (a) Transmission and (b) Back-reflection Laue methods
(Ref.: B.D. Cullity, Elements of x-ray Diffraction)

4.3.2   The Rotating Crystal Method

In this method a single crystal is mounted with one of its crystallographic axes normal to the direction of travel of a monochromatic x-ray beam. A cylindrical x-ray film is placed around the specimen crystal. The crystal is rotated about the chosen direction which coincide with the axis of the film. Here, the diffracted beams are located on imaginary cones whose axes coincide with the axis of rotation, as shown in Fig. 4.7. Here, a particular set of atomic planes are rotated about one particular axis which is the chosen crystallographic axis about which the crystal is rotated. Thus, the sets of crystallographic planes which are almost perpendicular to rotation axis are not in a position to give diffraction spots since in all rotational positions, Bragg angle θ can never be reached by them. The sets of crystallographic planes which contain the direction of rotation are able to give diffraction spots corresponding to nearly all possible orders of reflection because these planes go through all incident angles with respect to the incident x-ray beam right from $0°$ to $90°$. The remaining sets of crystallographic planes in the crystal are in a situation intermediate between these two extremes noted above. It is clear that all sets of planes are not able to participate in diffraction processes at all possible Bragg angles. Thus, it can be observed here that the Bragg angle is variable not completely but partially for most of the sets of planes in the crystal.

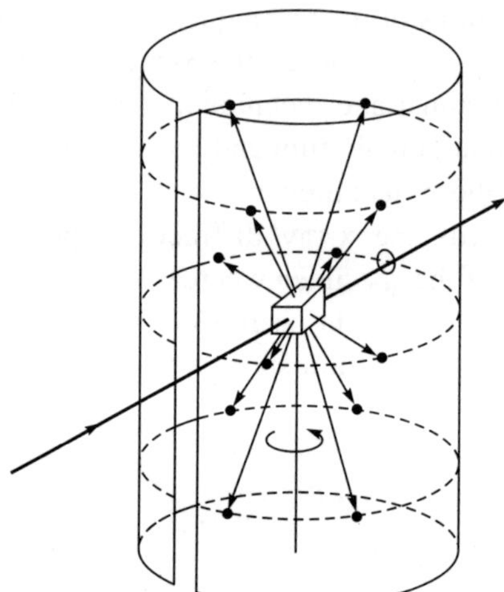

Fig. 4.7   Rotating-crystal method

4.3.3   The Powder Method

Here, the specimen is a polycrystalline material with a large number of tiny crystals oriented in all possible random directions with respect to the incident beam. The specimen can be reduced to powder in order to increase the number of crystals interacting with the incident x-ray beam. Thus, in this method all possible

Bragg angles for a given (hkl) plane can be encountered or in other words, all possible orders of (hkl) reflections can occur.

Fig. 4.8 shows a magnified cross-sectional view of the cylindrical powder specimen. An incident monochromatic x-ray beam is shown to strike the specimen at right angles to the cylindrical specimen axis. A set of (hkl) planes lying in one of the tiny crystals satisfies the Bragg's law and first order diffraction takes place in the direction shown at an angle 2θ with respect to the transmitted beam, where θ is the Bragg angle. If the set of (hkl) planes in Fig. 4.8 is rotated about the axis coincident with the incident beam direction, then it can be clearly visualized that the diffracted beams lie on the curved surface of a cone whose apex angle is four times the Bragg angle as shown. All these rotational orientation positions are available in the powder sample (without physically rotating the particular tiny crystal shown) as the incident beam encounters a large number of tiny crystals with all possible rotational Bragg angle positions mentioned above. So in the powder method, a given set of (hkl) planes produces a diffraction cone whose apex angle is 4θ. The interplanar distance 'd' of this particular (hkl) plane can then be calculated making use of Bragg's law provided the wavelength of the monochromatic x-ray beam is known. Most of the applications of x-ray diffraction are based on the powder method.

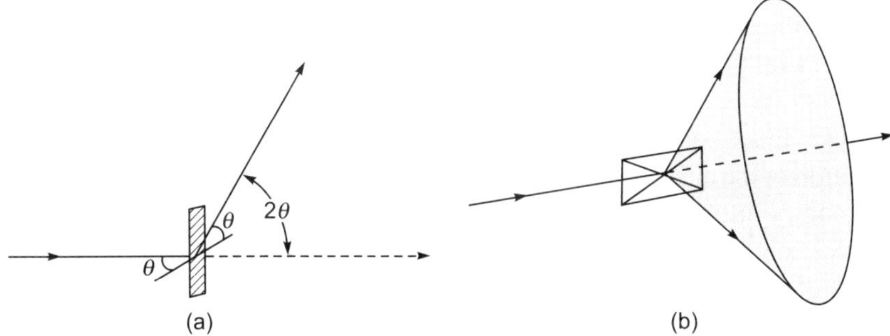

Fig. 4.8   Formation of a diffracted cone of radiation in the powder method (Ref.: B.D. Cullity, Elements of x-ray Diffraction)

4.4 DIFFRACTION INTENSITY

Bragg's law of diffraction merely refers to the direction of diffraction and does not refer to the intensity of the diffracted beam. The intensity of the diffracted beam depends on several factors like the atoms involved, the type of the crystal structure, the number and type of the reflecting planes involved, the geometry of the diffraction method used, the temperature during diffraction, etc., So it is possible that even though Bragg's law is satisfied, the diffraction intensity may be negligible or zero if the factors mentioned are not favourable. In such conditions the diffracted beam may not practically exist. In other words, it is not necessary that diffraction should occur whenever Bragg's law is satisfied. But, if Bragg's law is not satisfied, the diffraction can never occur. It is in this context that Bragg's law is said to be a negative law.

4.4.1 Polarization Factor and Electron Scattering

It has been noted earlier, that an x-ray beam is an electromagnetic wave with a sinusoidally varying electric field. Whenever, such an x-ray beam encounters a bound electron (shell electron other than valence or free electron) of an atom, the sinusoidally varying electric field of the x-ray beam sets this electron into oscillating motion about its mean position. The frequency of oscillation is exactly the same as the frequency of incident x-ray beam. Any oscillating (continuously accelerating and decelerating) charged particle emits the electromagnetic wave. In other words, the bound electron scatters the x-ray beam in all directions and the wavelength scattered is exactly equal to the wavelength of the incident beam, since, the frequencies of the incident beam and the oscillation of electron are the same. This scattering is referred to as coherent scattering. J. J. Thomson found that the intensity 'I' of the beam scattered by a single electron of charge 'e' and mass 'm', at a distance 'r' from the electron, is given by

$$I = I_0. \ e^4 \ \mathrm{Sin}^2 \ \alpha/(r^2.m^2.c^4). \tag{4.6}$$

where I_0 is the intensity of the incident beam, c is the velocity of visible light, A is a constant (10^{-14}) and α is the angle between the scattering direction and the direction of acceleration of the electron. Fig. 4.9 illustrates the geometry of coherent scattering by a bound electron. In this figure, the X axis is the direction of travel of the incident beam and OX represents the transmitted beam. The point O refers to the point where the electron is situated at the time of interaction with the x-ray beam. The electric vector 'E' of the x-ray beam is shown to lie in a random direction in

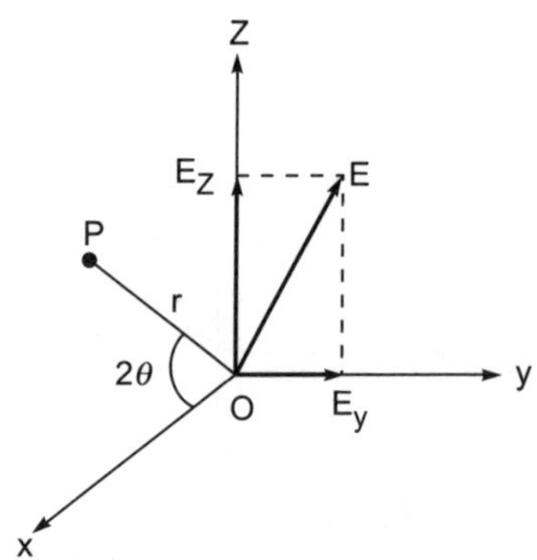

Fig. 4.9 Coherent scattering by a bound (shell) electron

the YOZ plane perpendicular to the direction of travel of the incident beam. The electric vector 'E' can be resolved as shown in the figure into vectors E_y and E_z in y and z axis directions which are at 90° to each other. It can be seen that $E^2 = E_y^2 + E_z^2$ and $E_y = E_z$ on the average as the direction of 'E' is perfectly random.

So,
$$E = E_y^2 + E_z^2 = E_y^2 + E_y^2 = 2 \ E_y^2$$

or
$$E_y^2 = E_z^2 = 1/2 \ E^2$$

The intensity of an x-ray beam is proportional to the square of its electric vector since this vector is a measure of the amplitude of the wave and since the intensity of a wave is proportional to the square of its amplitude.

Thus, $I_{OY} = I_{OZ} = 1/2\ I_O$, where I_{OY} and I_{OZ} are the intensities of the x-ray beam resolved in Y and Z direction respectively. The component I_{OY} of the incident beam accelerates the electron in the direction OY and the I_{OZ} accelerates the electron in the direction OZ. Let 'P' be a point at a distance 'r' from O in the XOZ plane and lying in the direction OP making an angle of 2θ with the transmitted beam direction OZ. The scattered beam intensity at P due to the intensity component I_{OY} of the incident beam by the electron at O is given by

$$I_{PY} = I_{OY}.e^4/(r^2.m^2.c^4)\ Sin^2(90°)\ .$$
$$= I_{OY}.e^4/(r^2.m^2.c^4)\ .\ A \quad \text{since, } Sin\ 90° = 0.$$

The scattered beam intensity at P due to the intensity component I_{OZ} of the incident beam by the electron at O is given by

$$I_{PZ} = I_{OZ}.e^4/(r^2.m^2.c^4)\ Sin^2(90° - 2\theta)\ .$$
$$= I_{OZ}.e^4/(r^2.m^2.c^4)\ Cos^2(2\theta)\ .$$

The total scattered intensity at P 'I_P' by the electron at O is then given by the summation the two intensities I_{PY} and I_{PZ}.

So,
$$I_P = I_{PY} + I_{PZ}$$
$$= e^4/(r^2.m^2.c^4)\ [I_{OY} + I_{OZ}\ Cos^2(2\theta)]\ .$$

But as seen earlier, $I_{OY} = I_{OZ} = I_O/2$.
Substituting this in the above equation,

$$I_P = e^4/(r^2.m^2.c^4).I_O/2.[1 + Cos^2(2\theta)]\ . \tag{4.7}$$

This is the Thomson equation for the scattering of the x-ray beam by a single electron. If the values of the constants are inserted in the equation, it can be seen that the intensity scattered is a very small fraction of the incident intensity. The absolute intensities of the scattered x-ray beam can be calculated by making use of the equation 4.7. The measurement and calculation of absolute intensities are difficult in practice. For most of the practical purposes, it is sufficient to know the relative intensities.

In the equation 4.7, all terms are constant for a given diffraction experiment except for the term $[1 + Cos^2(2\theta)]/2$. This term is referred to as the polarization factor. This factor becomes 1 when the angle 2θ is 0 or 180°. When the angle 2θ is 90°, the polarization factor reduces to 1/2. Thus, the scattered beam is stronger in both forward and backward direction than in a direction at right angles to the incident beam.

So far the discussion has been centered on Thomson scattering which is coherent scattering. But, there is another type of scattering phenomenon discovered by A. H. Compton and hence termed 'Compton scattering' which can occur when a valence or free electron encounters an x-ray beam. This can be understood by considering x-ray beam as a stream of x-ray photons of energy 'hv' where 'h' is Planck's constant and 'v' is their frequency. As shown in Fig. 4.10, the incident x-ray photon having energy 'hv_1' strikes the electrons and an elastic collision occurs between the x-ray photon and the free electron. The free electron is knocked out of the path and the x-ray photon is scattered at an angle 2θ as shown. In the process, some part of the incident energy of the x-ray photon is used up in providing kinetic energy to the free electron to move out of the path. Thus, the energy of the x-ray photon is now reduced to 'hv_2', where the new frequency v_2 is slightly less than the incident frequency v_1. Thus, the resulting wavelength λ_2 of the scattered beam is slightly greater than incident wavelength λ_1. The magnitude of the change in the wavelength is given by:

$$\lambda_2 - \lambda_1 = 0.03243\ (1 - \cos 2\theta) \tag{4.8}$$

Thus, the scattered beam, not only has a different wavelength compared to the incident one, but also there is no fixed phase relationship between the two. Thus, it is called the incoherent scattering or Compton modified scattering. This scattering cannot take part in diffraction because it cannot produce interference enhancement effect due to non-coherency. However, this contributes to the background radiation.

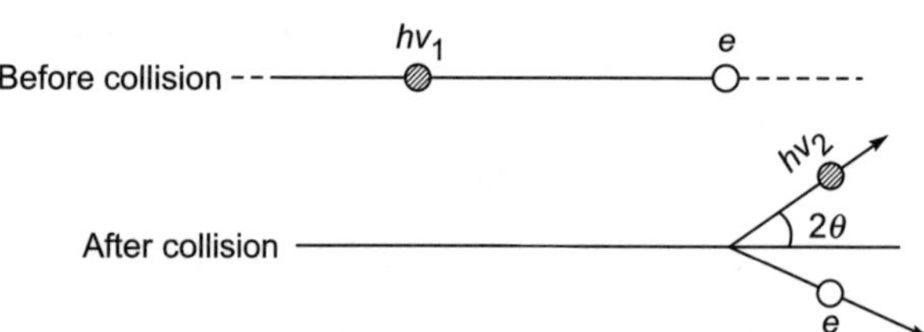

Fig. 4.10   Elastic collision of photon and electron

4.4.2   Structure Factor

When an x-ray beam strikes an atom, coherent (Thomson) scattering takes place from all the bound electrons of the atom in accordance with the Thomson equation. Nucleus of the atom is not able to take part in scattering since it cannot be made to oscillate to an appreciable extent due to its extremely large mass relative to an electron. Thomson equation shows that the scattering intensity is inversely proportional to the square of the mass of the scattering particle and the intensity scattered by nucleus is therefore negligible due to its large mass. In other words, scattering by the atom is the sum total of scattering by all the electrons present in the atom.

Fig. 4.11 shows the scattering by electrons in atom at a specific angle 2θ with respect to the transmitted x-ray beam. It is clear from the figure that the path difference between the two scattered rays from the two electrons shown is given by (CB-AD). This value being small compared to the wavelength of x-rays used for the purpose of diffraction, destructive interference takes place, the extent of

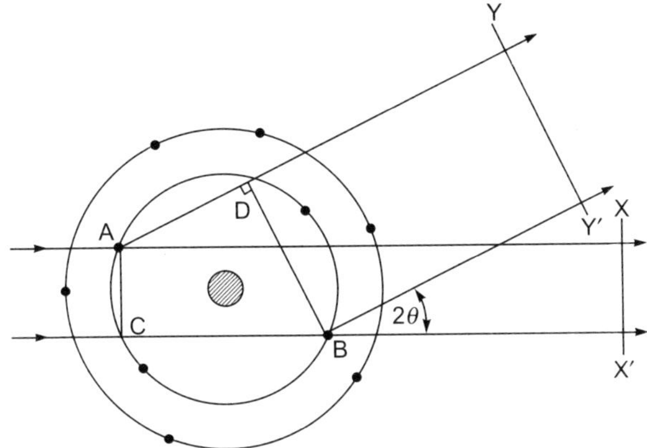

Fig. 4.11   X-ray scattering by an atom
(Ref.: B.D. Cullity, Elements of x-ray Diffraction)

which depends on the angle 2θ. The scattered rays in the forward direction ($2\theta = 0$) are completely in phase with each other whereas the rays scattered in all other directions are out of phase to different extents. As the angle 2θ increases, the path difference increases between the scattered rays and the intensity consequently decreases, because of increased "out-of-phaseness". The atomic scattering factor 'f' is defined as the ratio of the amplitude of the x-ray wave scattered by an atom (i.e., by all the electrons in the atom) to that scattered by a single electron:

$$f = \frac{\text{amplitude of the x - ray wave scattered by an atom}}{\text{amplitude of the x - ray wave scattered by one electron}}$$

As explained above, the factor $f = Z$ for an atom scattering in the forward direction ($2\theta = 0$) as Z is the atomic number indicating the number of electrons in a given atom. As 2θ increases, the factor 'f' keeps on decreasing because of increasing amount of destructive interference. Further, as the wavelength 'λ', of the incident beam, decreases, the out-of-phaseness (with respect to λ) of the scattered beams increases, for the same path difference (i.e., for a fixed 2θ) and consequently, the factor 'f' decreases. The variation of the atomic scattering factor 'f' of copper with 'Sin θ/λ' is shown in Fig. 4.12. It is seen here that as 'Sin θ/λ' increases the atomic scattering factor decreases as explained above. Table 4.1 gives the calculated values of the atomic scattering factors for various elemental atoms.

Diffraction by a unit cell of a crystal structure is the resultant of the waves scattered by all the atoms of the unit cell. This resultant wave can be obtained by summing up of the waves scattered by atoms present in the unit cell with due respect to their individual positions and the amplitudes. It has been already noted that the direction in which the diffraction takes place should satisfy Bragg's law. Fig. 4.13 shows the geometry of diffraction from a (h00) plane in an orthorhombic unit cell. The path difference (MC+CN), between rays 1′ and 2′

Table 4.1 Atomic scattering factors

$\frac{\sin\theta}{\lambda}\,(A^{-1})$	0.0	0.1	0.2	0.3	0.4	0.5	0.6	0.7	0.8	0.9	1.0	1.1	1.2
H	1	0.81	0.48	0.25	0.13	0.07	0.04	0.03	0.02	0.01	0.00	0.00	
He	2	1.88	1.46	1.05	0.75	0.52	0.35	0.24	0.18	0.14	0.11	0.09	
Li^+	2	1.96	1.8	1.5	1.3	1.0	0.8	0.6	0.5	0.4	0.3	0.3	
Li	3	2.2	1.8	1.5	1.3	1.0	0.8	0.6	0.5	0.4	0.3	0.3	
Be^{+2}	2	2.0	1.9	1.7	1.6	1.4	1.2	1.0	0.9	0.7	0.6	0.5	
Be	4	2.9	1.9	1.7	1.6	1.4	1.2	1.0	0.9	0.7	0.6	0.5	
B^{+3}	2	1.99	1.9	1.8	1.7	1.6	1.4	1.3	1.2	1.0	0.9	0.7	
B	5	3.5	2.4	1.9	1.7	1.5	1.4	1.2	1.2	1.0	0.9	0.7	
C	6	4.6	3.0	2.2	1.9	1.7	1.6	1.4	1.3	1.16	1.0	0.9	
N^{+5}	2	2.0	2.0	1.9	1.9	1.8	1.7	1.6	1.5	1.4	1.3	1.16	
N^{+3}	4	3.7	3.0	2.4	2.0	1.8	1.66	1.56	1.49	1.39	1.28	1.17	
N	7	5.8	4.2	3.0	2.3	1.9	1.65	1.54	1.49	1.39	1.29	1.17	
O	8	7.1	5.3	3.9	2.9	2.2	1.8	1.6	1.1	1.4	1.35	1.26	
O^{-2}	10	8.0	5.5	3.8	2.7	2.1	1.8	1.5	1.5	1.4	1.35	1.26	
F	9	7.8	6.2	4.45	3.35	2.65	2.15	1.9	1.7	1.6	1.5	1.35	
F^-	10	8.7	6.7	4.8	3.5	2.8	2.2	1.9	1.7	1.55	1.5	1.35	
Ne	10	9.3	7.5	5.8	4.4	3.4	2.65	2.2	1.9	1.65	1.55	1.5	
Na^+	10	9.5	8.2	6.7	5.25	4.05	3.2	2.65	2.25	1.95	1.75	1.6	
Na	11	9.65	8.2	6.7	5.25	4.05	3.2	2.65	2.25	1.95	1.75	1.6	
Mg^{+2}	10	9.75	8.6	7.25	5.95	4.8	3.85	3.15	2.55	2.2	2.0	1.8	
Mg	12	10.5	8.6	7.25	5.95	4.8	3.85	3.15	2.55	2.2	2.0	1.8	
Al^{+3}	10	9.7	8.9	7.8	6.65	5.5	4.45	3.65	3.1	2.65	2.3	2.0	
Al	13	11.0	8.95	7.75	6.6	5.5	4.5	3.7	3.1	2.65	2.3	2.0	
Si^{+4}	10	9.75	9.15	8.25	7.15	6.05	5.05	4.2	3.4	2.95	2.6	2.3	
Si	14	11.35	9.4	8.2	7.15	6.1	5.1	4.2	3.4	2.95	2.6	2.3	
P^{+5}	10	9.8	9.25	8.45	7.5	6.55	5.65	4.8	4.05	3.4	3.0	2.6	
P	15	12.4	10.0	8.45	7.45	6.5	5.65	4.8	4.05	3.4	3.0	2.6	
P^{-3}	18	12.7	9.8	8.4	7.45	6.5	5.65	4.85	4.05	3.4	3.0	2.6	
S^{+6}	10	9.85	9.4	8.7	7.85	6.85	6.05	5.25	4.5	3.9	3.35	2.9	
S	16	13.6	10.7	8.95	7.85	6.85	6.0	5.25	4.5	3.9	3.35	2.9	
S^{-2}	18	14.3	10.7	8.9	7.85	6.85	6.0	5.25	4.5	3.9	3.35	2.9	
Cl	17	14.6	11.3	9.25	8.05	7.25	6.5	5.75	5.05	4.4	3.85	3.35	
Cl^-	18	15.2	11.5	9.3	8.05	7.25	6.5	5.75	5.05	4.4	3.85	3.35	
A	18	15.9	12.6	10.4	8.7	7.8	7.0	6.2	5.4	4.7	4.1	3.6	
K^+	18	16.5	13.3	10.8	8.85	7.75	7.05	6.44	5.9	5.3	4.8	4.2	
K	19	16.5	13.3	10.8	9.2	7.9	6.7	5.9	5.2	4.6	4.2	3.7	3.3
Ca^{+2}	18	16.8	14.0	11.5	9.3	8.1	7.35	6.7	6.2	5.7	5.1	4.6	
Ca	20	17.5	14.1	11.4	9.7	8.4	7.3	6.3	5.6	4.9	4.5	4.0	3.6
Sc^{+3}	18	16.7	14.0	11.4	9.4	8.3	7.6	6.9	6.4	5.8	5.35	4.85	
Sc	21	18.4	14.9	12.1	10.3	8.9	7.7	6.7	5.9	5.3	4.7	4.3	3.9
Ti^{+4}	18	17.0	14.4	11.9	9.9	8.5	7.85	7.3	6.7	6.15	5.65	5.05	
Ti	22	19.3	15.7	12.8	10.9	9.5	8.2	7.2	6.3	5.6	5.0	4.6	4.2
V	23	20.2	16.6	13.5	11.5	10.1	8.7	7.6	6.7	5.9	5.3	4.9	4.4
Cr	24	21.1	17.4	14.2	12.1	10.6	9.2	8.0	7.1	6.3	5.7	5.1	4.6
Mn	25	22.1	18.2	14.9	12.7	11.1	9.7	8.4	7.5	6.6	6.0	5.4	4.9

$\frac{\sin\theta}{\lambda}$ (A^{-1})	0.0	0.1	0.2	0.3	0.4	0.5	0.6	0.7	0.8	0.9	1.0	1.1	1.2
Fe	26	23.1	18.9	15.6	13.3	11.6	10.2	8.9	7.9	7.0	6.3	5.7	5.2
Co	27	24.1	19.8	16.4	14.0	12.1	10.7	9.3	8.3	7.3	6.7	6.0	5.5
Ni	28	25.0	20.7	17.2	14.6	12.7	11.2	9.8	8.7	7.7	7.0	6.3	5.8
Cu	29	25.9	21.6	17.9	15.2	13.3	11.7	10.2	9.1	8.1	7.3	6.6	6.0
Zn	30	26.8	22.4	18.6	15.8	13.9	12.2	10.7	9.6	8.5	7.6	6.9	6.3
Ga	31	27.8	23.3	19.3	16.5	14.5	12.7	11.2	10.0	8.9	7.9	7.3	6.7
Ge	32	28.8	24.1	20.0	17.1	15.0	13.2	11.6	10.4	9.3	8.3	7.6	7.0
As	33	29.7	25.0	20.8	17.7	15.6	13.8	12.1	10.8	9.7	8.7	7.9	7.3
Se	34	30.6	25.8	21.5	18.3	16.1	14.3	12.6	11.2	10.0	9.0	8.2	7.5
Br	35	31.6	26.6	22.3	18.9	16.7	14.8	13.1	11.7	10.4	9.4	8.6	7.8
Kr	36	32.5	27.4	23.0	19.5	17.3	15.3	13.6	12.1	10.8	9.8	8.9	8.1
Rb^+	36	33.6	28.7	24.6	21.4	18.9	16.7	14.6	12.8	11.2	9.9	8.9	
Rb	37	33.5	28.2	23.8	20.2	17.9	15.9	14.1	12.5	11.2	10.2	9.2	8.4
Sr	38	34.4	29.0	24.5	20.8	18.4	16.4	14.6	12.9	11.6	10.5	9.5	8.7
Y	39	35.4	29.9	25.3	21.5	19.0	17.0	15.1	13.4	12.0	10.9	9.9	9.0
Zr	40	36.3	30.8	26.0	22.1	19.7	17.5	15.6	13.8	12.4	11.2	10.2	9.3
Nb	41	37.3	31.7	26.8	22.8	20.2	18.1	16.0	14.3	12.8	11.6	10.6	9.7
Mo	42	38.2	32.6	27.6	23.5	20.8	18.6	16.5	14.8	13.2	12.0	10.9	10.0
Tc	43	39.1	33.4	28.3	24.1	21.3	19.1	17.0	15.2	13.6	12.3	11.3	10.3
Ru	44	40.0	34.3	29.1	24.7	21.9	19.6	17.5	15.6	14.1	12.7	11.6	10.6
Rh	45	41.0	35.1	29.9	25.4	22.5	20.2	18.0	16.1	14.5	13.1	12.0	11.0
Pd	46	41.9	36.0	30.7	26.2	23.1	20.8	18.5	16.6	14.9	13.6	12.3	11.3
Ag	47	42.8	36.9	31.5	26.9	23.8	21.3	19.0	17.1	15.3	14.0	12.7	11.7
Cd	48	43.7	37.7	32.2	27.5	24.4	21.8	19.6	17.6	15.7	14.3	13.0	12.0
In	49	44.7	38.6	33.0	28.1	25.0	22.4	20.1	18.0	16.2	14.7	13.4	12.3
Sn	50	45.7	39.5	33.8	28.7	25.6	22.9	20.6	18.5	16.6	15.1	13.7	12.7
Sb	51	46.7	40.4	34.6	29.5	26.3	23.5	21.1	19.0	17.0	15.5	14.1	13.0
Te	52	47.7	41.3	35.4	30.3	26.9	24.0	21.7	19.5	17.5	16.0	14.5	13.3
I	53	48.6	42.1	36.1	31.0	27.5	24.6	22.2	20.0	17.9	16.4	14.8	13.6
Xe	54	49.6	43.0	36.8	31.6	28.0	25.2	22.7	20.4	18.4	16.7	15.2	13.9
Cs	55	50.7	43.8	37.6	32.4	28.7	25.8	23.2	20.8	18.8	17.0	15.6	14.5
Ba	56	51.7	44.7	38.4	33.1	29.3	26.4	23.7	21.3	19.2	17.4	16.0	14.7
La	57	52.6	45.6	39.3	33.8	29.8	26.9	24.3	21.9	19.7	17.9	16.4	15.0
Ce	58	53.6	46.5	40.1	34.5	30.4	27.4	24.8	22.4	20.2	18.4	16.6	15.3
Pr	59	54.5	47.4	40.9	35.2	31.1	28.0	25.4	22.9	20.6	18.8	17.1	15.7
Nd	60	55.4	48.3	41.6	35.9	31.8	28.6	25.9	23.4	21.1	19.2	17.5	16.1
Pm	61	56.4	49.1	42.4	36.6	32.4	29.2	26.4	23.9	21.5	19.6	17.9	16.4
Sm	62	57.3	50.0	43.2	37.3	32.9	29.8	26.9	24.4	22.0	20.0	18.3	16.8
Eu	63	58.3	50.9	44.0	38.1	33.5	30.4	27.5	24.9	22.4	20.4	18.7	17.1
Gd	64	59.3	51.7	44.8	38.8	34.1	31.0	28.1	25.4	22.9	20.8	19.1	17.5
Tb	65	60.2	52.6	45.7	39.6	34.7	31.6	28.6	25.9	23.4	21.2	19.5	17.9
Dy	66	61.1	53.6	46.5	40.4	35.4	32.2	29.2	26.3	23.9	21.6	19.9	18.3
Ho	67	62.1	54.5	47.3	41.1	36.1	32.7	29.7	26.8	24.3	22.0	20.3	18.6
Er	68	63.0	55.3	48.1	41.7	36.7	33.3	30.2	27.3	24.7	22.4	20.7	18.9
Tm	69	64.0	56.2	48.9	42.4	37.4	33.9	30.8	27.9	25.2	22.9	21.0	19.3

Contd.

$\dfrac{\sin\theta}{\lambda}$ (A^{-1})	0.0	0.1	0.2	0.3	0.4	0.5	0.6	0.7	0.8	0.9	1.0	1.1	1.2
Yb	70	64.9	57.0	49.7	43.2	38.0	34.4	31.3	28.4	25.7	23.3	21.4	19.7
Lu	71	65.9	57.8	50.4	43.9	38.7	35.0	31.8	28.9	26.2	23.8	21.8	20.0
Hf	72	66.8	58.6	51.2	44.5	39.3	35.6	32.3	29.3	26.7	24.2	22.3	20.4
Ta	73	67.8	59.5	52.0	45.3	39.9	36.2	32.9	29.8	27.1	24.7	22.6	20.9
W	74	68.8	60.4	52.8	46.1	40.5	36.8	33.5	30.4	27.6	25.2	23.0	21.3
Re	75	69.8	61.3	53.6	46.8	41.1	37.4	34.0	30.9	28.1	25.6	23.4	21.6
Os	76	70.8	62.2	54.4	47.5	41.7	38.0	34.6	31.4	28.6	26.0	23.9	22.0
Ir	77	71.7	63.1	55.3	48.2	42.4	38.6	35.1	32.0	29.0	26.5	24.3	22.3
Pt	78	72.6	64.0	56.2	48.9	43.1	39.2	35.6	32.5	29.5	27.0	24.7	22.7
Au	79	73.6	65.0	57.0	49.7	43.8	39.8	36.2	33.1	30.0	27.4	25.1	23.1
Hg	80	74.6	65.9	57.9	50.5	44.4	40.5	36.8	33.6	30.6	27.8	25.6	23.6
Tl	81	75.5	66.7	58.7	51.2	45.0	41.1	37.4	34.1	31.1	28.3	26.0	24.1
Pb	82	76.5	67.5	59.5	51.9	45.7	41.6	37.9	34.6	31.5	28.8	26.4	24.5
Bi	83	77.5	68.4	60.4	52.7	46.4	42.2	38.5	35.1	32.0	29.2	26.8	24.8
Po	84	78.4	69.4	61.3	53.5	47.1	42.8	39.1	35.6	32.6	29.7	27.2	25.2
At	85	79.4	70.3	62.1	54.2	47.7	43.4	39.6	36.2	33.1	30.1	27.6	25.6
Rn	86	80.3	71.3	63.0	55.1	48.4	44.0	40.2	36.8	33.5	30.5	28.0	26.0
Fr	87	81.3	72.2	63.8	55.8	49.1	44.5	40.7	37.3	34.0	31.0	28.4	26.4
Ra	88	82.2	73.2	64.6	56.5	49.8	45.1	41.3	37.8	34.6	31.5	28.8	26.7
Ac	89	83.2	74.1	65.5	57.3	50.4	45.8	41.8	38.3	35.1	32.0	29.2	27.1
Th	90	84.1	75.1	66.3	58.1	51.1	46.5	42.4	38.8	35.5	32.4	29.6	27.5
Pa	91	85.1	76.0	67.1	58.8	51.7	47.1	43.0	39.3	36.0	32.8	30.1	27.9
U	92	86.0	76.9	67.9	59.6	52.4	47.7	43.5	39.8	36.5	33.3	30.6	28.3
Np	93	87	78	69	60	53	48	44	40	37	34	31	29
Pu	94	88	79	69	61	54	49	44	41	38	34	31	29
Am	95	89	79	70	62	55	50	45	42	38	35	32	30
Cm	96	90	80	71	62	55	50	46	42	39	35	32	30
Bk	97	91	81	72	63	56	51	46	43	39	36	33	30
Cf	98	92	82	73	64	57	52	47	43	40	36	33	31
	99	93	83	74	65	57	52	48	44	40	37	34	31
	100	94	84	75	66	58	53	48	44	41	37	34	31

From X-Ray Diffraction by Polycrystalline Materials, edited by H.S. Peiser, H.P. Rooksby, and A.J.C. Wilson (The Institute of Physics, London, 1955).

scattered from the atoms A and C respectively, is given by λ, the wavelength of the incident x-rays, i.e., $2d_{h00}\,Sin\theta = \lambda$ from Bragg's Law, where θ is the Bragg angle. If A is the origin, from the definition of the Miller indices the interplanar distance, $d_{h00}(AC) = a/h$ where 'a' is the lattice parameter of the orthorhombic unit cell in the x direction. An atom B present in the unit cell at a distance 'x' from A along AC (x axis) affects the (h00) reflection considered. [For example, if 'x' = AC/2, then the path difference between 1′ and 3′ is equal to one half of the path difference between 1′ and 2′ by geometry. Since, the path difference between 1′ and 2′ is 'λ', the path difference between 1′ and 3′ is $\lambda/2$. Thus, there would be destructive interference and the intensity of diffracted beam would be zero. So, it can be

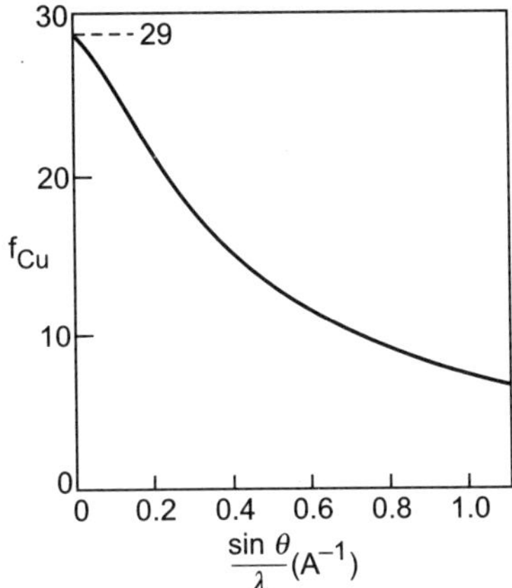

Fig. 4.12 The atomic scattering factor of copper
(Ref.: B.D. Cullity, Elements of X-ray Diffraction)

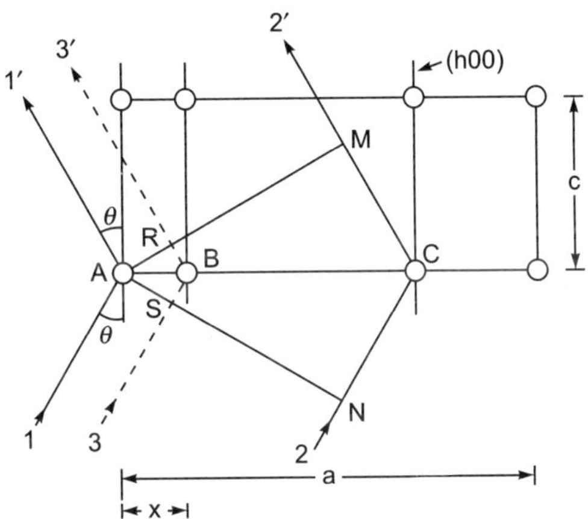

Fig. 4.13 The effect of atom position on the phase
difference between diffracted rays
(Ref.: B.D. Cullity, Elements of X-ray Diffraction)

visualized that the position of the atom B in the structure decides the intensity of the diffracted beam]. From the Fig. 4.13, it can be seen that the path difference between 1′ and 2′ is equal to λ. But AB = x and AC = a/h.

Therefore, the path difference between 1′ and 3′ $= \dfrac{x}{a/h}\,\lambda$.

This path difference can be expressed as phase angle difference ϕ by the relation, path difference $\times 2\pi/\lambda$ = phase difference ϕ. Since, a path difference of λ is equivalent to a phase difference of 2π.

The phase difference between 1′ and 3′ $= \phi_{1'3'} = \dfrac{x}{a/h} \times 2\pi$

$$= 2\pi \, \frac{h\,x}{a}$$

But x/a is the fractional coordinate position 'u' of atom B.

$$\therefore \qquad\qquad \phi_{1'3'} = 2\,\pi\,hu.$$

The same argument can be extended to a general hkl reflection in three dimensions from an atom B located at fractional indices positions u, v, w. Fractional indices v and w are defined as v = y/b and w = z/c, where x, y, z are coordinates of B; and b and c are lattice parameters in y and z directions respectively. The phase difference in such a case, between the wave scattered by atom B and that by atom A, for the hkl reflection, is given by

$$\phi = 2\pi(hu + kv + lw) \qquad\qquad (4.9)$$

This is a general relation applicable to a unit cell of any crystal structure.

The scattered rays from A and B, not only may differ in phase, but also in amplitude if atom B is different from atom A. The amplitudes of the waves scattered by A and B are related to their atomic scattering factors as seen earlier.

Thus, the diffracted beam from a unit cell is made up of the sum total of waves scattered by all the atoms in the unit cell. The waves scattered by the atoms in the unit cell may have different phase angles and different amplitudes depending on their fractional coordinates (uvw) and scattering factor (f). So, a summation of waves having different amplitudes and phase angles have to be carried out in order to determine the diffracted wave from a unit cell. To do this, a convenient analytical method would be to consider each scattered wave as a vector in the complex plane as shown in Fig. 4.14. A complex plane is a plane containing an axis of real numbers 1, 2, 3, etc. and an axis of imaginary numbers i (that is = $\sqrt{-1}$), 2i, 3i, etc. as shown. To find an expression for the wave vector, the same can be plotted in the complex plane as shown in Fig. 4.14. The length of the vector 'V' can be drawn to represent the amplitude value of the wave and the angle ϕ, the phase angle of it. The analytical expression for the wave can now be written as (V cos ϕ + iV sin ϕ). This expression is a complex number, i.e., a summation of a real number V cos ϕ and an imaginary number, iV sin ϕ, since i = imaginary number, $\sqrt{-1}$. From the power series expansions of cos x, sin x and e^{ix}, it can be observed that

$$\cos x + i \sin x = e^{ix}$$

$$\therefore \qquad\qquad V \cos \phi + iV \sin \phi = V\,(\cos \phi + i \sin \phi) = V\,.\,e^{i\varphi}.$$

So, any scattered wave with an amplitude V and a phase angle Φ can be expressed as a complex exponential function $V.e^{i\varphi}$. The amplitude of the wave scattered by an atom is proportional to its atomic scattering factor 'f' and for all practical purpose, 'f' can be used in calculations as the relative amplitude of the scattered wave from the atom considered. The phase angle of scattering by an

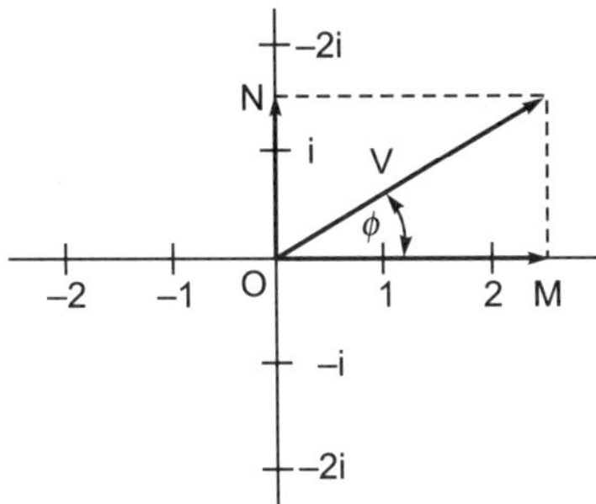

Fig. 4.14   A wave vector in the complex plane

atom in the unit cell at fractional indices position u, v, w is given by $\Phi = 2\pi(hu+kv+lw)$ as deduced earlier. The wave scattered by this atom can now be considered a vector $f.e^{2\pi i(hu+kv+lw)}$, since here $V = f$ and $\Phi = 2\pi(hu+kv+lw)$.

As discussed earlier, the hkl reflection from the unit cell is a summation of all the waves scattered by all the atoms in the unit cell. This resultant wave from a unit cell is called the structure factor 'F'.

$$F = f_1.e^{\,2\pi i(hu_1+kv_1+lw_1)} + f_2.e^{2\pi i(hu_2+kv_2+lw_2)} + \ldots\ldots\ldots$$

where f_1, f_2, etc. are the atomic scattering factors of atoms with fractional indices positions $(u_1\ v_1\ w_1)$ and $(u_2\ v_2\ w_2)$, etc. respectively. This can be rewritten as

$$F_{hkl} = \sum_{1}^{n} f_n.e^{\,2\pi i(hu_n+kv_n+lw_n)} \tag{4.10}$$

where 'n' is the number of atoms present in the unit cell.

The absolute value of F, i.e., $|F|$ gives the amplitude of the resultant wave. $|F|$ can be defined as

$$|F| = \frac{\text{amplitude of the wave scattered by all atoms of unit cell}}{\text{amplitude of the x-ray wave scattered by one electron}}$$

The diffracted beam intensity is proportional to $|F|^2$ since the intensity of a wave is proportional to the square of its amplitude.

The structure factor is a very important factor controlling the intensity of the diffracted beam from a given crystal structure. If this factor is favourable the intensity of the diffracted beam can be high and if unfavourable, the intensity can be zero. Under situation where the intensity of diffraction reduces to zero, the particular hkl reflection may not occur even if the Bragg's law is satisfied. This can be clearly understood by working out the structure factors for some simple crystal structures.

(i) Consider the case of a cubic unit cell having eight atoms at the eight corners of the unit cell which is a simple cubic unit cell. In this case there is only one atom per unit cell since the contribution of each of the 8 corner atoms to the unit cell is only 1/8 [i.e., $8 \times 1/8 = 1$]. This atom can conveniently be chosen to be present at the origin. So, the fractional indices position for this atom is (000). Making use of the equation 4.10, the structure factor

$$F_{hkl} = f.e^{\,2\pi i(h.0\, +\, k.0\, +\, l.0)}$$

where   f = atomic scattering factor of the atoms considered and
hkl = Miller indices of the reflecting plane.

Therefore, $F_{hkl} = f.\,e^0 = f.$

Intensity of any hkl reflection is proportional to the square of its amplitude which in turn is proportional to the mod of the structure factor:

So, $I \propto \left|F\right|^2 \propto f^2.$

So intensity for all the hkl reflections are the same in this structure and it is independent of hkl values.

(ii) In a body centered cubic cell, there are two atoms per cell and the fractional indices for one can be selected as cube corner atom position (000) and for the other body centered atom as (1/2, 1/2, 1/2). If both of these atoms are of the same kind with atomic scattering factor f, then

$$F_{hkl} = f.e^{2\pi i(h.0\, +\, k.0\, +\, l.0)} + f.e^{2\pi i(h.1/2\, +\, k.1/2\, +\, l.1/2)}$$
$$= f.[1 + e^{\,\pi i(h+k+l)}].$$

But, it is known that

$$e^{\,\pi i} = \cos \pi + i \sin \pi \quad [\text{since } e^{ix} = \cos x + i.\sin x]$$
$$= -1 + 0 = -1.$$

In general, it can be observed that

$$e^{\,\pi i.n} = (-1)^n \text{ where n is an integer.}$$

So, for a bcc structure

$$F_{hkl} = f\,\{1 + e^{\,\pi i(h+k+l)}\}$$
$$= f\,\{1 + 1\}$$
$$= 2f, \quad \text{if (h+k+l) is an even integer.}$$

and $\left|F\right|^2 = 4f^2.$

$$F_{hkl} = f\,\{1 - 1\}$$
$$= 0 \quad \text{if (h+k+l) is an odd integer.}$$

In other words, intensities of diffracted beams from (100), (111), (210), (320) and other planes, where the indices add up to odd integers, are all equal to zero. And the intensities of diffracted beams from all other planes are proportional to $4f^2$.

(iii) In a base centered cell, shown in Fig. 4.15, the contribution of the two base centered atoms is half each or totally one to the unit cell. Thus, a total of two atoms per unit cell exists in a base centered structure considering the contribution of one atom by eight corner atoms. The fractional indices of the positions for these can be conveniently taken as 000 and 1/2 1/2 0. Thus, the structure factor can be written as (if all the atoms are of the same kind with atomic scattering factor f),

$$F_{hkl} = f.e^{2\pi i(h.0 + k.0 + l.0)} + f.e^{2\pi i(h.1/2 + k.1/2 + l.0)}$$
$$= f.[1 + e^{\pi i(h+k)}].$$

It can be seen that if h and k both are odd or both are even integers (that is, h and k are unmixed indices), then the term (h+k) becomes an even integer in which case

$$F = 2f, \text{ or intensity, } I \alpha 4f^2$$

In case of mixed indices, i.e., when one of the indices (h or k) is odd and the other is even, then the term (h+k) becomes odd, and $F = f.(1-1) = 0$, or the intensity $I = 0$.

Further, it can be seen that the intensity is independent of l value.

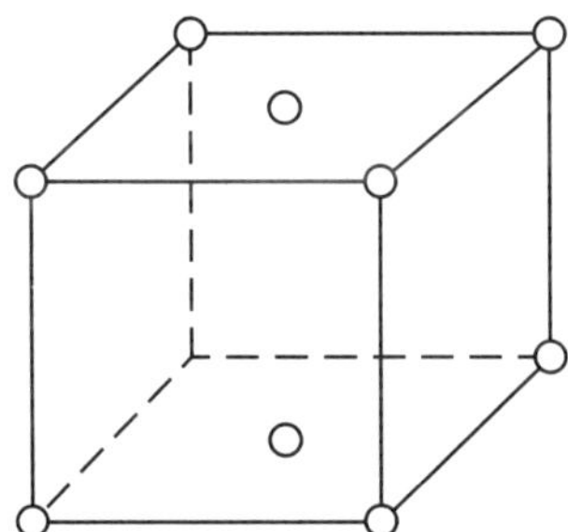

Fig. 4.15 Base centered cell

(iv) For a face centered cell, the number of atoms per unit cell is four, considering the contribution of one atom from eight corner atoms and three atoms from six face centered atoms. The fractional indices for the positions of these atoms can be taken as (000), (½ ½ 0), (½ 0 ½) and (0 ½ ½). If all the atoms in the structure are of the same kind with atomic scattering factor f, then the structure factor can be written as

$$F = f.e^{2\pi i(h.0 + k.0 + l.0)} + f.e^{2\pi i(h.1/2 + k.1/2 + l.0)}$$
$$= f.e^{2\pi i(h.1/2 + k.0 + l.1/2)} + f.e^{2\pi i(h.0 + k.1/2 + l.1/2)}$$
$$F = f.\{1 + e^{\pi i(h+k)} + e^{\pi i(k+l)} + e^{\pi i(l+h)}\}$$

It can be observed that if the reflecting plane indices hkl are all odd or all even (i.e., unmixed) integers, then the terms (h+k), (k+l) and (l+h) are all even integers and the structure factor

$$F = f\,(1+1+1+1) = 4f$$

and the diffraction intensity $I \propto |F|^2 = 16f^2$.

Further, if hkl are mixed indices, it can be seen that

$$F = 0 \text{ and the diffraction intensity} = 0.$$

(v) Consider the CsCl structure wherein Cs^+ ions are present at the cube corners (eight of them in all) of the unit cell and one Cl^- ion at the center of the body of the cube. Thus, the structure has one Cs^+ ion and one Cl^- ion per unit cell. The fractional indices of the position of the Cs ion can be taken as 000 and the those of the Cl ion can be taken as 1/2 1/2 1/2. Thus, the structure factor

$$F = f_{Cs}.\ e^{2\pi i(h.0 + k.0 + l.0)} + f_{Cl}.\ e^{2\pi i(h.1/2 + k.1/2 + l.1/2)}$$

where f_{Cs} is the scattering factor of Cs^+ ion
and f_{Cl} is the scattering factor of Cl^- ion.

Simplifying,

$$F = f_{Cs} + f_{Cl}.\ e^{\pi i(h+k+l)}$$
$$= f_{Cs} + f_{Cl}, \text{ if } (h+k+l) \text{ is an even integer,}$$
$$= f_{Cs} - f_{Cl}, \text{ if } (h+k+l) \text{ is an odd integer.}$$

It can be noted from the above that though this structure resembles the body centered one referred to in section (ii), but unlike it, the reflections are present even when (h+k+l) is an odd integer. But the intensities of these planes, the indices of which sum up to an odd value, are very low since the intensities for these planes are proportional to $|F|^2$ which in turn is equal to $(f_{Cs} - f_{Cl})^2$. So, here all hkl reflections are present like a simple structure considered in section (i).

(vi) The NaCl structure is cubic with 4 Na^+ ions at positions with fractional indices (000), (½ ½ 0), (0 ½ ½) and (½ 0 ½) and 4 Cl^- ions at (½ ½ ½), (0 0 ½), (½ 0 0) and (0 ½ 0).

These fractional indices positions can be referred to as 000 face centering translations for Na and 1/2 1/2 1/2 face centering translations for Cl ions. If the scattering factors of Na and Cl are f_{Na} and f_{Cl} respectively,

$$F = f_{Na}.\{1 + e^{\pi i(h+k)} + e^{\pi i(k+l)} + e^{\pi i(l+h)}\}$$
$$+ f_{Cl}.\{ e^{\pi i(h+k+l)} + e^{\pi i(l)} + e^{\pi i(h)} + e^{\pi i(k)}\}$$

This can be rewritten as

$$F = f_{Na}.\{1 + e^{\pi i(h+k)} + e^{\pi i(k+l)} + e^{\pi i(l+h)}\}$$
$$+ f_{Cl}.e^{\pi i(h+k+l)}.\{1 + e^{\pi i(-h-k)} + e^{\pi i(-k-l)} + e^{\pi i(-l-h)}\}$$

Since $e^{n\pi i} = e^{-n\pi i}$, the above expression can be simplified as

$$F = \{f_{Na} + f_{Cl}.e^{\pi i(h+k+l)}\}.[1 + e^{\pi i(h+k)} + e^{\pi i(k+l)} + e^{\pi i(l+h)}]$$

It is clear that the terms in the second bracket are the same as those obtained in the case of face centered cell discussed in (iv). Thus,

$$F = 0 \text{ for mixed hkl indices.}$$

For the case of unmixed indices the second bracket reduces to 4.

So, $F = 4.\{f_{Na} + f_{Cl}.e^{\pi i(h+k+l)}\}$ for unmixed indices hkl.

Further, for the unmixed indices case, when (h+k+l) is an even integer,

$$F = 4\,\{f_{Na} + f_{Cl}\} \text{ and the intensity I } \alpha\ 16\{f_{Na} + f_{Cl}\}^2$$

and when (h+k+l) is odd integer,

$$F = 4\,\{f_{Na} - f_{Cl}\} \text{ and the intensity I } \alpha\ 16\{f_{Na} - f_{Cl}\}^2$$

This structure is similar to face centered cell, but the intensities of reflections from some of the unmixed indices planes {where (h+k+l) is an odd integer} are decreased and some others {where (h+k+l) is an even integer} are increased.

(vii) Consider the cubic unit cell of diamond in which there are eight carbon atoms per unit cell. Let f_C be the scattering factor of the carbon atom. The fractional indices of the positions of carbon atoms are

$$000,\ 1/2\ 1/2\ 0,\ 0\ 1/2\ 1/2,\ 1/2\ 0\ 1/2,\ \text{and}$$

$$1/4\ 1/4\ 1/4,\ 3/4\ 3/4\ 1/4,\ 1/4\ 3/4\ 3/4,\ 3/4\ 1/4\ 3/4.$$

So,
$$F = f_C.\{1 + e^{\pi i(h+k)} + e^{\pi i(k+l)} + e^{\pi i(l+h)}\} + f_C.\{e^{\pi i(h+k+l)/2} + e^{\pi i(3h+3k+l)/2} + e^{\pi i(h+3k+3l)/2} + e^{\pi i(3h+k+3l)/2}\}$$

or, rewriting,

$$F = f_C.\{1 + e^{\pi i(h+k)} + e^{\pi i(k+l)} + e^{\pi i(l+h)}\}$$
$$+ f_C.\,e^{\pi i(h+k+l)/2}.\{1 + e^{\pi i(h+k)} + e^{\pi i(k+l)} + e^{\pi i(l+h)}\}$$

Taking the terms of the entire bracket {} as a common factor,

$$F = f_C.\{1 + e^{\pi i(h+k+l)/2}\}.\{1 + e^{\pi i(h+k)} + e^{\pi i(k+l)} + e^{\pi i(l+h)}\}$$

The second bracket {} reduces to zero under mixed hkl indices and to 4 under unmixed hkl indices.

Thus, F = 0 when hkl are mixed indices and the intensities of corresponding reflections are zero,

and $\quad F = 4f_C.\{1 + e^{\pi i(h+k+l)/2}\}$ when hkl are unmixed indices.

The exponent term has (h+k+l)/2 as a factor of π, which can be a fraction. To overcome this difficulty, another approach can be used. The intensity of a wave, $Ve^{i\varphi}$, is proportional to the square of the amplitude V, and can be obtained by multiplying the wave vector by its complex conjugate. The complex conjugate of the wave vector $A.e^{i\varphi}$ is $A.e^{-i\varphi}$, which is obtained by replacing i by –i.

i.e.,
$$Ve^{i\varphi} \times Ae^{-i\varphi} = V^2$$

or, wave vector $\times$ its complex conjugate $= V^2$

If this approach is used in the present case,

$$|F_{hkl}|^2 = 4\ f_C.\{1 + e^{\pi i(h+k+l)/2}\} \times 4f_C.\{1 + e^{-\pi i(h+k+l)/2}\}$$
$$= 16\ f_C^2.\{1 + e^{\pi i(h+k+l)/2} + 1 + e^{-\pi i(h+k+l)/2}\}$$

But since $e^{ix} + e^{-ix} = 2\cos x$,

$$|F_{hkl}|^2 = 16f_C^2.\{2 + 2\cos \pi/2.(h+k+l)\}$$
$$= 32f_C^2.\{1 + \cos \pi/2.(h+k+l)\}$$
$$= 32f_C^2 \text{ when } (h+k+l) \text{ is an odd integer,}$$
$$= 64f_C^2 \text{ when } (h+k+l) \text{ is an even multiple of two and}$$
$$= 0 \text{ when } (h+k+l) \text{ is an odd multiple of two.}$$

(viii) Zinc blende form of ZnS a structure almost similar to diamond wherein in place of eight carbon atoms, there are four Zn atoms at 000 face centering translations and four S atoms at 1/4 1/4 1/4 face centering translations.

From the example of the previous structure, the structure factor can be written as

$$F = f_{Zn}.\{1 + e^{\pi i(h+k)} + e^{\pi i(k+l)} + e^{\pi i(l+h)}$$
$$+ f_S.\ e^{\pi i(h+k+l)/2}.\{1 + e^{\pi i(h+k)} + e^{\pi i(k+l)} + e^{\pi i(l+h)}\}$$

Taking the terms of the entire bracket $\{\ \}$ as a common factor,

$$F = \{f_{zn} + f_S.\ e^{\pi i(h+k+l)/2}\}.\{1 + e^{\pi i(h+k)} + e^{\pi i(k+l)} + e^{\pi i(l+h)}\}.$$

The second bracket $\{\ \}$ reduces to zero under mixed hkl indices and to 4 under unmixed hkl indices.

Thus, $F = 0$ when hkl are mixed indices and the intensities of corresponding reflections are zero,

and $$F = 4.\{f_{zn} + f_S.\ e^{\pi i(h+k+l)/2}\} \text{ when hkl are unmixed indices.}$$

$|F|^2$ is obtained by multiplication of this expression by its complex conjugate:

$$|F|^2 = 16.\{f_{zn} + f_S.\ e^{\pi i(h+k+l)/2}\}.\{f_{zn} + f_S.\ e^{\pi -i(h+k+l)/2}\}$$

On term by term multiplication,

$$|F|^2 = 16.\{ f_{zn}^2 + f_s^2 + 2f_{zn}.f_s.\cos \pi/2(h+k+l)\}$$

On simplification,

$$|F|^2 = 16.\{f_{zn}^2 + f_s^2 \} \text{ when } (h+k+l) \text{ is an odd integer,}$$

$$|F|^2 = 16.\{f_{zn} - f_s\}^2 \text{ when } (h+k+l) \text{ is an odd multiple of two,}$$

$$|F|^2 = 16.\{f_{zn} + f_s\}^2 \text{ when } (h+k+l) \text{ is an even multiple of two.}$$

(ix) A close packed hexagonal (cph or hcp) structure may be considered as one having two atoms located at positions with the fractional indices (000) and (1/3 2/3 1/2).

The structure factor for this cell

$$F = f.e^{2\pi i(0)} + f.e^{2\pi i(h/3 + 2k/3 + 1/2)}$$
$$F = f.\{1 + e^{2\pi i[(h+2k)/3 + 1/2]}\}$$

Put $\left[\dfrac{(h + 2k)}{3} + \dfrac{1}{2}\right] = p$ for convenience,

$$F = f.\{1 + e^{2\pi ip}\}.$$

It can be readily observed that p may take fractional values and this expression remains complex. Multiplying by its complex conjugate:

$$|F|^2 = f^2 .\{1 + e^{2\pi ip}\}.\{1 + e^{-2\pi ip}\}$$

$$|F|^2 = f^2 .\{2 + e^{2\pi ip} + e^{-2\pi ip}\}$$

Converting this into a cosine expression as done in (vii),

$$|F|^2 = f^2 .\{2 + 2 \cos 2\pi p\}$$
$$= f^2 .\{2 + 2 (2 \cos^2 \pi p - 1)\}$$
$$= f^2 .\{ 4 \cos^2 \pi p\}$$
$$= 4f^2 \cos^2 \pi .\left\{\dfrac{(h + 2k)}{3} + \dfrac{1}{2}\right\}$$
$$= 0 \text{ when } (h + 2k) \text{ is a multiple of 3 and l is odd.}$$

Thus, it is these missing reflections, like 11.1, 11.3, 22.1, 22.3, which indicate close packing in a hexagonal structure. From the expression above, it can be seen that if (h + 2k) is a multiple of three and l is even, then

$$\left\{\dfrac{(h + 2k)}{3} + \dfrac{1}{2}\right\} = n \text{ where n is an integer;}$$

$$\cos \pi n = + \text{ or } - 1 \text{ and } \cos^2 \pi n = 1,$$

$$\text{so } |F|^2 = 4f^2.$$

When all possible values of h, k and l are considered, the results may be summarized as follows:

$h + 2k$	l	$|F|^2$
3n	odd	0
3n	even	$4f^2$
3n ± 1	odd	$3f^2$
3n ± 1	even	f^2

It is clear from the above discussions that the structure factor is affected not by the shape and the size of the unit cell but by the atomic positions only. In other words, any of the above examples (i), (ii), (iii) and (iv) could have been any of the seven crystal systems from cubic to triclinic and still have the same structure factors discussed respectively. Table 4.2 shows the Miller indices of the planes and the corresponding values of $|F|^2$ for various structures. The effect of structure factor on the intensity is independent of the method of diffraction used. Some factors, which affect the intensity, are dependent on the experimental method of diffraction used and will be pointed out in respective sections. As the powder method is the most common method of x-ray diffraction, the remaining factors, which affect the intensity of diffraction, are discussed with reference to the powder method.

4.4.3 Multiplicity Factor

This is a factor (p) which refers to the number of planes of a given Miller indices present in the structure of interest having identical plane spacings. Similar to structure factor, this factor is also independent of the diffraction method used. But this factor is relevant and useful only in rotating crystal and powder methods. In a cubic crystal, the planes (100), (010) and (001) have identical plane spacings. Counting ($\bar{1}$00), (0$\bar{1}$0) and (00$\bar{1}$) planes together with the above three make six planes (of {100} family) having identical plane spacings. Thus, the multiplicity factor for {100} planes is six. It should be noted, however, that the multiplicity factor 'p' depends on the crystal system considered. For example, in a tetragonal system, the planes (100) and (010) have the same interplanar spacing equal to the lattice parameter 'a' of the concerned tetragonal lattice, and the plane (001) has a different spacing which is equal to the lattice parameter 'c'. Thus, for this system, the multiplicity factor for {100} is four and that for {001} is two. As a further example, for an orthorhombic structure, the multiplicity factors for {100}, {010} and {001} are each equal to two since their plane spacings are equal to lattice

Table 4.2 Miller indices of atomic planes and $|F|^2$ values for cubic crystals

$h^2+k^2+l^2$	hkl	*simple*	*/body centered*	*/face centered*	*/diamond*
1	100	f^2	-	-	-
2	110	f^2	$4f^2$	-	-
3	111	f^2	-	$16f^2$	$32f^2$
4	200	f^2	$4f^2$	$16f^2$	-
5	210	f^2	-	-	-
6	211	f^2	$4f^2$	-	-
8	220	f^2	$4f^2$	$16f^2$	$64f^2$
9	300, 221	f^2	-	-	-
10	310	f^2	$4f^2$	-	-
11	311	f^2	-	$16f^2$	$32f^2$
12	222	f^2	$4f^2$	$16f^2$	-
13	320	f^2	-	-	-
14	321	f^2	$4f^2$	-	-
16	400	f^2	$4f^2$	$16f^2$	$64f^2$
17	410, 322	f^2	-	-	-
18	411, 330	f^2	$4f^2$	-	-
19	331	f^2	-	$16f^2$	$32f^2$
20	420	f^2	$4f^2$	$16f^2$	-
21	421	f^2	-	-	-
22	332	f^2	$4f^2$	-	-
24	422	f^2	$4f^2$	$16f$	$64f^2$
25	430, 500	f^2	-	-	-
26	510, 431	f^2	$4f^2$	-	-
27	511, 333	f^2	-	$16f^2$	$32f^2$
29	520, 432	f^2	-	-	-
30	521	f^2	$4f^2$	-	-
32	440	f^2	$4f^2$	$16f^2$	$64f^2$
33	522, 441	f^2	-	-	-
34	530, 433	f^2	$4f^2$	-	-
35	531	f^2	-	$16f^2$	$32f^2$
36	600, 442	f^2	$4f^2$	$16f^2$	-
37	610	f^2	-	-	-
38	611, 532	f^2	$4f^2$	-	-
40	620	f^2	$4f^2$	$16f^2$	$64f^2$

parameters 'a', 'b' and 'c' respectively. The multiplicity factors for different {hkl} planes are tabulated under Table 4.3 for various crystal systems.

Table 4.3   Usual multiplicity factors for power diffraction methods

Structure							
Cubic	:	hkl: 48	hhl: 24	0kl: 24	0kk: 12	hhh: 8	00l: 6
Tetragonal	:	hkl: 16	hhl: 8	0kl: 8	hk0: 8	hh0: 4	0k0: 4	00l: 2
Orthorhombic	:	hkl: 8	0kl: 4	h0l: 4	hk0: 4	h00: 2	0k0: 2	00l: 2
	:						
Monoclinic	:	hkl: 4	h0l: 2	0k0: 2			
Triclinic	:	hkl: 2					

4.4.4   Lorentz Factor

This factor arises from the geometry of diffraction in a given diffraction technique. This factor for the powder method is determined as follows:

The intensity versus 2θ values for a particular diffraction line resemble the curve shown in Fig. 4.16 in this technique. This shows that the diffraction not only occurs at the exact Bragg angle θ_B, but also at angles close to it on either side. The area under the curve is termed as 'integrated intensity' of diffraction. This integrated intensity not only influences the eye estimation of the darkening on the diffraction pattern but also is a measure of the total diffraction energy. The area under the curve can be approximated by $I_{max}.B$ where I_{max} is the maximum intensity (Fig. 4.16) of the diffraction peak and B is the width of the diffraction line at half the maximum intensity level.

The breadth B of a diffraction line is a function of the thickness or size of the crystals in a polycrystalline powder sample. Fig. 4.17 shows a situation of diffraction where the thickness 't' of a given crystal is given by $(m-1).d$ where m is the total number of diffracting (hkl) planes available with interplanar distance d. Thus, if the diffraction is taking place at angle θ_B, then the path difference

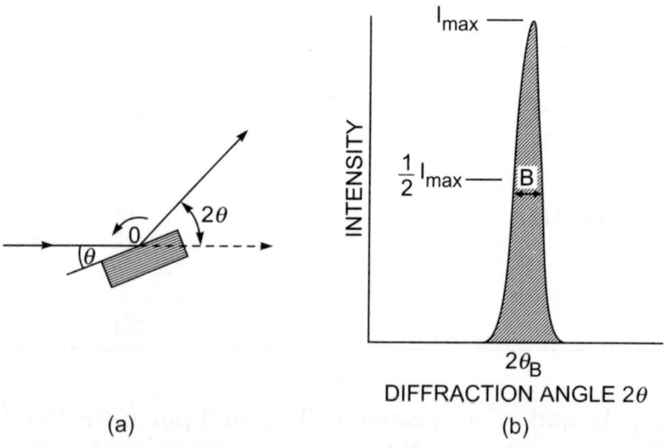

Fig. 4.16   Diffraction by a crystal rotated through the Bragg angle
(Ref.: B.D. Cullity, Elements of x-ray Diffraction)

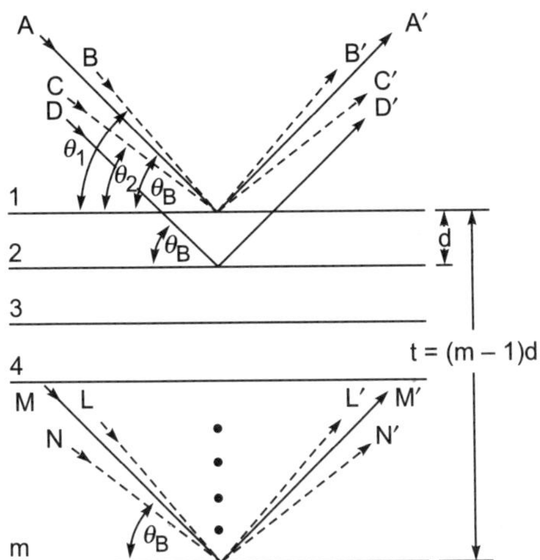

Fig. 4.17 Effect of crystal size on diffraction
(Ref.: B.D. Cullity, Elements of X-ray Diffraction)

between the rays scattered (at θ_B) from the top plane and the plane number 2 shown in the Fig. 4.17 should be λ and that from the top plane and the plane number 3 is 2λ and so on where λ is the wavelength of the monochromatic x-ray beam used for diffraction. Extending this argument, the path difference between the rays scattered from the top plane and the 'm'th plane at an angle θ_B should be $(m-1)$. λ. Under such a circumstance, all the scattered rays, 1′, 2′, 3′, ….. and m′ from the top plane, 2nd plane, 3rd plane, …… and 'm'th plane respectively, interfere with each other constructively producing a strong diffracted beam at θ_B.

But, if the incident beam does not conform to the ideal situation of being strictly parallel, there can be certain divergent or convergent rays. In Fig. 4.17 a pencil of rays, B, is incident on the crystal at an angle θ_1 which is slightly more than the Bragg angle θ_B. The path difference, between the beams scattered at this angle θ_1 from the first and the second plane, is slightly more than λ as can be observed from the geometry. Let the path difference in this case be $\lambda+\delta$ where δ is a small fraction of λ. Similarly, the path difference between the beams scattered at this angle θ_1 from the first and the third plane is $2(\lambda+\delta)$ by geometry. Thus, the path difference between the beams scattered at this angle θ_1 from the first and the last 'm'th plane is $(m-1)$. $(\lambda+\delta)$. Suppose the angle θ_1 is such that the extra path difference '$(m-1)$. δ' happens to be equal to λ, then there would be destructive interference between the scattered ray from the first plane and that from the plane lying at a distance 't/2' below the surface (where 't' is the thickness of the crystal perpendicular to the surface plane) since, the path difference concerned is $\lambda/2$. Then the scattered rays from the first plane would cancel the scattered ray from the plane lying at a depth t/2. If this argument is extended further, it can be seen that scattered rays from the top half of the crystal would cancel those from the bottom half of the crystal. No diffracted beam occurs at this scattering angle θ_1.

At incident angles lesser than θ_1, some amount of diffraction would occur since the cancellation of the scattered beams is not complete. But at incident angles more than this θ_1, the cancellation always occurs completely and no diffracted beam results. This is why the intensity versus 2θ curve shows a maximum at the exact Bragg angle θ_B and the intensity decreases as 2θ increases till it finally reduces to zero at $2\theta_1$ as discussed (Fig. 4.16). Similar argument holds for angles smaller than θ_B. In Fig. 4.16 as the angle of diffraction decreases from $2\theta_B$, the diffraction intensity decreases and reaches zero at $2\theta_2$. Or, θ_2 is such a scattering angle (slightly less than θ_B) that the scattered rays from the top half of the crystal cancel those from the bottom half. From the figure it can be noted that the width of the diffraction line 'B' can be approximated to $(2\theta_1 - 2\theta_2)/2$ or $(\theta_1 - \theta_2)$.

At the incident angle θ_1 where complete cancellation of scattered rays occur, the path difference between the scattered rays from the first and the 'm'th plane is given by $2t \sin \theta_1$. But this path difference should be equal to $(m-1)\lambda + \lambda$ as discussed in the previous paragraph. Or,

$$(m - 1)\lambda + \lambda = 2t.\sin \theta_1 \text{ , or } m\lambda = 2t.\sin \theta_1 \tag{4.11}$$

By similar means it can be shown that for the incident angle θ_2 (which is slightly less than θ_B),

$$(m - 1)\lambda - \lambda = 2t.\sin \theta_2, \text{ or } (m-2)\lambda = 2t.\sin \theta_2 \tag{4.12}$$

Subtracting equation 4.12 from 4.11,

$$2\lambda = 2t.\sin \theta_1 - 2t.\sin \theta_2 \text{ or,}$$
$$\lambda = t.(\sin \theta_1 - \sin \theta_2),$$

or,
$$\lambda = t \cdot 2 \cdot \cos \left[\frac{(\theta_1 + \theta_2)}{2} \right] \cdot \sin \left[\frac{(\theta_1 - \theta_2)}{2} \right]$$

Here, it can be noted that $(\theta_1 + \theta_2)/2$ is approximately equal to θ_B and $(\theta_1 - \theta_2)$ is approximately equal to 'B' as seen earlier. Noting further that $\sin[(\theta_1 - \theta_2)/2]$ is equal to approximately $[(\theta_1 - \theta_2)/2]$ since $(\theta_1 - \theta_2)$ is a very small angle, the above equation can be rewritten as

$$\lambda = 2t.\cos \theta_B . \frac{B}{2} = t.\cos \theta_B.B \text{ or,}$$

$$t = \lambda/(\cos \theta_B.B).$$

A more rigorous treatment shows that

$$t = \frac{0.9\lambda}{(\cos \theta_B \cdot B)} \tag{4.13}$$

This equation is known as the Scherrer formula and can be used for the calculation of the particle size, t, of very tiny crystals by measurement of the diffraction line width 'B'. The equation also shows that the width 'B' is proportional to $(1/\cos \theta_B)$ for given 'λ' and 't' values.

So far, it has been shown that the diffracted beam width 'B' is dependent on the size of the individual crystal particles. What affects the intensity of diffraction (I_{max}) at the exact Bragg angle θ_B? For examining this, Fig. 4.18 a may be studied. The incident X-ray beam and the diffracted beam at the exact Bragg angle is shown in this figure. If the crystal is rotated as shown by dotted lines to a position making an angle θ_1 with the incident beam, some intensity is still diffracted in the direction of the original diffraction angle $2\theta_B$ shown. In this situation, the intensity maximum 'I_{max}' is a summation of the diffraction intensity at exact Bragg angle θ_B and the intensities of diffraction at all angles of incidence slightly more than θ_B provided all these diffracted beams are at the diffraction angle $2\theta_B$. So, the value of I_{max} depends on angular range of $\Delta\theta$, of crystal rotation shown in Fig. 4.18, over which the intensity of diffraction at $2\theta_B$ is appreciable. The situation of diffraction at the limiting angle θ_1, upto which the diffraction intensity at $2\theta_B$ is

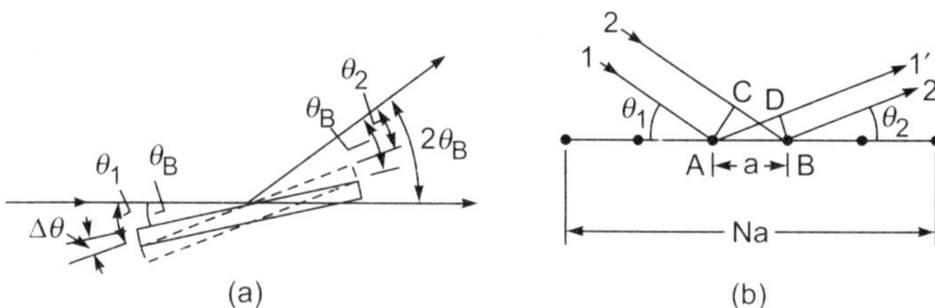

Fig. 4.18 Scattering in a fixed direction during crystal rotation
(Ref.: B.D. Cullity, Elements of x-ray diffraction)

appreciable, is shown in Fig. 4.18(b). It should be noted, however, that this physical rotation of the reflecting plane may not occur in powder method but all such rotational positions of the reflecting plane (hkl) could be available in different crystallites in a powder sample because of their random orientations. Fig. 4.18(b) shows the situation of an incident beam making an angle θ_1 ($\theta_B + \Delta\theta$) with the concerned reflecting plane (hkl). It should be noted that the reflected beam makes an angle θ_2 ($\theta_B - \Delta\theta$) when the incident beam is making an angle of θ_1 with the reflecting plane. Let 'a' be the interatomic spacing as shown in figure and 'N' be the number of atoms involved in scattering. The path difference between scattered rays 1' and 2' shown in Fig. 4.18(b) is given by

$$(AD - CB) = a.\cos\,\theta_2 - a.\cos\,\theta_1$$
$$= a\{\cos(\theta_B - \Delta\theta) - \cos(\theta_B + \Delta\theta)\}.$$

By expanding the cosine terms and noting $\sin\Delta\theta$ is approximately equal to $\Delta\theta$ as $\Delta\theta$ is small,

$$(AD - CB) = 2a.\Delta\theta.\sin\,\theta_B$$

and the path difference between the rays scattered by end atoms shown is $N.2a.\,\Delta\theta.\sin\,\theta_B$.

The conditions for zero diffracted intensity is that this path difference N.2a. $\Delta\theta.\sin\theta_B$ must equal λ. (This condition is almost analogous to the situation while deriving the Scherrer formula).

Therefore, $N.2a.\ \Delta\theta.\sin\theta_B = \lambda$

or,

$$\Delta\theta = \frac{\lambda}{(N\cdot 2a\cdot\sin\theta_B)}.$$

This equation shows the angular range $\Delta\theta$ over which the diffracted intensity is appreciable (which in turn determines the intensity I_{max}) is proportional to $1/\sin\theta_B$.

Thus, the width 'B' of the diffraction curve is proportional to $1/\cos\theta_B$ and the maximum intensity I_{max} is proportional to $1/\sin\theta_B$ and the product '$I_{max}.B$' is, therefore, proportional to '$(1/\sin\theta_B).(1/\cos\theta_B)$' or to $(1/\sin 2\theta_B)$ This is the first geometrical factor.

A second geometrical factor is concerned with the number of powder particles oriented at or near the Bragg angle of a given (hkl) plane. Fig. 4.19 shows a reference sphere of radius 'r' drawn around the powder specimen at 'O'. 'ON' is the normal to the reflecting set of planes (hkl) shown in the figure. Let $\Delta\theta$ represent the range of angles near the Bragg angle over which the diffracted intensity is appreciable for the particular (hkl) set of planes considered. The particles or crystallites which are able to reflect are only those which have their normals lying in a band, on the reference sphere, of width r.$\Delta\theta$. As the powder particles are oriented at random in the powder specimen taken, the reference sphere surface contains evenly distributed ends of (hkl) plane normals. The fraction of particles favourably oriented so as to cause hkl reflection is simply given by the ratio of the area of the band of width r.$\Delta\theta$ to the area of the spherical surface. This ratio is given by

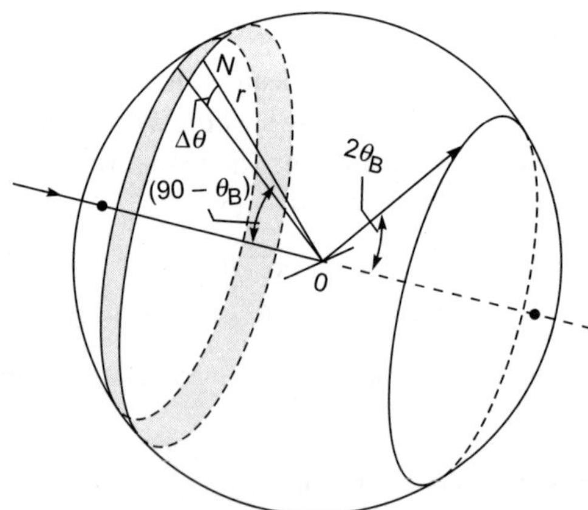

Fig. 4.19 The distribution of plane normals for a particular cone of reflected rays (Ref.: B.D. Cullity, Elements of x-ray Diffraction)

$$\frac{r \cdot \Delta\theta\, 2\pi \cdot \sin(90 - \theta_B)}{4\pi r^2} = \frac{\Delta\theta \cdot \cos\theta_B}{2}$$

Thus, it is seen that the fraction of particles favourably oriented, so as to cause hkl reflection, is proportional to $\cos\theta_B$.

A third geometrical factor arises from the consideration that the integrated intensity per unit length of the diffraction line is generally measured in any diffraction experiment rather than the total diffraction intensity which is clear from Fig. 4.20. It can be seen from the figure that the intensity which is recorded on the film (of a powder camera) or by an intensity counter (as in a diffractometer) is always for a constant portion of the diffraction ring irrespective of its size. If 'R' is the radius of the Debye-Scherrer camera or of the diffractometer circle, as the case may be, then the circumference of any diffraction ring is $2\pi R.\sin\theta_B$. Then the relative integrated intensity per unit length of the diffraction line is proportional to $1/(2\pi R.\sin\theta_B)$ or to $1/\sin\theta_B$ as $2\pi R$ is constant for a given diffraction experiment.

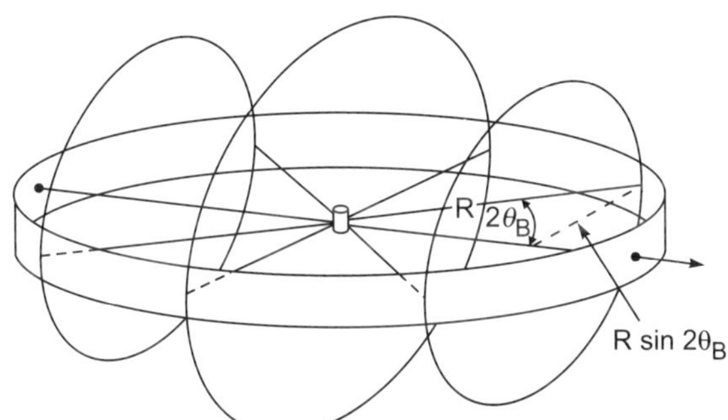

Fig. 4.20 Intersection of cones of diffracted rays with Debye-Scherrer film (Ref.: B.D. Cullity, Elements of x-ray Diffraction)

All these factors mentioned above can be put together and the combined factor is called the Lorentz factor.

So, the Lorentz factor $= \left(\dfrac{1}{\sin 2\theta_B}\right) \cdot \cos\theta_B \cdot \left(\dfrac{1}{\sin 2\theta_B}\right)$

$$= \frac{\cos\theta_B}{\sin^2 2\theta_B} = \frac{1}{4\sin^2\theta_B \cdot \cos\theta_B}.$$

Since, the polarization factor $\{(1 + \cos^2 2\theta_B)/2\}$ is also related to the Bragg angle like the Lorentz factor, these two can be combined together and can be called the Lorentz-Polarization factor.

The combined factor $= \dfrac{\{1 + \cos^2 2\theta_B)/2\}}{(4\sin^2\theta_B \cdot \cos\theta_B)}$

Neglecting the constant term 1/8,

$$\text{the Lorentz-Polarization factor} = \frac{1 + \cos^2 2\theta_B}{\sin^2 \theta_B \cdot \cos \theta_B}$$

Table 4.4 gives the values of these factors for different θ_B.

4.4.5 Absorption Factor

This factor has the effect of decreasing the intensity of both the incident and the diffracted x-ray beams in proportion to the lengths traversed through the specimen. Generally, it is found that as the Bragg angle θ_B increases, the absorption decreases for a cylindrical powder specimen in a Debye-Scherrer(D-S) camera. For a flat specimen used in a diffractometer, the absorption factor is practically independent of θ_B.

4.4.6 Temperature Factor

Temperature has an effect of increasing the thermal vibrations of atoms about their mean positions in a crystal lattice. The amplitude of vibration increases as the temperature increases. Thus, the positions of the atomic planes are no more fixed and they can be thought of as lying anywhere in a range corresponding to the range of displacement of atoms. And the diffraction from these vibrating planes would not be perfect when compared to that from the planes whose positions would have been fixed. If 'x' is the average displacement of the atoms from their mean positions in a plane of interplanar distance 'd', the ratio 'x/d' increases as the temperature increases. Further, as θ increases, 'd' increases for a constant λ and consequently 'x/d' decreases. The value of 'x/d' decides the intensity of the diffracted beam. So, it can be easily seen that as θ increases, the temperature factor decreases. But the absorption factor increases as θ increases, as observed in section 4.4.5. Thus, for powder diffraction work using the D-S technique, the absorption and temperature factors are working against each other and to a first approximation, can be thought of as cancelling each other.

The extent of thermal vibrations of atoms is a function of not only temperature but also the strength of the atomic bond existing in the crystal structure of the material concerned. As the bond strength is reflected in the melting point, materials having low melting points (i.e., low bond strengths) show greater 'x' values than those with high melting points. Thus, materials with low melting points show high 'x/d' ratios and consequently high absorption.

Another effect of temperature is the temperature diffuse scattering. This scattering is the general coherent scattering in all directions which contribute to the background of the diffraction pattern and its intensity increases slightly with the diffraction angle 2θ.

Table 4.4   Lorentz-Polarization factor $\left(\dfrac{1 + \cos^2 2\theta}{\sin^2 \theta \cos \theta} \right)$

$\theta°$	0.0	0.1	0.2	0.3	0.4	0.5	0.6	0.7	0.8	0.9
2	.9	1486	1354	1239	1138	1048	968.9	898.3	835.1	778.4
3	7.2	680.9	638.8	600.5	565.6	533.6	504.3	477.3	452.3	429.3
4	8.0	388.2	369.9	352.7	336.8	321.9	308.0	294.9	282.6	271.1
5	0.3	250.1	240.5	231.4	222.9	214.7	207.1	199.8	192.9	186.3
6	0.1	174.2	168.5	163.1	158.0	153.1	148.4	144.0	139.7	135.6
7	1.7	128.0	124.4	120.9	117.6	114.4	111.4	108.5	105.6	102.9
8	0.3	97.80	95.37	93.03	90.78	88.60	86.51	84.48	82.52	80.63
9	8.79	77.02	75.31	73.66	72.05	70.49	68.99	67.53	66.12	64.74
10	8.41	62.12	60.87	59.65	58.46	57.32	56.20	55.11	54.06	53.03
11	2.04	51.06	50.12	49.19	48.30	47.43	46.58	45.75	44.94	44.16
12	8.39	42.64	41.91	41.20	40.50	39.82	39.16	38.51	37.88	37.27
13	6.67	36.08	35.50	34.94	34.39	33.85	33.33	32.81	32.31	31.82
14	1.34	30.87	30.41	29.96	29.51	29.08	28.66	28.24	27.83	27.44
15	7.05	26.66	26.29	25.92	25.56	25.21	24.86	24.52	24.19	23.86
16	8.54	23.23	22.92	22.61	22.32	22.02	21.74	21.46	21.18	20.91
17	0.64	20.38	20.12	19.87	19.62	19.38	19.14	18.90	18.67	18.44
18	8.22	18.00	17.78	17.57	17.36	17.15	16.95	16.75	16.56	16.36
19	6.17	15.99	15.80	15.62	15.45	15.27	15.10	14.93	14.76	14.60
20	4.44	14.28	14.12	13.97	13.81	13.66	13.52	13.37	13.23	13.09
21	2.95	12.81	12.68	12.54	12.41	12.28	12.15	12.03	11.91	11.78
22	1.66	11.54	11.43	11.31	11.20	11.09	10.98	10.87	10.76	10.65
23	0.55	10.45	10.35	10.24	10.15	10.05	9.951	9.857	9.763	9.671
24	9.579	9.489	9.400	9.313	9.226	9.141	9.057	8.973	8.891	8.810
25	8.730	8.651	8.573	8.496	8.420	8.345	8.271	8.198	8.126	8.054
26	7.984	7.915	7.846	7.778	7.711	7.645	7.580	7.515	7.452	7.389
27	7.327	7.266	7.205	7.145	7.086	7.027	6.969	6.912	6.856	6.800
28	6.745	6.692	6.637	6.584	6.532	6.480	6.429	6.379	6.329	6.279
29	6.230	6.183	6.135	6.088	6.042	5.995	5.950	5.905	5.861	5.817
30	5.774	5.731	5.688	5.647	5.605	5.564	5.524	5.484	5.445	5.406
31	5.367	5.329	5.292	5.254	5.218	5.181	5.145	5.110	5.075	5.040
32	5.006	4.972	4.939	4.906	4.873	4.841	4.809	4.777	4.746	4.715
33	4.685	4.655	4.625	4.595	4.566	4.538	4.509	4.481	4.453	4.426
34	4.399	4.372	4.346	4.320	4.294	4.268	4.243	4.218	4.193	3.169
35	4.145	4.121	4.097	4.074	4.052	4.029	4.006	3.984	3.962	3.941
36	3.919	3.898	3.877	3.857	3.836	3.816	3.797	3.777	3.758	3.739
37	3.720	3.701	3.683	3.665	3.647	3.629	3.612	3.594	3.577	3.561
38	3.544	3.527	3.513	3.497	3.481	3.465	3.449	3.434	3.419	3.404
39	3.389	3.375	3.361	3.347	3.333	3.320	3.306	3.293	3.280	3.268
40	3.255	3.242	3.230	3.218	3.206	3.194	3.183	3.171	3.160	3.149
41	3.138	3.127	3.117	3.106	3.096	3.086	3.076	3.067	3.057	3.048
42	3.038	3.029	3.020	3.012	3.003	2.994	2.986	2.978	2.970	2.962
43	2.954	2.946	2.939	2.932	2.925	2.918	3.911	2.904	2.897	2.891
44	2.884	2.878	2.872	2.866	2.860	2.855	2.849	2.844	2.838	2.833

Contd.

Table 4.4 Contd.

θ°	0.0	0.1	0.2	0.3	0.4	0.5	0.6	0.7	0.8	0.9
45	2.828	2.824	2.819	2.814	2.810	2.805	2.801	2.797	2.793	2.789
46	2.785	2.782	2.778	2.775	2.772	2.769	2.766	2.763	2.760	2.757
47	2.755	2.752	2.750	2.748	2.746	2.744	2.742	2.740	2.738	2.737
48	2.736	2.735	2.733	2.732	2.731	2.730	2.730	2.729	2.729	2.728
49	2.728	2.728	2.728	2.728	2.728	2.728	2.729	2.729	2.730	2.730
50	2.731	2.732	2.733	2.734	2.735	2.737	2.738	2.740	2.741	2.743
51	2.745	2.747	2.749	2.751	2.753	2.755	2.758	2.760	2.763	2.766
52	2.769	2.772	2.775	2.778	2.782	2.785	2.788	2.792	2.795	2.799
53	2.803	2.807	2.811	2.815	2.820	2.824	2.828	2.833	2.838	2.843
54	2.848	2.853	2.858	2.863	2.868	2.874	2.879	2.885	2.890	2.896
55	2.902	2.908	2.914	2.921	2.927	2.933	2.940	2.946	2.953	2.960
56	2.967	2.974	2.981	2.988	2.996	3.004	3.011	3.019	3.026	3.034
57	3.042	3.050	3.059	3.067	3.075	3.084	3.092	3.101	3.110	3.119
58	3.128	3.137	3.147	3.156	3.166	3.175	3.185	3.195	3.205	3.215
59	3.225	3.235	3.246	3.256	3.267	3.278	3.289	3.300	3.311	3.322
60	3.333	3.345	3.356	3.368	3.380	3.392	3.404	3.416	3.429	3.441
61	3.454	3.466	3.479	3.492	3.505	3.518	3.532	3.545	3.559	3.573
62	3.587	3.601	3.615	3.629	3.643	3.658	3.673	3.688	3.703	3.718
63	3.733	3.749	3.764	3.780	3.796	3.812	3.828	3.844	3.861	3.878
64	3.894	3.911	3.928	3.946	3.963	3.980	3.998	4.016	4.034	4.052
65	4.071	4.090	4.108	4.127	4.147	4.166	4.185	4.205	4.225	4.245
66	4.265	4.285	4.306	4.327	4.348	4.369	4.390	4.412	4.434	4.456
67	4.478	4.500	4.523	4.546	4.569	4.592	4.616	4.640	4.664	4.688
68	4.712	4.737	4.762	4.787	4.812	4.838	4.864	4.890	4.916	4.943
69	4.970	4.997	5.024	5.052	5.080	5.109	5.137	5.166	5.195	5.224
70	5.254	5.284	5.315	5.345	5.376	5.408	5.440	5.471	5.504	5.536
71	5.569	5.602	5.636	5.670	5.705	5.740	5.775	5.810	5.846	5.883
72	5.919	5.956	5.994	6.032	6.071	6.109	6.149	6.189	6.229	6.270
73	6.311	6.352	6.394	6.437	6.480	6.524	6.568	6.613	6.658	6.703
74	6.750	6.797	6.844	6.892	6.941	6.991	7.041	7.091	7.142	7.194
75	7.247	7.300	7.354	7.409	7.465	7.521	7.578	7.636	7.694	7.753
76	7.813	7.874	7.936	7.999	8.063	8.128	8.193	8.259	8.327	8.395
77	8.465	8.536	8.607	8.680	8.754	8.829	8.905	8.982	9.061	9.142
78	9.223	9.305	9.389	9.474	9.561	9.649	9.739	9.831	9.924	10.02
79	10.12	10.21	10.31	10.41	10.52	10.62	10.73	10.84	10.95	11.06
80	11.18	11.30	11.42	11.54	11.67	11.80	11.93	12.06	12.20	12.34
81	12.48	12.63	12.78	12.93	13.08	13.24	13.40	13.57	13.74	13.92
82	14.10	14.28	14.47	14.66	14.86	15.07	15.28	15.49	15.71	15.94
83	16.17	16.41	16.66	16.91	17.17	17.44	17.72	18.01	18.31	18.61
84	18.93	19.25	19.59	19.94	20.30	20.68	21.07	21.47	21.89	22.32
85	22.77	23.24	23.73	24.24	24.78	25.34	25.92	26.52	27.16	27.83
86	28.53	29.27	30.04	30.86	31.73	32.64	33.60	34.63	35.72	36.88
87	38.11	39.43	40.84	42.36	44.00	45.76	47.68	49.76	52.02	54.50

From The Interpretation of X-Ray Diffraction Photographs, by N. F. M. Henry, H. Lipson and W. A. Wooster (Macmillan, London, 1951).

4.4.7 General Equation for the Intensity

The total diffracted intensity can now be written combining all the factors discussed so far. In the D-S camera the diffracted intensity (relative) of powder pattern lines can be expressed as

$$I = |F|^2 \cdot p \cdot \frac{1 + \cos^2 2\theta_B}{\sin^2 \theta_B \cdot \cos \theta_B} \qquad (4.14)$$

It should be noted that in this equation the temperature and absorption factors are neglected and this equation is applicable specifically to the D-S technique. Further, this equation gives the relative integrated intensity values and not the absolute intensity values. However, in most of the applications, the calculation of the absolute intensity values is difficult, if not impossible, and not necessary.

As an example of intensity calculations, the calculation of intensities in a diffraction pattern of iron made with Cr-Kα radiation is done below:

Table 4.5

1	2	3	4	5	6	7	8	9	
Line	*hkl*	θ	*(sin θ)/λ*	f_{Fe}	$|F|^2$	*p*	*L-P*	*cal*	*prac*
1	110	34.5	0.5664	17.3	1197	12	4.27	61305	str
2	200	53.2	0.8010	14.5	841	6	2.81	14194	med
3	211	78.8	0.9810	12.7	645	24	9.85	154076	vstr

Explanation:

Column 1: Indicates the line numbers in the increasing order of θ values.

Column 2: As iron has a body centered cubic structure, the structure factor shows that the reflection takes place only for those planes for which 'h+k+l' is an even number (section 4.4.2). The hkl values for which reflection is possible can be written down in the increasing order in this column.

Column 3: In a cubic crystal, the interplanar spacing 'd' is given by

$$d^2 = \frac{a^2}{h^2 + k^2 + l^2} ,$$

where 'a' is the lattice parameter.

From Bragg's law,

$$d = \frac{\lambda}{(2 \cdot \sin \theta)} \quad \text{or,}$$

$$d^2 = \frac{\lambda^2}{4 \cdot \sin^2 \theta} = \frac{a^2}{h^2 + k^2 + l^2} \quad \text{or,}$$

$$\sin^2 \theta = \lambda^2 \cdot \frac{(h^2 + k^2 + l^2)}{(4a^2)}$$

Here, λ is the wavelength of Cr Kα line i.e., 0.229nm

and 'a' is the lattice parameter of iron i.e., 0.286nm

With the help of sin θ values, θ can be calculated.

Column 5: From Table 4.1

Column 6: From Table 4.2

Column 7: From Table 4.3

Column 8: From Table 4.4

Column 9: The calculated value of the relative intensity is obtained by multiplying respective values from columns 6, 7 and 8. The visual estimation of the practical intensities are also given as strong, medium and very strong respectively for the first, second and the third lines, for the purpose of comparison, in the adjacent column.

SOLVED EXAMPLES

Example 4.1 Determine the position in terms of 2θ and the relative integrated intensity (R.I.I.) of the first eight lines of the diffraction pattern of silver (FCC with lattice parameter of 0.40856 nm) obtained using copper Kα radiation (of wavelength 0.15418 nm) on an x-ray diffractometer neglecting the temperature and absorption factors.

Solution: The following table shows how the calculation can be conducted systematically. It is known from the structure factor calculation that for FCC structures $(h^2+k^2+l^2)$ values of the reflecting planes will be increasing in the sequence 3, 4, 8, 11, 12, 16, 19, 20, 24 and so on. These values are entered in the third column.

Line	hkl	$h^2+k^2+l^2$	$sin^2\theta$	$sin\theta$	θ	2θ	$sin\theta/\lambda$	f_{Ag}	$\lvert F \rvert^2$	p	L-P	R.I.I
1.	111	3	0.1068	0.3268	19.1	38.2	0.212	36.3	21×10^3	08	16	2688
2.	200	4	0.1424	0.3774	22.2	44.4	0.245	34.2	19×10^3	06	11	1254
3.	220	8	0.2848	0.5337	32.3	64.6	0.346	29.6	14×10^3	12	05	840
4.	311	11	0.3916	0.6258	38.7	77.4	0.406	26.9	12×10^3	24	03	864
5.	222	12	0.4272	0.6536	40.8	81.6	0.424	26.1	11×10^3	08	03	264
6.	400	16	0.5696	0.7547	49.0	98.0	0.489	24.1	09×10^3	06	03	162
7.	331	19	0.6765	0.8225	55.3	110.6	0.533	23.0	08×10^3	24	03	576
8.	420	20	0.7121	0.8439	57.5	115.0	0.547	22.6	08×10^3	24	03	576

The rest of the calculations are done in a way similar to those done in example noted at the end of chapter 4. It is to be noted that the values in the R.I.I. column are only relative values. These values are all approximate but are useful for the purpose of comparison of line intensities.

EXERCISES

4.1 Determine the (hkl) and 2θ values of the first four lines on the powder patterns (using Cu Kα radiation) of substances with the following structures:

   (i) Simple cubic (a = 0.4 nm)

  (ii) Body centered cubic (a = 0.4 nm)

 (iii) Face centered cubic (a = 0.4 nm)

 (iv) Simple Tetragonal (a = 0.4 nm and c = 0.3 nm)

> [Ans: (i) (100)– 22.2, (110)– 31.6, (111)– 39.0, (200)– 45.3.
> (ii) (110)– 31.6, (200)– 45.3, (211)– 56.3, (220)– 66.1.
> (iii) (111)– 39.0, (200)– 45.3, (220)– 66.1, (311)– 79.5.
> (iv) (100)–22.2, (001)– 29.8, (110)– 31.6, (011)– 37.5.]

4.2 Determine the positions in terms of 2θ and the R.I.I. of the first eight lines of the diffraction pattern of copper (FCC with a lattice parameter of 0.36153 nm) obtained using copper Kα radiation on an x-ray diffractometer neglecting the temperature and absorption factors.

> [Ans: 6.20, 2.86, 1.75, 2.16, 0.66, 0.47, 2.50, 2.96]

4.3 Determine the position in terms of 2θ and the R.I.I. of the first three lines of chromium (BCC with a lattice parameter of 0.28845 nm) obtained using chromium Kα radiation (0.22909 nm) neglecting the temperature and absorption factors.

> [Ans: 64, 16.5, 143]

5

CHAPTER

LAUE TECHNIQUES

5.1 INTRODUCTION

This was the first diffraction technique ever used. This technique is based on Von Laue's original experiment. Here, a beam of white radiations, as obtained from an x-ray tube, is made to fall on a single crystal and the diffraction pattern is recorded on a flat film kept perpendicular to the x-ray beam. A particular set of planes (with a given 'd' value) picks up a definite wavelength λ from the continuous spectrum and diffracts it in accordance with the Bragg's law. The white or continuous radiation required for the purpose may be obtained from an x-ray tube with a heavy metal target such as tungsten.

The single crystal used may be a large isolated one or a large enough crystal in a polycrystalline aggregate. If the crystal is a part of a polycrystalline aggregate, care should be taken to see that the diffraction only from the crystal of interest is recorded.

Laue spots are formed by overlapping diffractions of different orders. Recalling the equation 4.1, i.e., n λ = 2d' sin θ, it can be seen that the product of the order of reflection 'n' and the wavelength 'λ' should be a constant value 2d' sin θ. This can be obtained by decreasing the λ value (λ_1, λ_2, λ_3, λ_4, λ_5 and so on) as the order of reflection increases in the sequence 1, 2, 3, 4, 5, etc. respectively so that the corresponding products 1. λ_1, 2. λ_2, 3. λ_3, 4. λ_4, 5. λ_5 are all equal to 2d' sin θ. In other words, if the first order reflection from a certain set of planes (with a plane spacing of d') takes place with a wavelength λ_1, the second order reflection takes place with a wavelength $\lambda_2 = \lambda_1/2$ and similarly, the n'th order reflection takes place with a wavelength $\lambda_n = \lambda_1/n$ provided such a short wavelength exists in the incident beam (i.e., n can have values such that $\lambda_1/n > \lambda_{swl}$). Here, it should be noted that when the orientation of a given crystal is fixed with respect to the incident beam a change in the plane spacing will not cause any difference in the Laue pattern of spots. It follows that two different crystals, having identical crystal structures but with different lattice parameters, would record identical diffraction patterns provided their orientations with respect to the incident beam are similar.

5.2 THE TRANSMISSION LAUE TECHNIQUE

The camera setup used for transmission Laue technique is shown in Fig. 5.1. The white x-ray beam from the source is first made to pass through a collimator c so that a narrow incident beam of nearly parallel x-rays is produced by two pinholes arranged in succession in a line. This beam is diffracted by the single crystal 's' mounted on a crystal holder 'H'. The diffracted beams are caught by the film placed in a light-tight flat film-holder. The film holder is a metal backed film container having only an opaque paper in front of the film. The film can be of a size of 100mm × 125mm. The beam-stop 'B' is generally a copper disc of about 0.5 mm thickness fixed on the paper cover of the film, designed to prevent the transmitted beam (from the crystal) from striking the film and causing excessive blackening of the film. The Bragg angle θ with reference to a transmission Laue spot can be determined by the relation

$$\tan 2\theta = \frac{r_1}{D_1} \tag{5.1}$$

where 2θ = diffraction angle = twice the Bragg angle, r_1 = the distance of the Laue spot from the center of the film, and D_1 = specimen to film distance (usually 5 cm).

It is to be observed that as the tube voltage increases, the Laue spots become more intense, since, the intensity of the incident beam increases with the tube voltage and also higher order reflections would occur because of the appearance of lower wavelength radiations with increased voltage (as λ_{swl} decreases as tube voltage increases). There is another effect of importance which occurs on increasing the tube voltage. The Laue spots near the center of the film is caused by

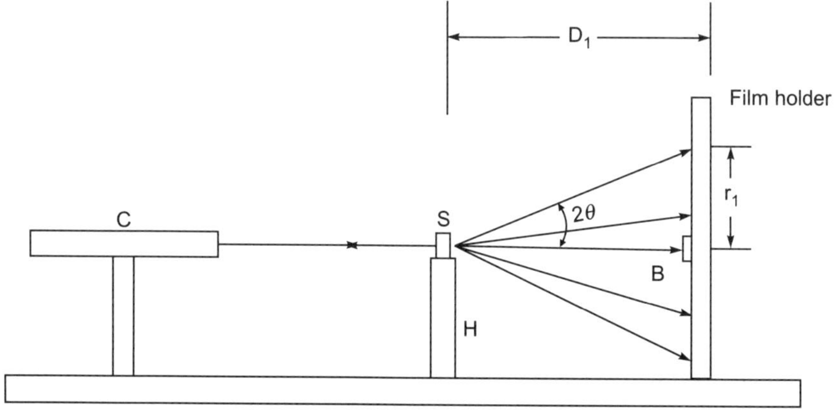

Fig. 5.1 Transmission Laue technique

planes which are inclined at very small angles to the incident beam which suggests that small wavelengths are necessary for the purpose of diffraction (since for a given d' value, if θ is small λ also correspondingly must be small to satisfy $n\lambda = 2d' \sin \theta$). Thus, as the voltage increases, shorter wavelength radiations are produced and planes inclined at very small angles to the incident beam are able to cause diffraction spots closer to the center of the film. In other words, as the voltage of the x-ray tube increases, additional diffraction spots appear closer to the center of the film.

The transmission Laue spots are generally seen to lie on certain imaginary curves like ellipses or hyperbolas as shown in Fig. 5.2. The spots which are lying on a particular curve are diffractions from planes belonging to a zone. This is because of the fact that the diffractions from the planes of a zone all lie on the surface of an imaginary cone whose axis is the zone axis of the zone concerned as shown in Fig. 5.2 and where the transmitted beam itself forms a part of the curved surface of the cone. Further, the angle of inclination of the zone axis to the transmitted beam, φ, is equal to the semi-apex angle of the cone.

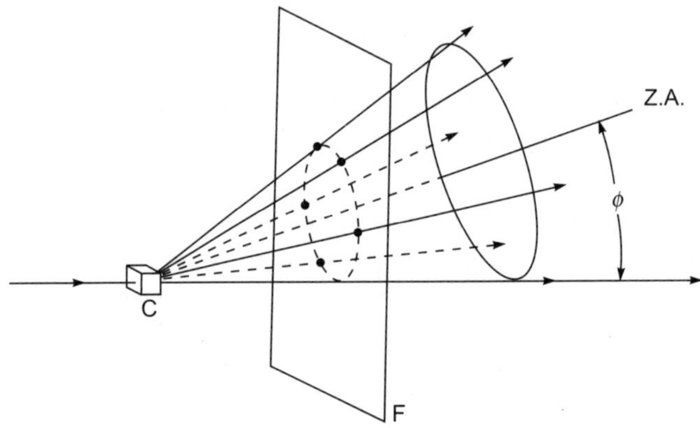

Fig. 5.2   Location of law spots in transmission

With the help of a stereographic projection, it can be easily shown that Laue reflections from planes belonging to a zone lie on the surface of a cone. In Fig. 5.3, the crystal is placed at the center of the reference sphere, the incident beam I enters at the left and the transmitted beam T leaves at the right. Let the zone axis be lying on the basic circle and the poles of the five planes belonging to this zone, viz., P_1, P_2, P_3, P_4, and P_5 be lying on the great circle shown. The corresponding diffracted beam positions are D_1, D_2, D_3, D_4 and D_5 respectively. It is already noted in section 4.2 that the incident beam, the normal to the reflecting plane, the diffracted beam and the transmitted beam are all co-planar. Thus, the poles I, P_1, D_1 and T on the stereographic projection must all lie on a great circle. The angle between I and P_1 (90-θ) should be the same as the angle between P_1 and D_1 (since, angle of incidence is equal to the angle of reflection). So, the position of D_1 on the

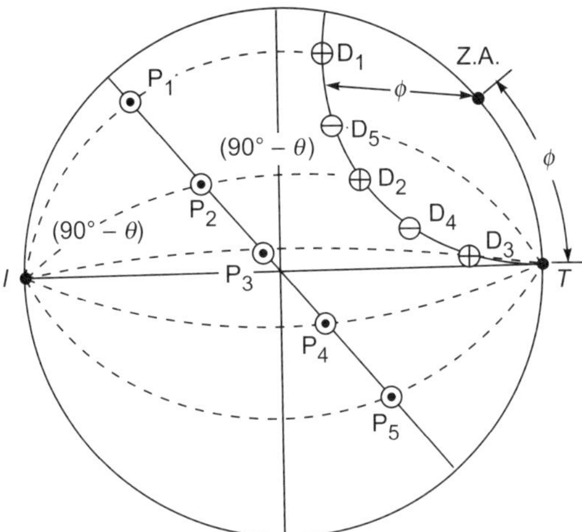

Fig. 5.3 Stereographic projection of transmission Laue method

stereographic projection can be located. Similarly, the other poles D_2, D_3, D_4 and D_5 are fixed. Further, it can be observed that the diffracted beams D_1, D_2, D_3, D_4 and D_5 all lie on a small circle passing through the transmitted beam. This small circle can be thought of as an intersection of a cone (whose axis is the zone axis) with the reference sphere.

5.3 BACK REFLECTION LAUE TECHNIQUE

Fig. 5.4 is a schematic sketch of a back reflection Laue camera arrangement. The cassette in this camera supports the film and the collimator. The film should have

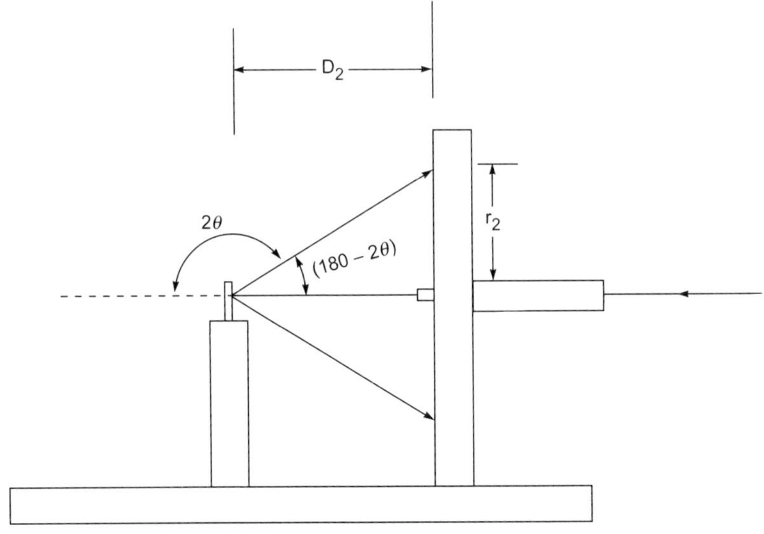

Fig. 5.4 Back reflection Laue technique

a hole to permit the collimator to project through it as shown. The relation between the Bragg angle θ for any diffraction spot and the distance of the spot from the center of the film 'r_2' is

$$\tan(180 - 2\theta) = \frac{r_2}{D_2} \qquad (5.2)$$

where D_2 is the specimen to film distance which is usually 3 cm. Back reflection Laue patterns may generally have diffraction spots as close to the center of the film as the size of the collimator permits. The spots closer to the center of the film are caused by high order reflections from planes almost perpendicular to the incident beam. From discussions made earlier, it is clear that these spots are caused by long wavelength incident radiations and thus decrease in voltage will not cause removal of spots from the center.

In back reflection Laue method as the film is placed between the crystal and the x-ray source, the beams diffracted from the crystal in a backward direction only are recorded. The array of diffraction spots are located on imaginary hyperbolas as shown in Fig. 5.5.

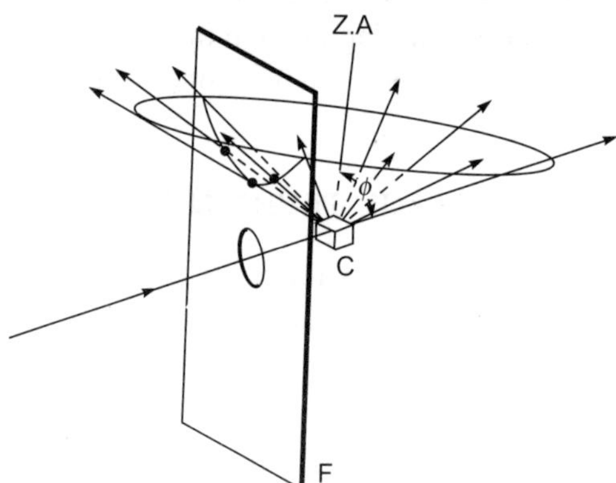

Fig. 5.5   Laue spots in back reflection

The exposure times for obtaining back reflection patterns are longer compared to those needed to get transmission patterns. The atomic scattering factor 'f' decreases as $(\sin\theta)/\lambda$ increases. This ratio $(\sin\theta)/\lambda$ is higher in back reflection technique compared to that in transmission technique. So, the atomic scattering factor 'f' is low in back reflection technique and hence the necessity for long exposure times. Further, the back reflection patterns have less contrast between the diffraction spots and the film background compared to the transmission Laue patterns. This is because the incoherent scattering intensity (which is responsible for the background darkening) is high at high $(\sin\theta)/\lambda$ values and further the coherent scattering intensity (which is responsible for diffraction spots) is low at high $(\sin\theta)/\lambda$ values.

Intensifying screens, generally used in radiography, may also be used here in order to increase the intensity of Laue spots. These intensifying screens are nothing but an active fluorescing material coated on an inert backing paper or cardboard. When such a screen is placed with its active face in contact with the film, the film is blackened not only by the incident x-ray beam but also by the visible light (generally blue) emitted by the fluorescing material. The active fluorescing material used in the screens may be calcium tungstate or zinc sulphide containing a trace of silver, the former is effective at wavelengths lesser than 0.05 nm and the latter can be used for longer wavelengths. These intensifying screens should not be used if the sharpness of the spots is an important criterion being studied as in deciphering the crystal distortion. The visible light emitted by the screens travels in all directions and makes the Laue spots to spread out to a larger area and makes the outlines of the spots hazy.

5.4 SPECIMENS AND SPECIMEN HOLDERS

The specimens in Laue methods are single crystals of the materials of interest. Transmission method requires specimens which allow diffracted beams to pass through them without too much of absorption. This means that the thickness of specimens must be restricted particularly if they have high absorption coefficients. In cases like copper, the thickness must be reduced, by some technique, to a few hundredths of a millimeter. But, if the specimen is too thin, the diffraction spots may have too low intensities to be recognized since the intensities of the spots depend on the volume of the material diffracting the x-ray beam. In the back reflection method such restriction is unnecessary and massive specimens may be examined since the diffracted beams originate only from a thin surface layer of the specimen and they need not travel through the specimen as the diffracted beam lies on the same side as the incident beam. But, this also indicates the drawback that the information which can be gathered from a back reflection pattern is representative of only a thin surface layer and not of the bulk specimen while the transmission pattern is representative of the specimen bulk.

Specimen holders are available in a large variety of types, each suited for a specific purpose. The simplest type consists of a fixed post to which the specimen can be attached with the help of plasticine. Sometimes, it may be necessary to fix the crystal in some specific orientation with respect to the incident x-ray beam. In such cases, it may be necessary to use a three-circle goniometer. This goniometer has three mutually perpendicular axes of rotation and when the crystal is mounted on it, the crystal can be oriented in any desired position with respect to the incident beam. In case of sheet specimens, it may be necessary to move the specimen in two directions at right angles to each other in order to obtain patterns from different locations on the sheet surface. For this purpose, a mechanical stage from a microscope can be converted into a specimen holder.

5.5　COLLIMATORS

Collimators are used in order to obtain a parallel beam of x-rays for the purposes of diffraction. A perfectly parallel beam of x-rays is almost impossible to obtain by any collimator, but it is attempted and the result is an almost parallel beam.

A simple pinhole collimator consists of two circular apertures or pinholes of small diameter 'd' separated by a large distance 's'. Suppose a point source 'S' is placed in line with the collimator axis, as shown in Fig. 5.6. Since, the source 'S' is small, the rays of the beam coming through the collimator are non-parallel and the beam is in the shape of a cone with an apex angle of divergence α. It can be noted by geometry that

$$\tan \frac{\alpha}{2} = \frac{d/2}{s_1}$$

where 's_1' is the distance of the source from the exit pinhole. Since, α is very small, $\tan \alpha/2$ can be equated to $\alpha/2$ and so the above equation can be rewritten as

$$\frac{\alpha}{2} = \frac{d/2}{s_1} \text{ or,}$$

$$\alpha = \frac{d}{s_1} \tag{5.3}$$

If it is necessary to have more nearly parallel beam, the angle α is to be decreased by decreasing the diameter 'd' of the exit pinhole. But this decrease in 'd' decreases the energy of the beam. It can be further noted that the entrance pinhole has no part to play in the collimator except to decrease the size of the incoming beam and can be omitted if the source is very small.

Usually x-ray tubes have focal spots of finite size and of rectangular shape. The x-ray beam issuing out of the x-ray tube depends on the take-off angle subtended

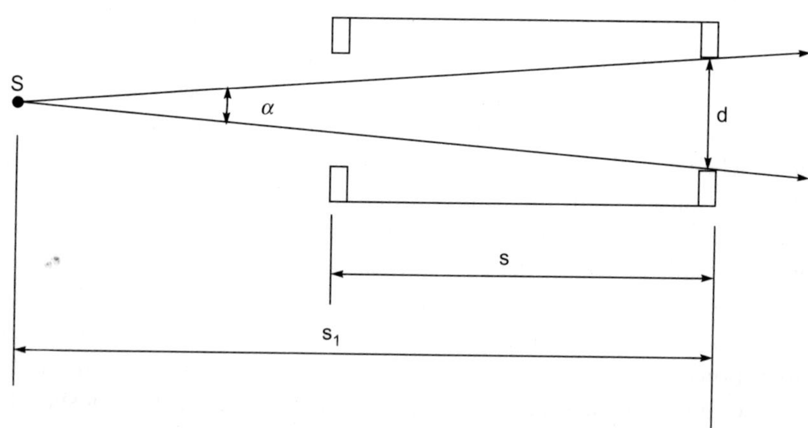

Fig. 5.6　Small source and pinhole collimator

at window by the focal spot. Generally, the take-off angle is quite small and the beam cross-section is usually a small square or a line as shown in Fig. 5.7(a) and (b) depending on the direction of projection. When the shape of the projected source is square of size 'a', the convergent rays from the edges of the source crossover at the center of the collimator and then diverge as shown in Fig. 5.8. The maximum divergence angle β can be calculated by

$$\tan \frac{\beta}{2} = \frac{d/2}{s_3/2}$$

where s_3 is the length of the collimator.

Since $\frac{\beta}{2}$ is a small angle, $\tan \frac{\beta}{2} = \frac{\beta}{2}$

So,

$$\frac{\beta}{2} = \frac{d}{s_3} \quad \text{or,}$$

$$\beta = \frac{2d}{s_3} \text{ in radians} \tag{5.4}$$

The x-ray beam coming out of the collimator contains not only divergent rays but also parallel and convergent rays. The maximum angle of convergence is given by

$$\tan \frac{\alpha}{2} = \frac{d/2}{s_2 + s_3}$$

where s_2 is collimator exit-hole to crystal distance.

Here again, as $\alpha/2$ is a small angle, the equation can be modified as

$$\frac{\alpha}{2} = \frac{d/2}{s_2 + s_3}$$

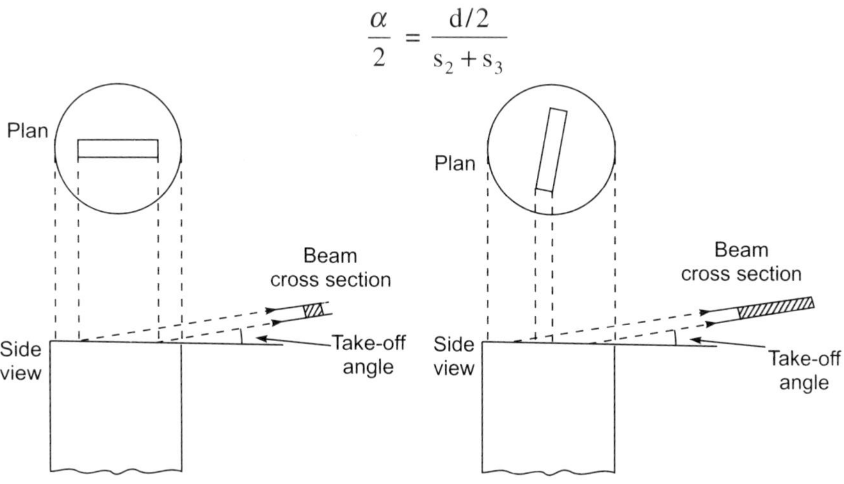

(a) Focal spot and issuing beam
cross section

(b) 90° rotational view of Fig. 5.7(a)

Fig. 5.7 (a) and (b)

or,
$$\alpha = \frac{d}{s_2 + s_3} \qquad (5.5)$$

It can also be noted that

$$\tan \beta = \frac{a/2}{s_1 - (s_3/2)} = \frac{d/2}{s_3/2}$$

or,
$$\frac{a}{2s_1 - s_3} = \frac{d}{s_3}$$

i.e.,
$$a = \frac{d(2s_1 - s_3)}{s_3} \qquad (5.6)$$

Generally, s_1 is about twice as large as s_3 which indicates that the condition of Fig. 5.8 is achieved when the pinhole diameters (d) are about one-third the size of the projected source 'a'. If the value of 'a' is smaller than that given by equation 5.6 then the conditions will be intermediate between those shown in Fig. 5.6 and 5.8. And as 'a' approaches zero, the maximum divergence angle decreases from the value given by the equation 5.4 to that given by 5.3. When 'a' exceeds the value given by the equation 5.6, there is unnecessary wastage of energy. Typical values of various parameters are : d = 0.5mm, s_3 = 5cm , s_2 = 3cm for which the equation 5.6 gives the β value 1.15° and $\alpha = 0.36°$. These angles may be reduced by decreasing the size 'd' of the pinholes, but it eventually results in decreasing the energy of the beam and consequently increases the exposure time required for obtaining the diffraction patterns.

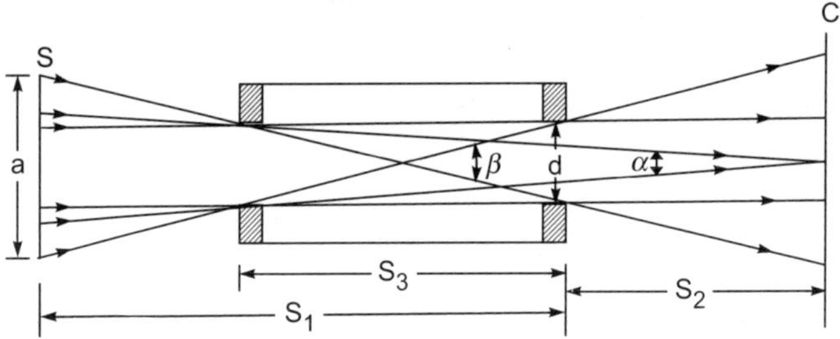

Fig. 5.8   Pinhole collimator and large source S = Source C = Crystal

5.6   LAUE SPOT SHAPES

The shapes of Laue spots will not be circular if the crystal is distorted. The topic of crystal distortion and its consequent effect on the shape of Laue spots will be discussed later in chapter 10. Here, the discussion is restricted to the shape of Laue spots obtained by the diffraction of undistorted crystals which depends on the divergence or convergence of the incident beam.

Assuming that the crystal is thin and larger than the cross-section of the incident beam which is divergent (usually so), a focusing action is present during diffraction using the transmission Laue technique. Fig. 5.9 shows the technique in which 'S' is a point source issuing a divergent polychromatic x-ray beam, the cross-section of which at any location is circular. 'C' is a crystal showing the inner reflecting planes and 'F' is the focal point of the diffracted beam. Assume that the reflecting plane at 'A' and at 'B' are very close to the Bragg angle on the

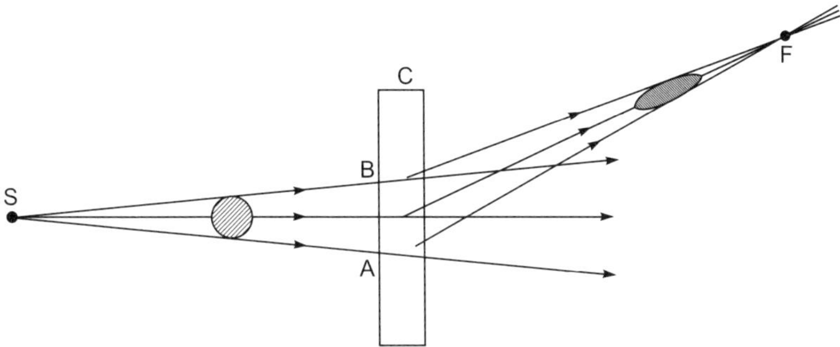

Fig. 5.9   Focusing of diffracted beam in the transmission Laue method. S = source, C = crystal, F = focal point

higher and the lower side respectively. Thus, the reflected rays from these planes tend to converge at a focal point F. This sort of convergence of the diffracted rays occurs only in the plane of the drawing. In a direction normal to the plane of the drawing, the diffracted rays would continue to diverge. Thus, the cross-section of the diffracted beam would be elliptical with the minor axis in the converging direction and the major axis in the diverging direction. The shape of the Laue spots as obtained in such a situation is shown in Fig. 5.10. Note that the Laue spots are generally of elliptical cross-section except the ones falling exactly at the center of the x-ray film (left-hand bottom corner in the figure). Spots of elliptical cross-section described above have their minor axes in radial direction from the center

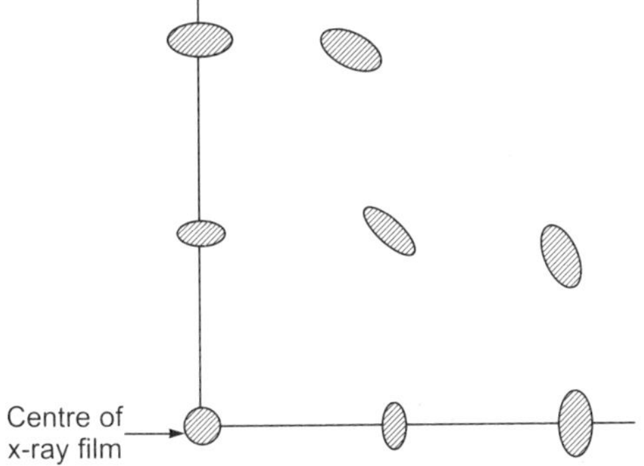

Fig. 5.10   Shape of transmission Laue spots as a function of position.
(Ref.: B.D. Cullity, Elements of x-ray Diffraction)

and the major axes along the circumferential direction. Further, as the distance of the spot from the center increases, the spot size increases and the intensity decreases.

In back reflection technique, focusing does not occur and the beam continues to diverge in all directions after diffraction. Here, the spots are elliptical in shape with their major axes along the radial directions from the center of the film.

SOLVED EXAMPLES

Example 5.1   A transmission Laue pattern of a copper crystal (lattice parameter = 0.36153 nm) is obtained on a film kept at a distance of 4 cm. The (100) planes of the crystal make an angle of 5° with the incident beam. Calculate the minimum tube voltage required to produce a 200 reflection. In addition, determine the location of the spot with reference to the center of the film.

Solution: From Bragg's law, $\lambda = 2d_{200}.\sin\theta$ for a 200 reflection.

And $\qquad\qquad d_{200} = d_{100}/2$. But d_{100} = lattice parameter = 0.3615 nm.

So, $\qquad\qquad \lambda = 0.3615.\sin(5°)/2 = 0.01575$ nm.

This wavelength must be available for 200 reflection to occur. It is known that for a given tube voltage, the minimum wavelength, λ_{swl}, can be calculated as

$\qquad \lambda_{swl}$ in nm = 1239.8/tube voltage or, rewriting,

$\qquad\qquad$ tube voltage = 1239.8/0.01575 = 78,730 Volts.

It is better to have a little higher voltage as the intensity of λ_{swl} is very low at the calculated voltage. If a little higher voltage is used, λ_{swl} will shift to lower wavelength values and the intensity of the wavelength 0.01575 nm will increase. Hence, high voltage of 80 kV or more may have to be used.

Further, the equation 5.1 relates the Bragg angle with the location of the diffraction spot on the film, i.e.,

$$\tan 2\theta = \frac{r_1}{D_1}. \text{ Here, } \theta \text{ is } 5° \text{ and } D_1 \text{ is } 4 \text{ cm.}$$

So, $\qquad \tan(10°) = \dfrac{r_1}{4}$, or, $r_1 = 4.\tan(10°) = 0.705$ cm.

Thus, the diffraction spot for 200 reflection would be located at 0.705 cm away from the center of the x-ray film.

EXERCISES

5.1   A transmission Laue pattern of a cubic crystal (lattice parameter = 0.3 nm) is obtained with a horizontal x-ray beam in the orientation mentioned below. The [001] axis of the crystal points along the beam away from the x-ray tube, [010] points vertically upwards and [100] is horizontal and parallel to the x-ray film. The film is kept 5 cm away from the crystal. Determine the wavelengths of the radiation diffracted from the $(01\bar{2})$ planes and the location of this diffraction spot on the x-ray film.

[Ans: 0.12 nm, 0.06 nm, 0.03 nm and so on if the incident beam contains these wavelengths. Location = 6.67 cm away from the center.]

5.2 A transmission Laue pattern of an iron crystal (lattice parameter = 0.286 nm) is obtained on a film kept at a distance of 5 cm using 30 kV tungsten radiation. How close to the center of the pattern can Laue spots be formed by reflecting planes (110) and (200) assuming that these planes are present in the desired orientation?

[Ans: 1.04 cm, 1.49 cm]

5.3 A back-reflection Laue pattern of an aluminium crystal (lattice parameter = 0.4049 nm) is obtained with the (111) planes making an angle of 85° with the incident x-ray beam. Assuming that wavelengths larger than 0.25 nm are too weak and are easily absorbed by air,

(i) What orders of (111) reflection are present if the tube voltage is 60 kV?

(ii) If the tube voltage is 50 kV, what orders are present?

[Ans: (i) second to twenty-second (ii) second to eighteenth.]

ROTATING CRYSTAL TECHNIQUE

6.1 EQUIPMENT

The equipment used for recording rotating crystal photographs consists of a cylindrical film holder, a collimation system, a goniometer for mounting the crystal and a small motor for rotating the goniometer head spindle as shown in Fig.6.1. The photographic film, normally a sheet 125mm by 175mm in size, is held tightly against the inside walls of the cylindrical film holder by suitable clips. The film should be placed inside an opaque holder to prevent its exposure to visible light. The schematic drawings of the collimator and the beam stop are as in Fig. 6.2. Generally, collimator consists of two pinholes kept at a certain distance apart

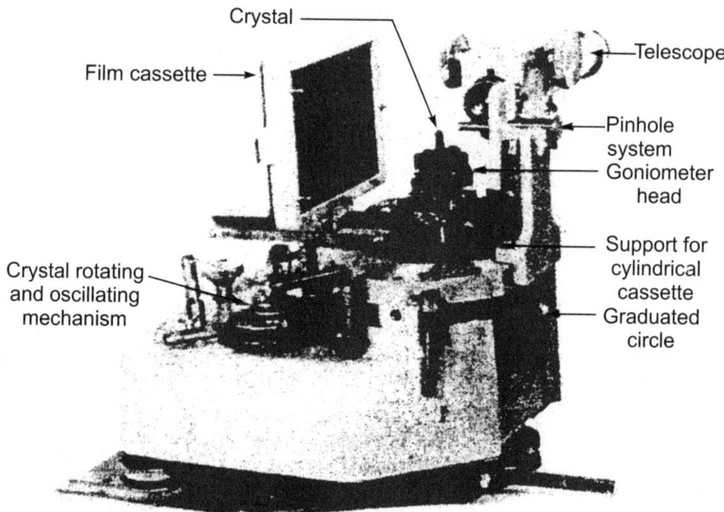

Fig. 6.1   An x-ray goniometer for Laue, oscillating-crystal, rotating-crystal, and powder methods. A flat-film cassette is shown (Unicam Instruments, Ltd., Cambridge, England.)
(Ref.: C. Barrett & T.B. Massalsiki, Structure of Metals, 3rd Edition, Chand & Co, New Delhi, 1968)

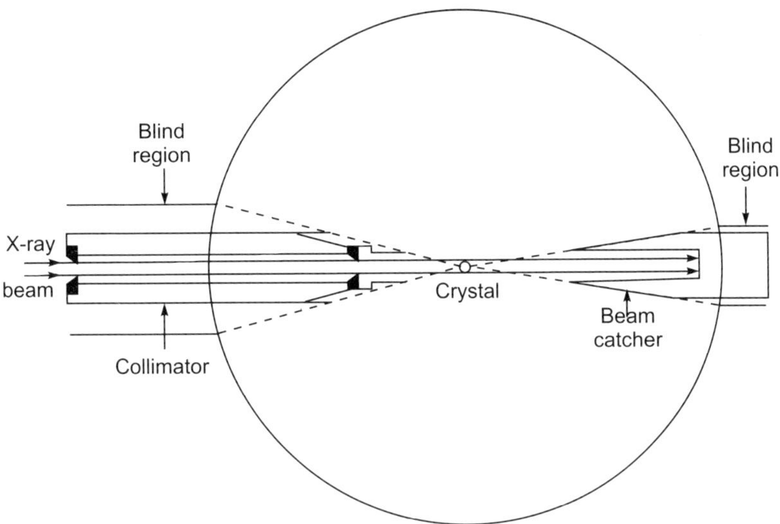

Fig. 6.2 Collimator, crystal and the beam stop

as discussed in Laue method in chapter 5. The exit pinhole scatters x-rays in all directions and these scattered rays, if not prevented from striking the film, can seriously increase the intensity of the background radiations obscuring the diffraction spots. So, the exit pinhole is 'guarded' by extending the collimator tube in a conical shape beyond the exit pinhole to prevent the scattered rays from affecting the film. The film stop consists of a 'guarded' lead (Pb) cup to absorb the transmitted beam. Another reason for extending the collimator and the beam stop tubes as close to the crystal as possible is to minimize the effect of scattering of the primary beam by air as it passes through the camera. Further, both these collimator and beam stop tubes are tapered in a conical fashion in order to minimize their interference with the low angle and high angle diffracted beams.

6.2 MOUNTING AND ADJUSTMENT OF CRYSTAL

The crystal size is limited usually to below half a millimeter in diameter in order to ensure a uniformly exposed diffraction photograph. This size also assures that the crystal is completely surrounded by the incident beam during the exposure. The amount of extraneous material used to hold the crystal should be kept to a minimum. For this reason a thin glass capillary is preferred to a solid fiber for holding the crystal in the path of the x-ray beam. Some kind of glue, which does not react chemically with the crystal, could be used to affix the crystal to the glass capillary. At the other end the capillary must be attached to the goniometer head through a brass pin as illustrated in Fig.6.3. A small set screw serves to lock the brass pin in position. When a crystal has one or more well developed faces, at least one of them should be made parallel to a goniometer arc. This is usually done by rotating the brass pin while the crystal is viewed through a binocular microscope. The presence of such well-defined faces then permits the proper alignment of the crystal rotation axis on an optical goniometer. After the crystal is correctly aligned the goniometer is mounted on the camera, s spindle and the crystal is viewed directly through the collimator for centering the crystal. The centering adjustments are carried out by means of two mutually orthogonal translations built into the goniometer head. When the crystal is properly centered, it appears to remain stationary while the spindle is rotated.

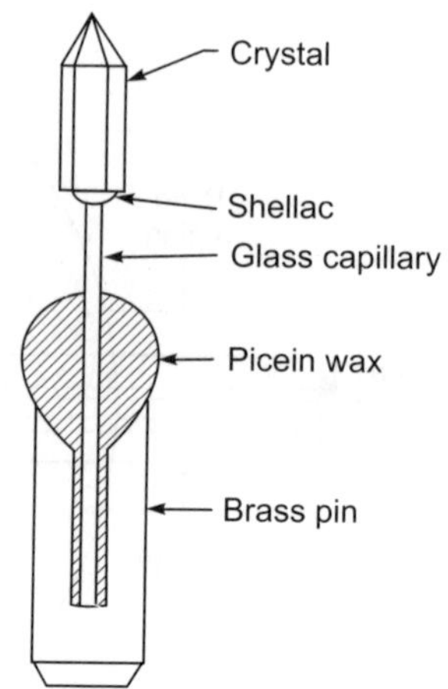

Fig. 6.3   Brass bin with crystal attached

6.3   THE RECIPROCAL LATTICE

6.3.1   Concept

The reciprocal lattice concept is indispensable for the interpretation of the rotating crystal diffraction patterns because of the convenient way in which the points in reciprocal space represent the orientations and spacings of the crystal planes. Further, the reciprocal lattice also indicates the conditions under which any plane will reflect and the position at which any reflection will be found on a rotating crystal pattern.

6.3.2   Introduction

Most of the diffraction phenomena could be explained by the simple Bragg's law. But there are many diffraction effects, which cannot be explained by Bragg's law like diffraction at non-Bragg angles. The explanation for these phenomena needs a more general theory and the reciprocal lattice provides the basis for such a theory. German physicist, Ewald, in 1921, introduced the concept of reciprocal lattice and since then it has become a very important tool in explaining certain complex diffraction effects.

6.3.3   Vector Algebra

A certain basics of vector algebra is essential to understand the concept of reciprocal lattice particularly the vector multiplication. The convention followed here for representing a vector is to write the alphabet denoting the vector in capitals and the small letter of the same alphabet is to be treated as the magnitude of the vector. The scalar product of the two vectors A and B , written as A . B is a scalar quantity and it is expressed as

$$A . B = ab . \cos \alpha \qquad (6.1)$$

The equation shows that the scalar product (also called dot product) of two vectors is a scalar quantity which is equal in magnitude to the product of the absolute values of the two vectors and the cosine of the angle α between them. Further, the scalar product of the sums or differences of vectors is found by term by term multiplication:

$$(A + B) . (C - D) = (A . C) - (A . D) + (B . C) - (B . D)$$

Further, the order of multiplication is not important.

i.e., $\qquad\qquad\qquad A . B = B . A$

The vector product of two vectors A and B written as A X B (called cross product) is a vector C at right angles to the plane containing A and B and equal in magnitude to the product of the absolute values of the two vectors and the sine of the angle α between them, i.e.,

$$C = A \times B \text{ and } c = a . b . \sin \alpha \qquad (6.2)$$

The magnitude of the vector C is the area of the parallelogram constructed on A and B as in Fig. 6.4. The direction of C is that in which a right handed screw would move if rotated in such a way as to bring A to B. Thus, it can be seen that

$$A \times B = - (B \times A).$$

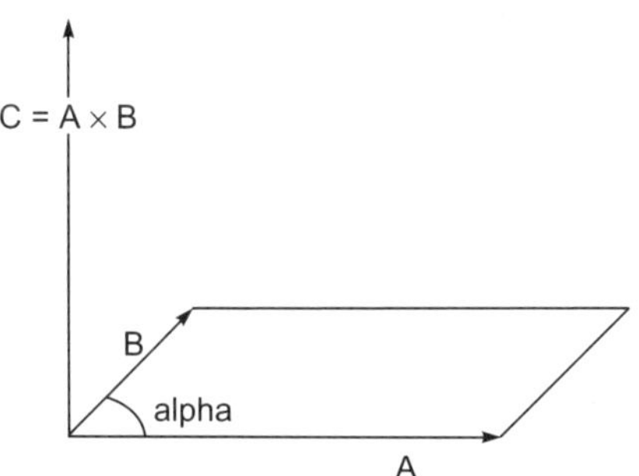

Fig. 6.4   Vector product

6.3.4 Construction and Properties of Reciprocal Lattice

It is possible to construct a reciprocal lattice for any given crystal lattice. Let a crystal lattice be defined by the lattice vectors A_1, A_2 and A_3. Then the corresponding reciprocal lattice has a unit cell made of vectors B_1, B_2 and B_3 where B_1, B_2 and B_3 are defined as :

$$B_1 = (A_2 \times A_3) / v \qquad (6.3)$$
$$B_2 = (A_3 \times A_1) / v \qquad (6.4)$$
$$B_3 = (A_1 \times A_2) / v \qquad (6.5)$$

where 'v' is the volume of the unit cell of the crystal lattice.

Consider a triclinic unit cell as shown in Fig. 6.5. The reciprocal lattice vector B_3 of this cell according to equation 6.5 is normal to the plane 'oxzy' (or (001) planes of the crystal lattice) containing the crystal lattice vectors A_1 and A_2. The magnitude of the vector B_3 (i.e., b_3) is given by the magnitude of the vector expressed by the equation 6.5,

i.e.,
$$b_3 = \frac{|A_1 \times A_2|}{\text{Volume of the crystal lattice unit cell}}$$

$$= \frac{\text{Are of the parallelogram oxzy}}{(\text{Area of the parallelogram oxzy}) \cdot (\text{height om})}$$

$$= 1 \,/\, om = 1 \,/\, d_{001}.$$

(since, om = plane spacing of (001) planes of crystal lattice)

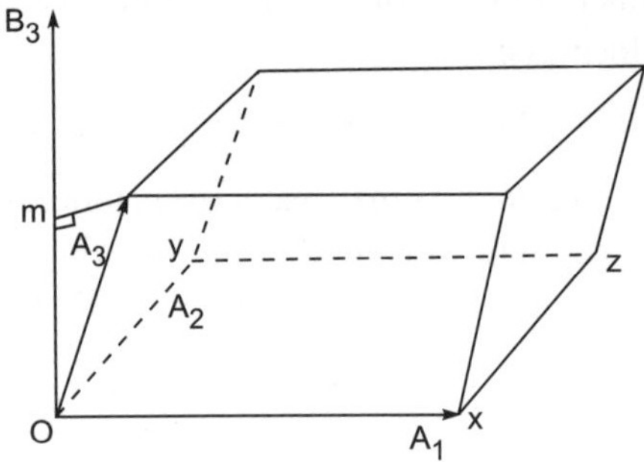

Fig. 6.5   Triclinic unit cell and the vector B_3

Similarly, it can be shown that the reciprocal lattice vectors B_1 and B_2 have their directions perpendicular to the (100) and the (010) planes of the crystal lattice respectively and their magnitudes are reciprocals of these plane spacings.

By extension of the above, similar relations can be found for all the planes of the crystal lattice. The reciprocal lattice can be constructed with unit cell vectors B_1, B_2 and B_3 and extending these unit cells in three dimensions. The lattice points of the reciprocal lattice points are labelled in terms of the basic vectors B_1, B_2 and B_3. For example, the point at the end of the vector B_1 is labelled as 100 and that at the end of B_2 is labelled as 010 and so on. The reciprocal lattice constructed in this fashion has the following properties:

1. A vector R_{hkl} drawn from the origin of the reciprocal lattice to any point having coordinates hkl is normal to the (hkl) plane in the crystal lattice. This vector R_{hkl} can be expressed in terms of the vectors B_1, B_2 and B_3 as

$$R_{hkl} = hB_1 + kB_2 + 1B_3$$

2. The dimension of the vector R_{hkl} is equal to the reciprocal of the spacing 'd' of (hkl) planes of the crystal lattice,

i.e., $$R_{hkl} = 1 / d_{hkl}.$$

It can be noted that the reciprocal lattice points represent sets of planes in the crystal lattice since both spacings and the orientations of the corresponding planes in the crystal lattice can be deduced from the above equations. Further, the reciprocal lattice is so called because many of its properties are reciprocal to the crystal lattice from which it is constructed.

6.3.5 Some Useful Relations

The dot product $B_3 . A_1$ or $B_3 . A_2$ are equal to zero since the reciprocal lattice vector B_3 is normal to both the crystal lattice vectors A_1 and A_2.

i.e., $$B_3 . A_1 = b_3 \times a_1 \cos \alpha = b_3 \times a_1 . 0 = 0$$
and, $$B_3 . A_2 = b_3 \times a_2 \cos \alpha = b_3 \times a_2 . 0 = 0$$

But the dot product $B_3 . A_3 = 1$,

since, $$B_3 . A_3 = b_3 \times (\text{projection of } A_3 \text{ on } B_3)$$
$$= 1/om \times (om)$$
$$= 1.$$

It follows , in general,

$$A_m . B_n = 1 \text{ if } m = n$$
$$= 0 \text{ if m is not equal to n.}$$

The relation between the planes of a zone and the zone axis, viz, hu+kv+lw = 0, can be proved with the concept of reciprocal lattice. The planes of a zone are all parallel to a line called zone axis. This means that the normals to all these planes of a zone must also be normals to the zone axis. If a plane (hkl) belongs to a zone whose zone axis is (uvw), then the normal to (hkl) plane, namely, the reciprocal lattice vector R_{hkl} must be normal to (uvw). The zone axis (uvw) can be expressed as a vector in the crystal lattice i.e.,

$$\text{zone axis} = uA_1 + vA_2 + wA_3.$$

Let R_{hkl} be the reciprocal lattice vector which is known to be perpendicular to the crystal lattice planes (hkl), i.e.,

$$R_{hkl} = hB_1 + kB_2 + lB_3.$$

It is known that R_{hkl} and the zone axis must be perpendicular to each other. Or, the dot product of these two vectors must be zero since the angle between these two vectors is 90°. In other words, $(uA_1 + vA_2 + wA_3) . (hB_1 + kB_2 + lB_3) = 0$

i.e., $$hu+kv+lw = 0 .$$

This is the relation used in chapter 2 while drawing standard projections.

6.4 DIFFRACTION AND RECIPROCAL LATTICE

Let us consider the effect of x-rays scattered by an atom at 'a' (p,q,r) on the x-rays scattered by the atom at the origin 'o' of the crystal lattice. Let the coordinates p,q,r be integers. The crystal lattice vector 'oa' can be expressed as

$$oa = pA_1 + qA_2 + rA_3.$$

Let the incident x-rays have a wavelength λ and be represented by the unit vector 'S_0' and the diffracted beam be represented by the unit vector 'S'. The vectors S_0, S and oa , in general, are not coplanar.

For the determination of conditions under which diffraction occurs, the phase difference between the rays scattered by the atom at 'o' and at 'a' must be found. The lines ou and ov in Fig.6.6 are the wave fronts perpendicular to the incident beam S_0 and the diffracted beam S respectively.

The path difference between the rays scattered by atoms at 'a' and 'o' is given by (from Fig.6.6),

$$\text{path difference} = ua + av$$
$$= om + on$$
$$= S_0 \cdot \text{vector oa} + (-S) \cdot \text{vector oa}$$
$$= -\text{vector oa} \cdot (S - S_0)$$

The phase difference f and the path difference are related by

$$f = (\text{path difference}) \times 2\pi / \lambda$$

Substituting for the path difference obtained

$$f = -2\pi \cdot \text{vector oa} \cdot (S - S_0) / \lambda$$

Let the vector $(S - S_0) / \lambda$ be expressed as:

$$(S - S_0) / \lambda = hB_1 + kB_2 + lB_3$$

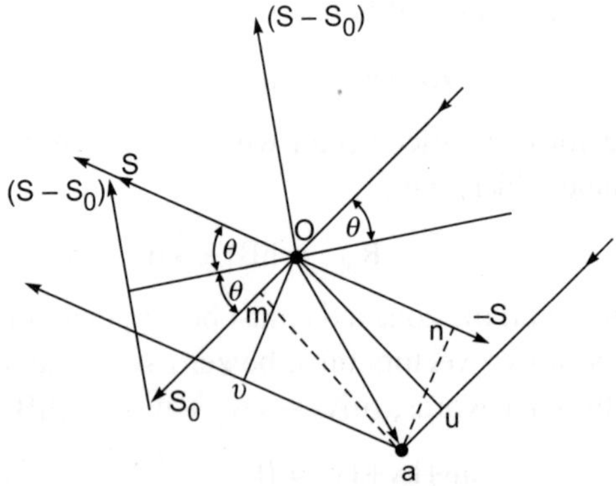

Fig. 6.6   X-ray scattering by atoms at O and a (After Guinier, X-ray Crystallographic Technology, Hilger & Watts, Ltd. London, 1952)

In other words, $hB_1+kB_2+1B_3$ is a vector in the reciprocal lattice. Now the phase difference is given by

$$f = -2\pi\,(hB_1+kB_2+1B_3)\,.\,(pA_1 + qA_2 + rA_3)$$
$$= -2\pi\,(hp+kq+1r)$$

Now a diffracted beam will be formed only if f is an integral multiple of 2π. This can happen only if h, k and l are integers as it is already assumed that p, q and r are integers. In other words, if diffraction is to occur, h, k, and l must be integers or $hB_1+kB_2+1B_3$ is a reciprocal lattice vector and the vector $(S - S_0)\,/\,\lambda$ must end on a reciprocal lattice point hk_1.

Or, $$(S - S_0)\,/\,\lambda = R_{hk1} = hB_1+kB_2+1B_3 \tag{6.6}$$

where h, k and l are integers.

Laue equations can be obtained from equation 6.6 by forming the dot product of each side of the equation with the three crystal lattice vectors A_1, A_2 and A_3 successively,

i.e., $$A_1\,.\,(S - S_0)\,/\,\lambda = A_1\,.\,(hB_1+kB_2+1B_3)$$
$$= h$$

similarly, $$A_2\,.\,(S - S_0)\,/\,\lambda = k$$

and $$A_3\,.\,(S - S_0)\,/\,\lambda = 1.$$

Or, $$A_1\,.\,(S - S_0) = h\lambda$$
$$A_2\,.\,(S - S_0) = k\lambda \tag{6.7}$$
$$A_3\,.\,(S - S_0) = 1\lambda$$

Equations 6.7 are the vector form of the equations derived by Von Laue in 1912 which must be satisfied in order that diffraction could occur.

Bragg's law can also be derived with the help of the reciprocal lattice concept as follows. From Fig. 6.6, the $(S - S_0)$ vector bisects the angle between the incident beam S_0 and the diffracted beam S. The diffracted beam S can be thus considered as being reflected from a set of planes perpendicular to the vector $(S - S_0)$. The equation 6.6 shows that the vector $(S-S_0)$ is parallel to the reciprocal lattice vector R_{hk1} which is in turn perpendicular to the crystal lattice plane (hk1). If θ is the angle between S or S_0 and these crystal lattice planes (hk1), then the magnitude of the vector $(S-S_0)$ is given by

$$S-S_0 = 2\sin\theta\ \text{(since, S and } S_0 \text{ are unit vectors)}$$

Therefore, $$2\sin\theta\,/\,\lambda = (s - s_0)\,/\,\lambda = |\,R_{hk1}\,|$$

But $$|\,R_{hk1}\,| = 1/d_{hk1}\ \text{from the reciprocal lattice concept,}$$

or, $$\lambda = 2\,d_{hk1}\sin\theta.$$

The conditions for diffraction expressed by equation 6.6 can be graphically represented by the 'Ewald construction' shown in Fig. 6.7. The vector $S_0\,/\,\lambda$ is drawn parallel to the incident beam and of length $1 / \lambda$ where λ is the wavelength of the incident x-ray beam. The terminal point O is taken as the origin of the reciprocal lattice. A sphere of radius $1 / \lambda$ is drawn about 'c' the initial point of the incident beam vector S_o / λ. The condition for diffraction from the (hk1) planes is that the reciprocal lattice point hk1 (the point 'p' in Fig. 6.7) touch the surface of the sphere. The direction of the diffracted beam vector S / λ is found by joining 'c'

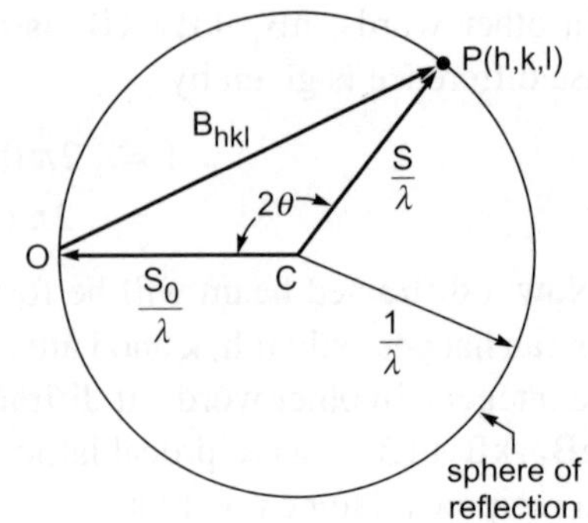

Fig. 6.7   The Ewald construction. Section through the sphere of reflection containing the incident and diffracted beam vectors.

with 'p'. When this is done, the vector 'op' equals both R_{hk1} and $(S - S_o) / \lambda$ thus, satisfying the equation 6.6. The sphere drawn is known as the sphere of reflection (Ewald sphere) because of the condition that the reflection occurs only if the reciprocal lattice point touches the surface of this sphere.

6.5   ROTATING CRYSTAL METHOD AND RECIPROCAL LATTICE

When monochromatic radiation is incident on a single crystal rotated about one of its axes, the diffracted beams lie on the surface of imaginary cones coaxial with the rotation axis. The schematic representation of how this reflection occurs can be shown by the Ewald construction. Rotation of a simple cubic crystal about the [001] axis is equivalent to rotation of its reciprocal lattice about the B_3 axis. Fig.6.8 shows a portion of the reciprocal lattice oriented in this fashion along with the sphere of reflection adjacent to it. All crystal planes having indices (hk1) are represented by points lying on a plane called the '1 = 1 layer' in the reciprocal lattice, normal to B_3. When the reciprocal lattice rotates, this plane cuts the sphere of reflection in the small circle shown, and any points on the '1 = 1 layer' which touch the surface of the sphere must touch it on this circle. Then all the diffracted beam vectors S/λ must end on this circle, which is similar to saying that the diffracted beams must lie on the surface of a cone. In this particular case, all the hk1 points shown intersect the surface of the sphere sometime during their rotation about the B_3 axis producing the diffracted beams shown in Fig.6.8. Similarly, hk0 and hk1 reflections would also be produced but these have not been drawn for the sake of clarity. Depending on the size of the reciprocal lattice and the size of the Ewald sphere, it is possible to have '1=2 layer' or higher layer reflections occurring.

Spots on a rotating crystal photograph are conveniently referred to a rectangular coordinate system that is related in a simple way to the cylindrical

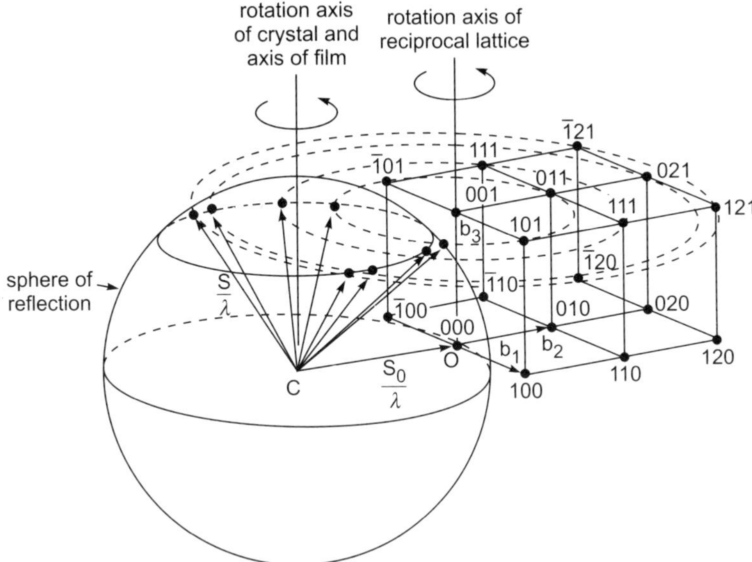

Fig. 6.8   Reciprocal-lattice treatment of rotating-crystal method

coordinates of a reflecting point in the reciprocal lattice. As shown in Fig. 6.9, the orientation of the reciprocal lattice vector B is specified by two mutually orthogonal vectors, viz., V along the vertical rotation axis and H in the horizontal plane containing the x-ray beam. The angle φ is the angle between the direct beam of x-rays and the plane containing the vectors V and H.

The important vector relation between these coordinates is

$$B = V + H$$

Similarly, their magnitudes are related by

$$b^2 = v^2 + h^2.$$

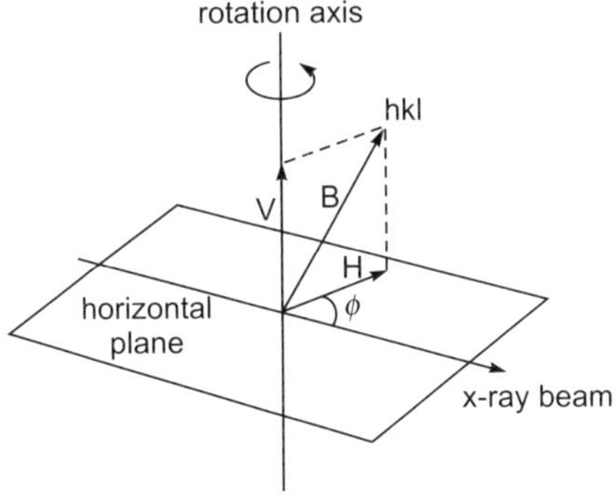

Fig. 6.9   Cylindrical coordinates for reciprocal lattice points

It follows from the above that points in the reciprocal lattice having a constant V value lie in a plane normal to the rotation axis. Similarly, points having a constant H value lie in a cylinder about the rotation axis. Fig.6.10 shows the intersection of both the plane (with constant V value) and the cylinder (with constant H value) with the Ewald sphere. The intersection curves of constant H value appear on a cylindrical film that surrounds the crystal's rotation axis in a way as shown on the

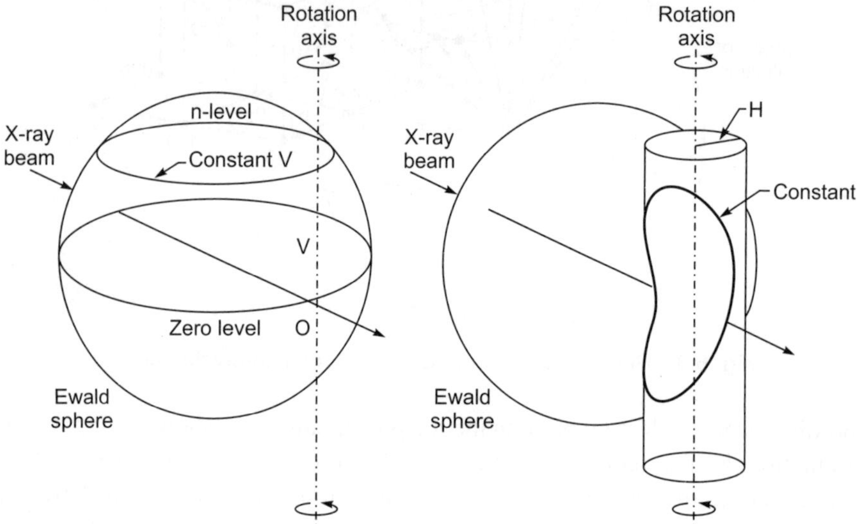

Fig. 6.10   Intersection of planes having constant V with Ewald sphere (left side) and cylinders having constant H (right side). (Ref.: M. J. Buerger, X-ray Crystallograhy, Wiley, New York, 1942).

Bernal chart reproduced in Fig.6.11. Each Bernal chart is prepared for a particular crystal-to-film distance. Such a chart can be superimposed directly over the rotating crystal diffraction pattern and the V and H coordinates of each spot can be read directly from the chart as shown in Fig. 6.12. Assuming that the crystal was mounted so as to rotate about the normal to the (001) planes, the value of d_{001} is given by

$$d_{001} = \lambda / V.$$

Accordingly, when the symmetry of the crystal is known, the dimension of the unit cell can be determined by rotating the crystal about each of the crystallographic axis and by using the above equation.

In the cubic system, all three cell edges are the same so that one rotation photograph about the cell edge 'a' is all that is needed. In the tetragonal and hexagonal systems, rotation photographs about the cell dimensions both 'c' and 'a' are necessary. In such cases as well as in case of more complex systems, it is better to index the reciprocal lattice spots and establish the crystal system and then proceed to determine the lattice parameters. Discussions on such procedures are elaborately dealt in other standard works.

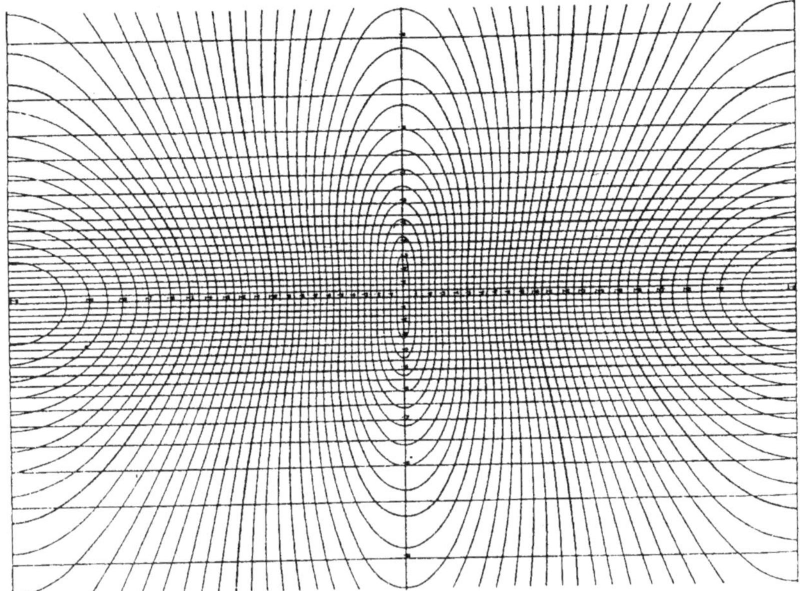

Fig. 6.11   Bernal chart (Ref: L.V. Azaroff, Elements of x-ray Crystallography, McGraw-Hill, New York, 1958)

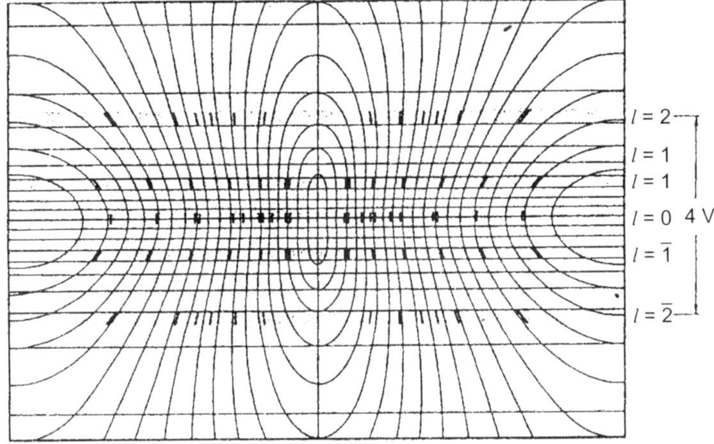

Fig. 6.12   Schematic view of a film overlaid on a Bernal chart. Note that the x-ray diffraction spots lie along rows of constant V and H (only a few possible reflections are indicated).

EXERCISES

6.1 Construct a reciprocal lattice based on

(i) a cubic crystal with a lattice parameter of 0.3 nm,

(ii) a tetragonal crystal with a = 0.3 nm and c = 0.4 nm.

7

CHAPTER

POWDER TECHNIQUES

7.1 INTRODUCTION

The powder method is the most commonly used method of x-ray diffraction and when properly employed can yield a lot of structural information about the powder sample. The method, devised first in 1916 by Debye and Scherrer in Germany and later in 1917 by Hull in USA, each independently, makes use of the diffraction of monochromatic x-rays by a powder specimen. The monochromatic x-ray beam can be obtained normally by β filtering of x-rays from an x-ray tube operated above the K-excitation voltage of a specific target material. The powder specimen refers to either powdered solid sample with a suitable binder to hold the powder particles together or a polycrystalline specimen in the form of a wire or sheet.

Depending on the positions of the photographic film and specimen, the three main powder techniques are the following:

(a) The Debye-Scherrer technique in which the film is placed on the inner surface of a cylindrical camera and specimen along the axis of the camera.

(b) The Focusing technique in which the film, specimen and x-ray source are all placed on the inner surface of a cylindrical camera.

(c) The Pinhole technique in which a flat photographic film is placed perpendicular to the incident x-ray beam and the specimen is placed either in front or at the back of the film.

In all these powder techniques, the diffracted beams lie on the surfaces of cones whose axes lie along the incident beam or its extension. Each cone of diffraction forms from a particular set of hkl plane satisfying the Bragg's law. In the pinhole technique, the entire diffraction ring is obtained on the photographic film whereas in Debye-Scherrer and focusing techniques, only a part of the diffraction ring is obtained since, the films used are in the form of narrow strips.

7.2   CHOICE OF RADIATION FOR POWDER TECHNIQUES

The choice of radiation for powder technique basically depends on the specimen being investigated whatever be the specific technique chosen. There are two important considerations in choosing the radiation.

1. The characteristic wavelength chosen should be longer than the K-absorption edge of the elements present in the specimen. If the chosen radiation is shorter than the K-absorption edge of any element of the specimen, K-fluorescent radiation of that element would be emitted which may badly fog the film masking the diffraction pattern.
2. The shorter the wavelength chosen, the larger the number of lines present on the diffraction pattern of a given crystalline material. It can be seen from Bragg law that as the wavelength decreases, the Bragg angle decreases for a given crystalline material, thereby increasing the number of lines for a given angular range of the camera.

The commonly employed monochromatic radiation for most of the experiments is copper $K\alpha$ with a wavelength of 0.15418 nm (weighted average of $K\alpha$ and $K\beta$ wavelengths). Depending on the specimen, other radiations (wavelengths) are also used like $Mo\text{-}K\alpha$, $Co\text{-}K\alpha$, $Fe\text{-}K\alpha$ and $Cr\text{-}K\alpha$. It should be noted that $Cu\text{-}K\alpha$ radiation is not suitable for irons and steels because its wavelength is shorter than the Fe-K-absorption edge (0.1734 nm).

7.3   BACKGROUND RADIATION

A powder diffraction pattern should have sharp diffraction lines on a dark background. There should be a good contrast the diffraction lines and the background on the pattern. For this it is necessary that the background intensity, on which the diffraction lines are superimposed, must have minimum intensity. This background intensity arises because of several causes noted below:

(a) Fluorescent radiation from the specimen

Even though the radiation selected has a wavelength longer than the K-absorption edge of the specimen used, there is a possibility of emission of fluorescent radiation by the specimen. This is because there are certain short wavelength components in the x-ray beam shorter than the K-absorption edge of the specimen even after the β-filtering. The β-filtering only minimizes the intensity of the $K\beta$ component of the radiation and incidentally reduces the intensities of other short wavelength components too. But, it does not completely eliminate these components. For example, imagine that a zinc specimen is being diffracted by $Cu\text{-}K\alpha$ radiation of wavelength 0.15418 nm from an x-ray tube operated at 40 kV. The radiation issuing out of the x-ray tube will have all the wavelengths right from a short wavelength limit of approximately 0.031 nm (i.e., $\lambda = 1239.8/40000 = 0.031$ nm). Even after β-filtering of this beam, the radiations still have the short wavelength components though with lesser intensities. These components of radiations of wavelengths between 0.031 and 0.128 nm (0.128 nm is the K-absorption edge of the specimen zinc) will cause fluorescent radiation from the specimen.

If it is not possible to choose a proper radiation with a wavelength longer than the K-absorption edge, proper filters can be used over the film so as to reduce the intensity of the fluorescent radiation reaching the film. Generally, in such cases, a considerably shorter wavelength radiation is to be used and the film covered with a suitable filter. The principle here is that the shorter the radiation, more would be the penetration through the filter and its consequent effect on the film and the fluorescent radiation being longer in wavelength is effectively absorbed by the filter. Further, it should be noted that longer wavelength fluorescent radiation like that of aluminium (0.834 nm) is easily absorbed by air and air can work as a efficient filter in cases of wavelengths more than about 0.3 nm.

(b) Diffraction of the continuous spectrum

A powder specimen is nothing but a collection of a large number of tiny crystals. Each crystal of the powder specimen can produce Laue patterns on the film due to the presence of continuous radiation in the incident beam. Thus, the amount of background radiation due to this cause can be quite substantial because of the superimposition of an extremely large number of Laue patterns though each Laue pattern may have low intensity spots.

(c) Diffuse scattering from the specimen may be due to:

(i) Incoherent diffused scattering which becomes more intense as the atomic number of the specimen decreases;

(ii) Coherent diffuse scattering caused either by heating referred as the temperature diffuse scattering or due to various types of imperfections (strain based or other) present in crystals.

(d) Diffraction and scattering from sources other than the specimen, e.g.,

(i) From the collimator and the beam-stop: this can be minimized by proper camera design.

(ii) From specimen-binder, specimen-holder, enclosure: the adhesive used to compact the powder specimens like gum, quickfix or araldite, the glass fibre to which the powder is attached, quartz tube in which the powder is enclosed can all contribute to the background radiation since all of these are amorphous substances. The quantity of these materials used should be kept to a minimum.

(iii) From air: This can be avoided by evacuating the camera or minimized by filling it with a light gas such as hydrogen or helium.

7.4   CRYSTAL MONOCHROMATORS

The most important contributors to the background are fluorescent radiations emitted by the specimen and diffraction from the continuous spectrum which are elaborated in a) and b) of the previous section 7.3. The cause discussed under c) cannot be remedied as it is inherently connected with the specimen itself. The cause d) can be taken care of and minimized by controlling the experimental setup as indicated elsewhere. The factors discussed under a) and b) above, can be completely eliminated if a strictly monochromatic beam is employed for diffraction, leading to diffraction patterns of extremely high clarity and contrast.

Strictly monochromatic radiations can be obtained using radiations from an x-ray tube provided the needed $K\alpha$ radiation is selectively diffracted by a set of planes of a known crystal at the required Bragg angle.

There are two types of crystal monochromators: i) unbent and ii) bent and cut. An unbent crystal can be used as a monochromator when it is set at the right Bragg angle for diffraction of the $K\alpha$ component of the incident beam so that it can diffract only that component. The diffracted beam, though of low intensity due to the highly inefficient diffraction process, is strictly monochromatic.

High intensity focused monochromatic radiation can be obtained by using bent and cut crystals either of reflecting type or of transmitting type. A focusing monochromator of reflection type is illustrated in Fig. 7.1. In the figure 'S' is a line source of x-rays located perpendicular to the plane of the drawing. AB is a crystal in the form of a rectangular plate with reflecting planes parallel to its surface. The crystal is elastically bent into a radius of curvature of 2R=CM. The face of the crystal is then cut away behind the dotted line to a radius R. Now, the rays diverging from the source 'S' will encounter the lattice planes at the same Bragg angle, since, the angles SDM, SCM and SEM are all equal to one another, being inscribed on the same arc SM, and have the value $(\pi/2)$-θ. When the angle of incidence is adjusted to that required for the diffraction of the $K\alpha$ component of the incident beam, a strong $K\alpha$ wavelength will be diffracted by the crystal and since the diffracted rays all originate on a circle passing through the source 'S', they will converge to a focus at 'F', located at a distance FC (same as SC). The value of Bragg angle θ required for the diffraction of a particular wavelength λ from planes of spacings 'd' can be calculated from the Bragg law, viz., $\lambda = 2d \sin \theta$. The source to crystal distance SC which is equal to CF is given by

$$SC = 2R \cos \left[\left(\frac{\pi}{2} \right) - \theta \right]$$

$$= 2R \sin \theta$$

$$= R. \frac{\lambda}{d} \left(\text{substituting } \frac{\lambda}{d} \text{ for } 2.\sin \theta \right).$$

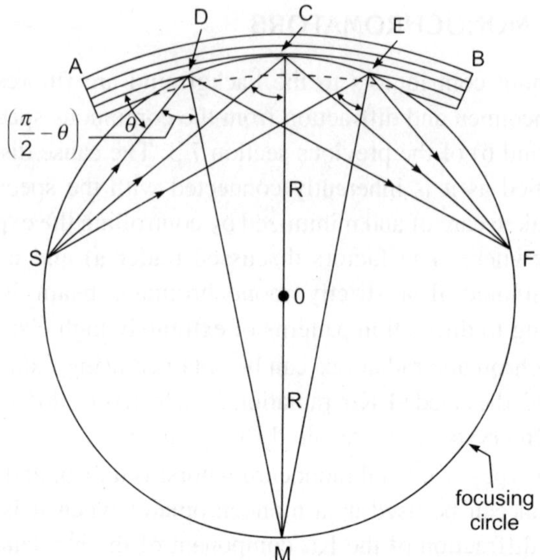

Fig. 7.1   Focusing monochromator (reflection type)

Since, there is focusing action in this bent and cut crystals, the diffracted beam is quite strong and exposure time can be considerably reduced using these monochromators.

7.5   MEASUREMENT OF LINE POSITION

Powder diffraction patterns are analyzed for the position of the diffraction lines and their intensities. The measurement of line intensities has been discussed in section 3.4.2. The measurement of positions of diffraction lines can be accomplished with the help of an illuminated ground glass screen over which a micrometer scale is fixed.

7.6   THE DEBYE-SCHERRER TECHNIQUE

7.6.1   General

The Debye-Scherrer camera consists of a cylindrical chamber with a light-tight cover, a collimator to admit and collimate the incident beam, a beam-stop to stop the transmitted beam, a film holder to hold the film tightly against the inner curved surface of the camera and a specimen holder for mounting the specimen at the camera axis.

The resolving power, in general, is the ability to distinguish very closely spaced objects as separate entities. In microscopes, the resolving power or resolution is the smallest distance of separation between two objects which can be clearly identified. In spectroscopy, the resolving power is given by $\lambda/\Delta\lambda$ where $\Delta\lambda$ is the difference between the two wavelengths close to each other and λ is their mean value.

Similarly, the resolving power in the crystal structure analysis may be taken as the ability to separate the diffraction lines from planes of very nearly the same spacing. Or, in other words, the resolution can be defined as d/Δd where d is the average plane spacing and Δd is the difference between the spacings of the two planes concerned.

Fig.7.2 illustrates the geometry of diffraction in the Debye-Scherrer camera. I_o is the incident beam striking the specimen at 'o'. The rays OD and OT are diffracted and transmitted beams respectively. The angle 2θ is the diffraction angle and 'S' is the distance measured on the film from the diffraction line 'D' to the transmitted beam 'T'. If 'R' is the radius of the camera, then,

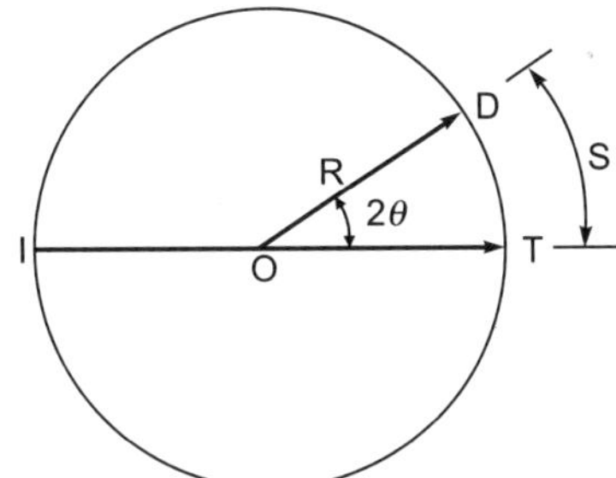

Fig. 7.2 Schematic section of D-S camera perpendicular to its axis

$$S = 2\theta.R \text{ and therefore, } \Delta S = R. \Delta 2\theta \qquad (7.1)$$

The equation 7.1 indicates that two planes of very nearly the same spacing will give rise to two diffracted beams separated by a small diffraction angle $\Delta 2\theta$ and the separation distance on the film would be ΔS. Differentiating the Bragg law with respect to 'd', one can obtain

$$0 = 2\left(d \cos\theta . \frac{d\theta}{dd} + \sin\theta . 1 \right)$$

or,
$$\frac{d\theta}{dd} = \frac{-\sin\theta}{d\cos\theta} = \frac{-\tan\theta}{d} \qquad (7.2)$$

But from 7.1 $d\theta = \dfrac{dS}{2R}$

Substituting this expression for $d\theta$ in 7.2,

$$\frac{dS}{2R} = -2\,dd.\frac{\tan\theta}{d}$$

or,   resolving power $\dfrac{d}{\Delta d} = -2R.\dfrac{\tan\theta}{\Delta S}$ (7.3)

where 'd' is the mean spacing of two planes and Δ d is the difference in their spacings and Δ S is the separation of two diffraction lines on the film. The equation 7.3 indicates that the resolving power is directly proportional to the radius of the camera and tan θ. Thus, as the camera size increases, the resolution increases but the intensity of the diffracted beam decreases since the beam has to travel through larger distances before registering on the film. In addition, the absorption by air also increases because the beam travels longer through air. Therefore, larger size cameras need longer exposure times for getting a good diffraction pattern. Generally, smaller diameter cameras are preferred unless a very high resolution pattern is required. Debye-Scherrer cameras with 57.3 mm diameter are commonly used and most suitable for general work. This 57.3 mm diameter camera has a convenient circumference of 180 mm corresponding to 360° which facilitates the calculation of the Bragg angle for any diffraction line. As shown in Fig. 7.2, in these cameras, 'S' in mm multiplied by 2 gives 2θ in degrees. Debye-Scherrer cameras with 60 mm or 120 mm diameter are also available. In these larger size cameras the decrease in diffracted intensity may be counteracted to some extent by evacuating the camera or filling it with a light gas such as helium or hydrogen during exposure.

The incident beam has to be collimated by a proper design of the pinhole system, particularly when weak diffracted beams are to be recorded. Similarly, a proper beam stop has to be utilized for stopping off the transmitted beam. Fig.7.3 shows the design of collimator and beam stop generally used in Debye-Scherrer cameras. The collimator has a 'guarded pinhole' assembly wherein the scattered radiation from the pinhole is prevented from striking the film by extending the collimator tube to a considerable distance beyond the exit pinhole as shown. The beam stop is generally a thick lead glass with a fluorescent screen on the inner side so as to facilitate the viewing of the transmitted beam with safety. Both the collimator and the beam stop have guard tubes which are tapered in the way shown in figure so as not to interfere with the low angle and high angle diffracted beams.

7.6.2 Specimen Preparation

In order to increase the chances of crystals which have all rotational orientation positions producing complete Debye diffraction cones as discussed in section 4.3.3, bulk specimens are converted into a fine powder by filing or by grinding in a small agate mortar. The fineness of the powder should be less than 50 μm in

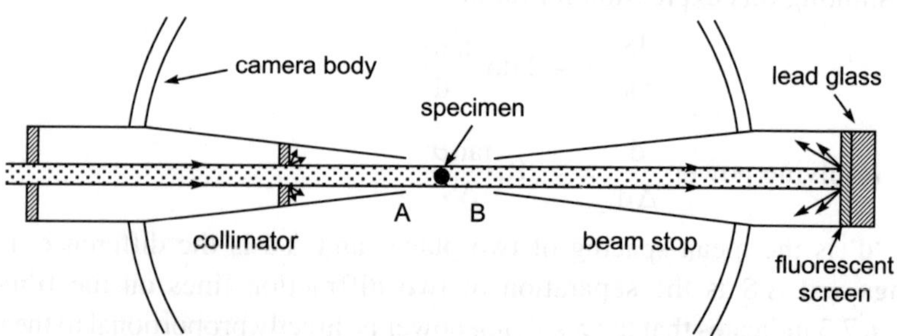

Fig. 7.3 Design of collimator and beam stop (schematic)

order to produce a continuous line pattern. The powder should then be annealed in evacuated glass or quartz capsules in order to minimize the effect of cold working.

Basically the Debye-Scherrer camera is so designed as to use polycrystalline specimens in the form of a thin rod of 0.5 mm in diameter or less and about 2 cm long. There are a number of ways of getting such a shape. One method is to mix the powder with an amorphous binder such as glue or araldite and roll into a rod shape. Another is to fill the powder into a thin tube of required size made of cellophane or a weakly absorbing amorphous material. Alternately, one can also use a thin glass fiber on which the powder can be made to adhere using a glue. In all these, it has to be remembered that the materials used as binders or supporters should be amorphous so that they will not cause additional diffraction lines of their own. Further, these materials should absorb the diffracted beams as little as possible. Polycrystalline wire specimens may be used directly, but the diffraction pattern must be analyzed keeping in mind the possibility of the existence of preferred orientation.

After the specimen has been prepared as discussed earlier, it can be mounted on the specimen holder coaxially with the camera. Care must be taken to see that when the specimen holder is rotated, the specimen rotates on its own axis without any wobbling. While the specimen is being diffracted, it is common practice to rotate the specimen by a small motor so that more powder particles will be oriented in diffraction positions.

7.6.3   Film Loading Methods

There are mainly three methods of film loading in a Debye-Scherrer camera as illustrated in Fig. 7.4 a), b) and c). The left hand figures show the opened out exposed film and the right side figures show the schematic cross-sectional views of the Debye-Scherrer camera for each of the methods. In both methods indicated by a) and b) referred to as symmetric methods, the film is placed symmetrically about the incident and transmitted beams and the method c) is a non-symmetrical method since both the ends of the film strip are present on the same side.

The method illustrated in Fig. 7.4 a) is termed as Bradley-Jay method where a hole is punched in the center of the film so that this strip may be slipped over the beam stop through the hole so that the film stays touching the inner curved surface of the camera. The transmitted beam leaves through the location of central hole of the film and the diffraction pattern is symmetrical with respect to the central hole. The Bragg angle θ for any diffraction line can be computed by measuring the distance 'U' using the relation : $U = 4\theta.R$
where 'R' is the radius of the Debye-Scherrer camera. During the process of developing the film, film shrinkage occurs. This shrinkage changes the distance 'U' measured on the film thereby making the θ calculation erroneous. This can be taken care of by slipping the ends of the film strips under metal knife-edges,

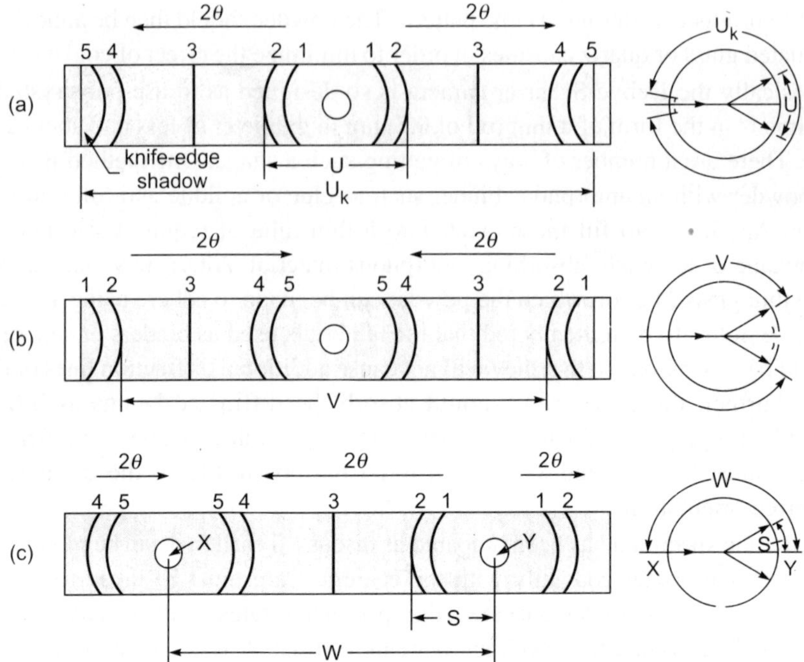

Fig. 7.4   Methods of film loading in Debye cameras. Corresponding lines have the same numbers in all films

specially provided for the purpose, which cast a sharp shadow near each end of the film. The distance 'U_k', between the knife-edges, and 'U', for a particular diffraction line, both will shrink in the same proportion during developing of the film. Thus, if the angular separation , $4\theta_k$, between the knife-edges of the camera is known, the Bragg angle θ for a particular set of diffraction lines can be calculated simply by $\theta/\theta_k = U/U_k$.

Fig. 7.4 b also shows a symmetric method, but, here, the incident beam enters at the center of the film and leaves at the ends of it. This method is known as Van-Arkel method. Here, the calculation of the Bragg angle θ can be done by using the relation, by geometry, $(2\pi - 4\theta)$. $R = V$. The shrinkage correction can be accomplished, here too, by the use of the knife-edges as discussed under Bradley-Jay method. It is well-known that high precision in lattice parameter determination can be obtained by making use of high angle diffraction lines. Better accuracy in high angle 'θ' measurements can be obtained by going in for Van-Arkel method since, on the high angle side, the film strip is continuous unlike in the Bradley-Jay method.

In the Straumani's method, illustrated in Fig. 7.4 c), the film strip has two holes, one for slipping over the beam stop and the other for the collimator. This method is called unsymmetrical because the diffraction line pairs would not be present symmetrically about the center of the film. Here, the knife-edges are not required for film shrinkage correction because the distance between the two holes in the film (i.e., the distance xy = w) corresponds to an angle of 180°. The Bragg angle θ corresponding to any distance 's' can be determined by the relation: $180°/2\theta = w/s$. In fact, even the knowledge of the camera diameter is not necessary here since, it is known from the loading technique that the distance between the two holes is directly equal to an angle of 180°.

7.6.4 High and Low Temperature Cameras

Many a time, in materials research, there is a need to go for high temperatures to determine phases present at such temperatures. Similarly, a low working temperature is also desired for similar reasons. In certain cases, it is possible to retain a high temperature phase on quenching to a low temperature. In these cases, the usual cameras can be used to identify the high temperature phase by quenching from such high temperatures and using the quenched specimens for the diffraction experiments.

Generally, all high temperature Debye cameras are equipped with a small furnace (usually electric resistance type) to heat the specimen and a thermocouple for temperature measurement both of which should not hinder the incident or diffracted beams. The film for recording the pattern has to be kept cool by cooling water circulation through the camera body or by keeping radiation shields (which should not interfere with the diffracted beams) between the furnace and the film. If the specimen is highly reactive at high temperatures, either the camera must be evacuated or filled with an inert gas or the specimen may be sealed in a thin walled silica tube. Further, the exposure times are generally longer here since the intensities of diffracted beams are low at high temperatures.

Cameras for low temperatures are equipped with an arrangement for passing a thin stream of coolant like liquid air over the specimen throughout the diffraction experiment. The diffraction pattern of the coolant also will be recorded but this can easily be distinguished from that of a crystalline solid. Diffraction pattern of a liquid will contain only one or two diffuse lines in contrast to the sharp lines of crystalline solids.

7.7 FOCUSING TECHNIQUES

7.7.1 Focusing Cameras: Principle

The diffracted rays originating from a large surface area of the specimen are all made to converge to one point on the x-ray film in all focusing cameras. The design of all these cameras is based on the geometrical concept that angles inscribed in a circle on a given arc 'SF' are all equal to one another and also equal to half the angle subtended at the center by the same arc 'SF' as shown in Fig. 7.5. In this figure, 'AB' is the powder specimen, 'S' and 'F' are the source of the x-

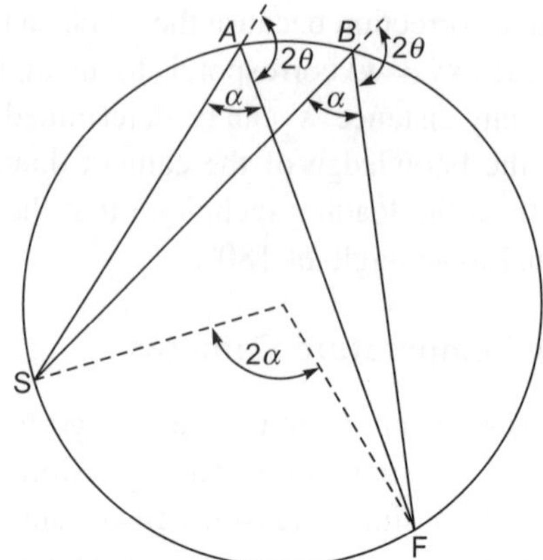

Fig. 7.5   Geometry of focusing cameras

ray beam and focused diffracted beam respectively. The rays diffracted at 'A' and 'B' both will be deviated through the same diffraction angle 2θ (=180° $-\alpha$) and will come to focus by the above mentioned geometry at F on the film placed along the circumference of the camera.

7.7.2   The Seemann-Bohlin Camera

Fig. 7.6 shows the geometry of Seemann-Bohlin camera with reference to the positions of the film MN, the specimen AB and the source slit S. The extended focal spot on the target T of the x-ray tube emits x-rays and such of those rays converging at the slit S will be able to enter the camera and then diverge and fall on to the specimen AB as the incident beam (divergent). If the x-ray tube has the line focal spot, then the focal spot itself could be used as the source in the camera replacing the slit. Whichever be the source, the divergent beam falling on the specimen AB will get diffracted from a particular set of (hkl) planes at diffraction angle 2θ and these diffracted rays would all converge to a line focus at F as shown in Fig. 7.6. The film strip ends are covered by knife-edges M and N which cast reference shadows on the film. It can be seen from Fig. 7.6 that

$$4\theta.R = \text{arc SABN} + U \qquad (7.4)$$

But this equation is rarely used in such cameras. Instead, a standard specimen, whose 'd' values are known, is used for drawing a calibrating curve of the film shrinkage corrected values of 'U' versus standard 'd' or 'θ' values after conducting diffraction experiments on it. This calibration curve can be used to determine the 'd' values or 'θ' values for an unknown specimen by measuring the 'U' values from its diffraction pattern.

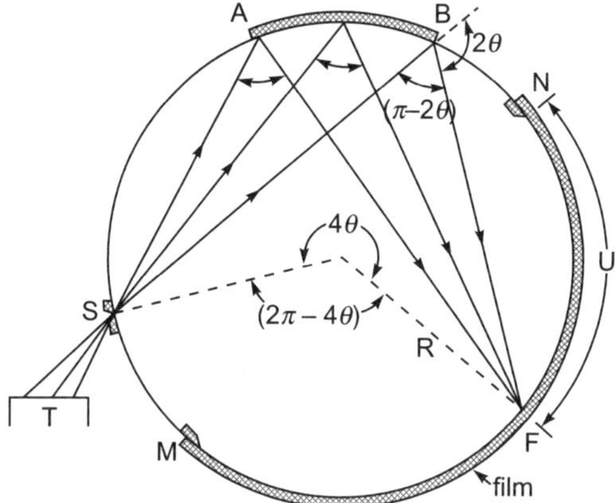

Fig. 7.6 Seemann-Bohlin focusing camera. Only one hkl reflection is shown

The resolving power of focusing cameras can be determined by differentiating the equation 7.4 and combining the result with the equation 7.2. Differentiating equation 7.4,

$$d\theta = \frac{dU}{4R}.$$

Combining with equation 7.2,

$$\frac{dU}{dd} = -4R.\frac{\tan\theta}{d}$$

or,

$$\frac{d}{\Delta d} = -4R.\frac{\tan\theta}{\Delta U} \tag{7.5}$$

It can be seen from equation 7.5 that the ability to separate diffraction lines from planes of almost similar spacing, is twice that of a Debye-Scherrer camera of the same radius. Further, the exposure time is much shorter because of larger area of diffracting specimen combined with the focusing action present in these cameras. These cameras are therefore convenient for the study of mixtures of phases where the diffraction patterns of phases overlap and the identification of closely spaced lines are of importance. Powder specimens for the diffraction work can be prepared by adhesive bonding of a thin layer of powder on to a piece of paper with the help of glue or petroleum jelly. The paper is then curved to fit the contour of the inside circumference of the camera and held in place by clamp attachments in the camera. This camera has a further advantage that bulk specimens can also be used by directly fastening the flattened surface to fit tangentially into the inner circumference of the camera. In this condition, the focusing action will not be perfect because the specimen surface is only tangential to the circumference and not exactly matching with the curvature of the camera. But, the advantage of the ability to conduct the diffraction experiment directly on a bulk specimen overrides the disadvantage of the decreased focusing action. The specimens prepared for microscopic examination can directly be used for diffraction purposes which is definitely a great advantage especially when the identification of a phase, examined by the microscopic technique, is required.

One disadvantage (if it is indeed one) of the Seemann-Bohlin camera is that the diffractions which can be registered on the film cover a limited range of 2θ values especially on the low angle side. It is particularly used for the case where a detailed study of certain 2θ ranges is required. Seemann-Bohlin cameras are available with designs for covering different 2θ ranges and some workers use a set of two or three cameras complementing one another to cover the entire range of 2θ values.

7.7.3   Back Reflection Focusing Cameras

Back reflection focusing cameras are mainly used for the study of back reflection lines (i.e., high angle diffraction lines) particularly for the purpose of the determination of the lattice parameter accurately. Fig. 7.7 illustrates the symmetrical back reflection focusing camera in which the slit 'S' is at the center of the film and the specimen AB is placed diametrically opposite to the slit. A means for oscillating the specimen through a few degrees about the camera axis is generally incorporated. The Bragg angle θ can be calculated by the relation

$$(4\pi - 8\theta).R = V \qquad\qquad (7.6)$$

where 'V' is the distance on the film between corresponding pair of diffraction lines lying on either side of the slit source 'S'. The resolution of this camera is same as that of the Seemann-Bohlin camera.

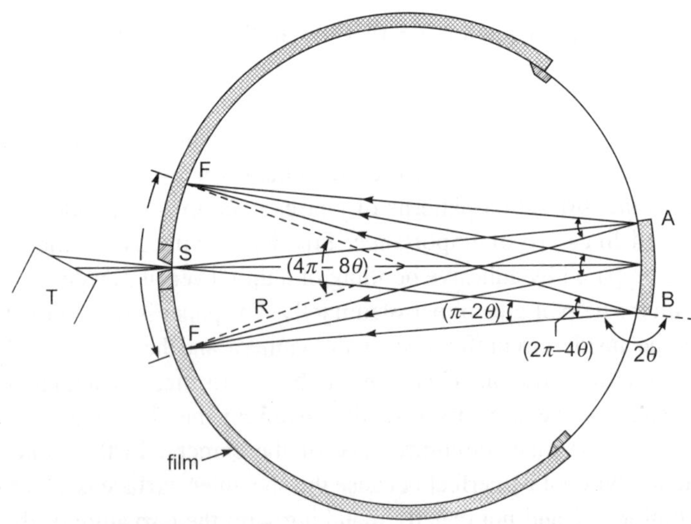

Fig. 7.7   Symmetrical back-reflection focusing camera. Only one hkl reflection is shown

7.8   THE PINHOLE TECHNIQUE

The technique involves the use of a Laue camera to study a polycrystalline specimen under diffraction by a monochromatic x-ray beam. Both the transmission and the back reflection Laue cameras can be used for the studies. Since, the x-ray film is flat and perpendicular to the x-ray beam, a particular (hkl) diffraction cone would intersect the x-ray film in a perfect circle called the Debye ring. Thus, the method has the advantage of studying the entire Debye ring and not just a part of it. This method has its own limitation in that the range of θ values that can be studied are rather narrow either on the high angle side or on the low as shown in Fig. 7.8. The expression for the determination of the Bragg angle θ can be obtained by the geometry of the diffraction. For example, for the transmission pinhole technique,

$$\tan 2\theta = \frac{U}{2D} \tag{7.7}$$

where U is the diameter of the Debye ring and D is the specimen to film distance. For the back reflection Laue technique, the expression is

$$\tan (\pi - 2\theta) = \frac{V}{2D} \tag{7.8}$$

where V is the diameter of the Debye ring obtained. The distance D is generally between 3-5 cm.

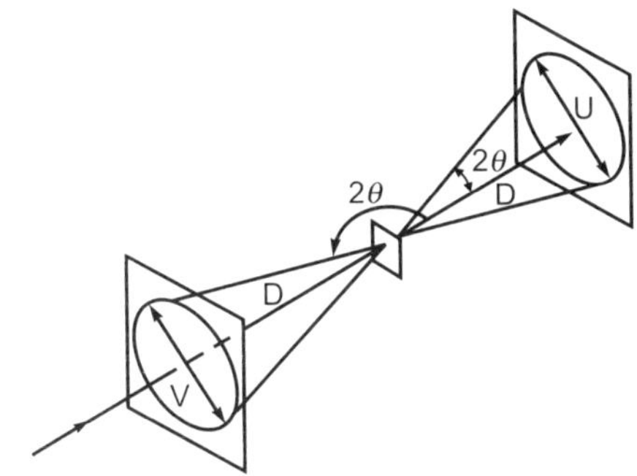

Fig. 7.8   Angular relationships in the pinhole method

Powder specimens for the technique may be prepared by spreading a little powder mixed with glue on to a thin paper or a glass slide. The distinct advantage of this method, however, lies in its ability to utilize bulk or massive specimens directly for the purposes of diffraction. In transmission method, since, the diffracted beam would be absorbed heavily by the specimen itself, the thickness of the bulk specimen which can be examined is highly limited (may be less than a hundred microns). But, here too, a slight alteration in the specimen mounting can help. As shown in Fig. 7.9 the specimen is kept in such a fashion that the incident

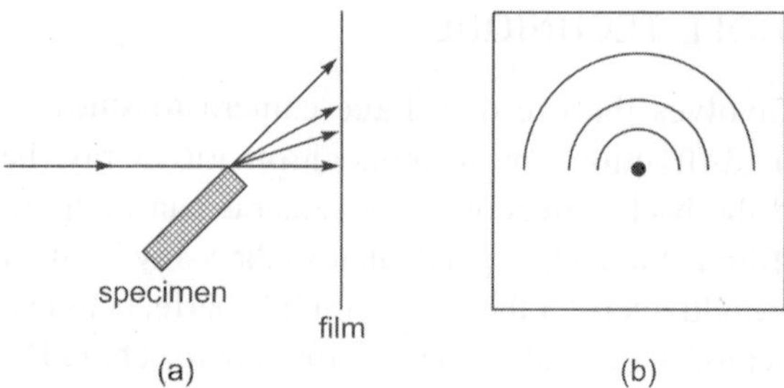

Fig. 7.9 Transmission pinhole method for thick specimens: (a) section through incident beam; (b) partial pattern obtained.

beam just makes contact with a top corner of the bulk specimen, so that the parts of the diffraction cones on the upper half are not reduced in intensity, whereas those on the lower half would have highly reduced intensity since these have to pass through thicker bulk of the specimen. Thus, a pattern as shown in Fig. 7.9(b) can be obtained which is good enough for most of the studies. The back reflection technique does not suffer from this drawback. Both these techniques of transmission pinhole and back reflection pinhole are very much utilized in the determination of preferred orientation, grain size, crystal perfection and precise lattice parameter. A precise knowledge of specimen to film distance is not necessary if a calibration of Debye ring diameters with θ values, made using a standard specimen, is available. Both rotation and oscillation of the film about the axis of the incident beam and the specimen about the incident beam axis can be practiced for increasing the number of reflections for producing smooth and continuous Debye rings.

EXERCISES

7.1 Derive an equation for the resolving power of a Debye-Scherrer camera for two very close wavelengths in terms of ΔS where S is defined by equation 7.1. Use this equation to determine the separation of Cu $K\alpha$ doublet ($K\alpha_1$ = 0.15443 nm, $K\alpha_2$ = 0.15405 nm) diffracted at Bragg angles of (i) 10 degrees (ii) 30 degrees (iii) 50 degrees (iv) 70 degrees and (v) 85 degrees in a standard (57.3 mm dia) D.S.camera.

[Ans: (i) 0.025 mm (ii) 0.081 mm (iii) 0.168 mm

(iv) 0.387 mm (v) 1.612 mm]

7.2 A transmission pinhole photograph of aluminium is to be obtained with Cu $K\alpha$ radiation using a film of size 50 by 60 mm. It is desired to have the first two Debye rings completely recorded on the film. What is the maximum specimen to film distance that can be recommended?

[Ans: 50 mm]

8

CHAPTER

X-RAY DIFFRACTOMETER

8.1 INTRODUCTION

For a long time, x-ray diffraction experiments for the study of crystalline materials were mainly confined to photographic recording in different types of cameras as discussed in chapters 5, 6 and 7. But, in recent years, x-ray diffractometers have become commonplace in research laboratories owing to certain typical merits of these over the photographic techniques. The present-day x-ray diffractometers are based mainly on the original x-ray spectrometer design developed by Friedman around the year 1943. An x-ray spectrometer can do basically two jobs, viz.,

1. measuring x-ray spectra by means of a crystal of known structure.
2. studying crystalline and non-crystalline materials by their diffraction patterns obtained by using x-rays of known wavelengths.

The instrument which can do job 1 referred to above should only be called x-ray spectrometer and the one which can do the job 2 should be called x-ray diffractometer, strictly speaking. Both these instruments have essentially the similar components.

8.2 ESSENTIALS OF X-RAY DIFFRACTOMETERS

The design of a diffractometer is based almost on the Debye-Scherrer camera. The film of the Debye-Scherrer camera is replaced by a movable counter, of any one of the types described in section 3.4.3, in a diffractometer. The movable counter measures the diffracted intensity directly and may be connected to a computerized output device which can give the intensity versus 2θ values on a chart.

The schematic sketch of a x-ray diffractometer is shown in Fig. 8.1 which illustrates the essential components. 'S' is the x-ray source which is nothing but the focal spot of the target 'T' in the x-ray tube shown. The divergent x-ray beams from the source 'S' passes through a collimator 'A' and strikes the flat

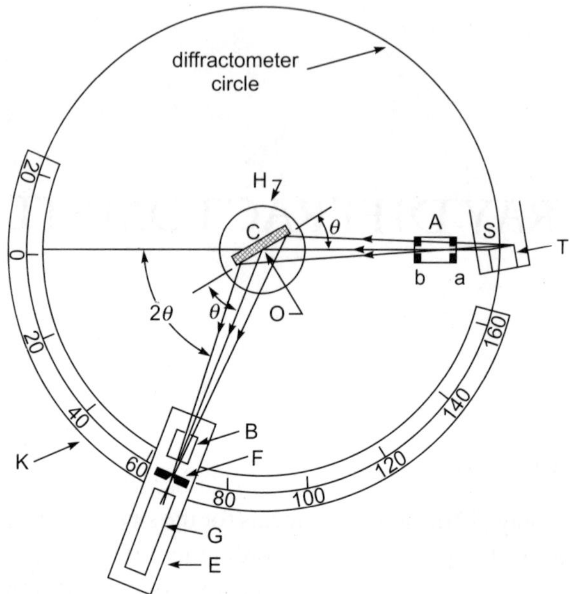

Fig. 8.1   X-ray diffractometer (schematic)

plate specimen 'C' kept on a specimen table 'H'. The diffracted beam after passing through the slit 'B' converges on the focal spot 'F' and enters the counter 'G' which is mounted on a carriage 'E'. The specimen table H can be rotated about an axis 'O' (called the diffractometer axis) perpendicular to the plane of the drawing. Likewise, the carriage E can also be rotated about the diffractometer axis 'O' and the angular position (2θ) of the carriage during such rotation can be read on a graduated scale 'K'. The counter carriage E and the specimen table H are mechanically so coupled that a rotation of the counter through '2x'. degrees makes the specimen automatically rotate through 'x' degrees. This arrangement is necessary to maintain the focusing conditions. The counter may be moved manually to any desired diffraction angle 2θ or alternately it may be driven by a motor at a desired constant angular velocity.

The diffraction intensity is obtained in a diffractometer by the measurement of the rate of production of pulses in counters. The pulse rate may be measured in two ways depending on the type of circuitry used and the counting method.

1. Continuous Method: Here, the pulse rate is converted into a steady current, the magnitude of which is dependent on the pulse rate and this current is measured by a counting rate meter calibrated to give counts or pulses per second. This technique gives a continuous indication of x-ray intensity. In this technique, the counter is set at $2\theta = 0$ and connected to the counting rate meter. The output of this circuit is fed into a fast acting automatic recorder. The counter is driven by a motor at a constant angular velocity through increasing value of 2θ until the entire range of 2θ values is scanned. As the scanning takes place, the counts per second versus the diffraction angle 2θ is plotted on a strip chart connected to the recorder.

2. Intermittent Counting Method: Here, the counter is connected to a scalar circuit which counts the pulses of current for a given time. If the pulse rate is required, the total number of pulses counted must be divided by the time spent in counting. The counter is set at a fixed 2θ value for a sufficiently long time so as to make an accurate counting. Then it is moved to a new 2θ position and the operation is repeated. The whole range of 2θ is covered in this fashion and the intensity at each 2θ value is calculated and then can be plotted in the form of a graph. Here, the determination of the profile of a single diffraction line would be a time consuming process since it requires measurements of intensity at regular intervals of $0.01°$ or better. So, this method is slower than the continuous method but yields more precise intensity measurements.

In diffractometer, the diffraction lines are recorded one after the other, whereas in powder camera, all diffractions are recorded simultaneously during the entire exposure time. Therefore, in diffractometer, the variation in the intensity of the incident x-ray beam will surely affect the diffraction line intensity which is being measured at a particular moment. So, the relative intensities of various diffraction lines obtained in a diffractometer can be compared only when the incident intensity remains constant throughout the measurement of all the diffraction lines of interest. Unless a voltage stabilizer is used, it may be difficult to maintain a constant voltage in a x-ray tube and consequently a constant intensity of the x-ray beam source from the x-ray tube. This difficulty is not experienced in powder cameras since, even when the incident x-ray intensity varies, it will affect simultaneously all the diffraction lines together and the relative intensities of diffraction lines will not be affected at all.

The specimen to be used in a diffractometer can be either a flat metal sheet or plate or wire or powder. When a flat sheet or wire is used, the fact that the preferred orientation affects the intensities of diffraction lines must be kept in mind. Flat sheet specimens can be directly used by placing it on the specimen table with the surface plane coinciding with the diffractometer axis as shown in Fig.8.1. Wires can be used directly by cementing a number of small lengths side by side to a glass plate and fixing this plate on to the specimen table such that the axis of the wires are at right angles to the diffractometer axis. Powder can be used by packing it in a square recess in a glass plate (if necessary with a binder) which can be mounted on the specimen table. Single crystal specimens also may be examined in a diffractometer by mounting this crystal on a three-circle goniometer. The diffractometer may also be used for measurements at high or low temperatures by surrounding the specimen with necessary heating or cooling unit. Such an arrangement is easier in a diffractometer, without hampering or blocking the incident or diffracted beams, than in powder cameras.

Another advantage of the diffractometer is its ability to quantitatively measure the line position and intensity both in a single operation. In powder cameras, quantitative measurement of the intensities of diffraction lines requires three steps viz., recording of the diffraction pattern on the film, making a microphotometer record of the film and conversion of the galvanometer deflection into intensities. Thus, the camera technique results in lower accuracy.

8.3 FOCUSING GEOMETRY AND X-RAY OPTICS

The diffractometer makes use of a diverging beam of x-rays for the purpose of producing a focusing geometry while working with a flat specimen, so that even a weak diffraction line could be measured accurately because of the increased intensity due to the focusing action. Fig. 8.2 shows the focusing geometry of x-ray diffractometers using flat specimens. For any given position of the counter, the receiving slit 'F' and the x-ray source 'S' are always on the diffractometer circle and, because of the mechanical coupling as discussed earlier, the face of the specimen is always tangent to a focusing circle passing through S and F. The focusing circle is not of a constant radius but decreases in size as the diffraction angle 2θ increases as indicated in Fig.8.2 (a) and (b). Perfect focusing requires that the specimen face be curved exactly to match the curvature of the focusing circle which is not practicable as the focusing circle curvature changes with the angle 2θ. The next best alternative of keeping the specimen face tangential to the focusing circle is practiced.

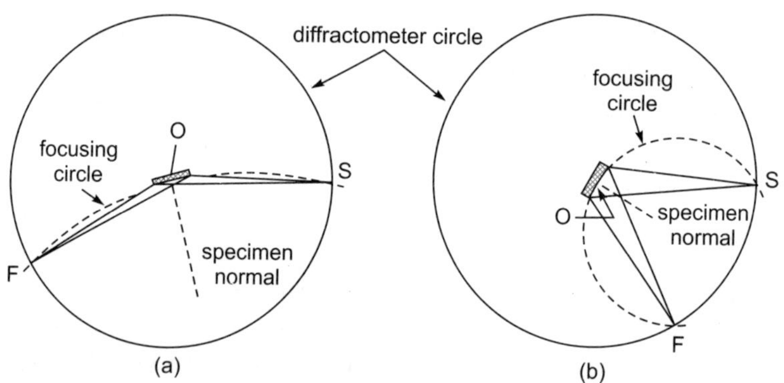

Fig. 8.2   Focusing geometry for flat specimens in (a) forward reflection and (b) back reflection

The line source 'S' extends considerably above and below the plane of drawing of Fig. 8.2 and emits radiation in all directions, but the focusing action requires that all the rays in the incident beam be parallel to the plane of the drawing. This condition is realized as closely as possible by passing the incident beam through a soller slit which contains closely spaced thin metal plates parallel to the plane of the diffractometer circle. These planes eliminate a large portion of the rays in the incident beam inclined to the plane of the diffractometer circle. Typical dimensions of a soller slit are the following: length of the plates - 32mm, thickness of plates - 0.05 mm, clearance between plates - 0.43 mm. The Fig.8.3 shows the arrangement of slits in a diffractometer. Slits 'a' and 'b' define the divergence of the incident beam in the plane of the diffractometer circle. The slits commonly available have divergence angles ranging from a fraction of degrees upto about 4. Similar type of soller slit is generally present also on the diffracted beam side as shown in Fig.8.3.

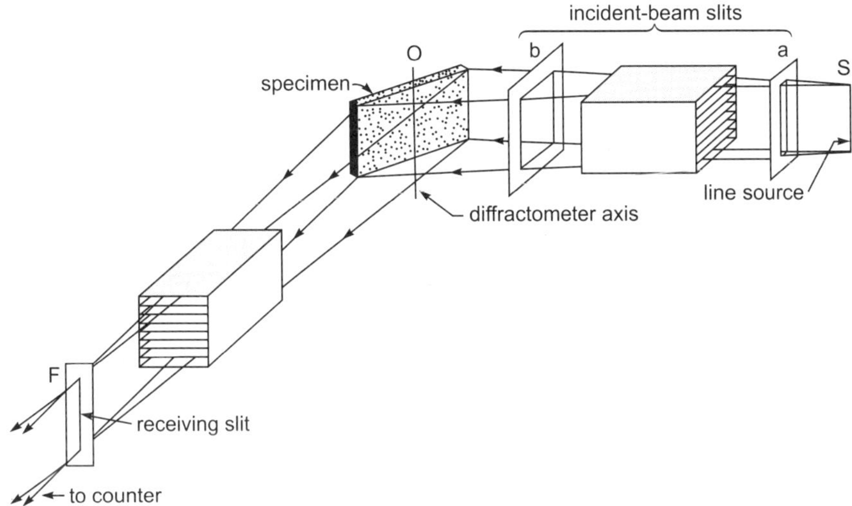

Fig. 8.3 Arrangement of slits in a diffractometer

Because of the focusing action and the relatively large radius of the diffractometer circle, about 15-20 cm in commercial instruments, a diffractometer can resolve very closely spaced diffraction lines (like the ones due to Cu-Kα doublets) provided all the components are properly aligned.

8.4 CALCULATION OF INTENSITIES OF DIFFRACTION LINES

General principles of calculation of intensities of diffraction lines are as discussed in section 4.4. The calculation is slightly modified to fit different geometries of specimens. If the specimen used is in the form of a thin rod, there is no focusing action and the incident beam slits are so chosen as to produce a thin, essentially parallel beam of x-rays. Then the geometry of the diffraction is exactly the same as that of Debye-Scherrer camera and the equation 4.14 can be applied, i.e.,

$$I = |F|^2.P. \frac{1 + \cos^2 2\theta_B}{\sin^2 \theta_B \cdot \cos \theta_B} \qquad (4.14)$$

If a flat specimen is used making equal angles with the incident and diffracted beams, it can be shown that the absorption factor is independent of θ. This independence of absorption factor with variation in θ is due to the balancing of two opposing effects. When θ is small, the area of the specimen irradiated is large but the effective depth of x-ray penetration is small; when θ is large , the area of the specimen irradiated is small but the penetration depth is large relatively. Thus, the effective irradiated volume remains constant and is independent of θ. The equation 4.14 can be used here too and if more precision is required, a temperature factor may be incorporated.

EXERCISES

8.1   Show that the resolving power of a diffractometer can be expressed by an equation similar to the equation 7.5 of chapter 7 where R is the radius of the focussing circle.

8.2   A powder specimen in the form of a rectangular plate has a width of 15 mm when measured in the plane of the diffractometer circle which has a radius of 150 mm. The width of the specimen is such that even at a lowest diffraction angle (2θ) of 10°, the specimen is able to fully cover the divergent incident beam issuing out of the primary slits. Determine the maximum divergence angle (measured in the plane of the diffractometer circle) of the incident x-ray beam.

[Hint: Use the geometry of diffraction in a diffractometer. Ans: 0.52]

DETERMINATION OF CRYSTAL STRUCTURE AND LATTICE PARAMETER

9.1 INTRODUCTION

The basic principles of structure determination have already been discussed in chapter 4. It has been shown that the crystal structure of a substance determines its diffraction pattern and the shape and the size of its unit cell determine the angular position of the diffraction lines. Further, the positions of the atoms within the unit cell decide the intensities of those diffraction lines. It should therefore be possible to go backwards and determine the crystal structure from the positions of the diffraction lines. The calculation of the positions of the diffraction lines from a known crystal structure is fairly easy and straightforward as done in chapter 4, but going backwards from diffraction lines to crystal structure is not so easy. Generally, the determination of the crystal structure is done by trial and error method in three major steps as noted below:

1. An assumption is made regarding the crystal system and Miller indices are assigned to each of the lines based on this assumption. This step is known as 'indexing the pattern'. Once this is done, the shape and the size of the unit cell can be calculated.

2. The number of atoms per unit cell can then be computed from the shape and the size of the unit cell, chemical composition and the measured density of the specimen.

3. The positions of the atoms within the unit cell can be deduced from the relative intensities of the diffraction lines.

The structure determination is complete only when all these three steps are accomplished. The third step is generally the most difficult in complex structures and for many substances, the structure determination is incomplete due to the inherent difficulty of this step. But, for most of the applications, the knowledge derived from the first two steps is sufficient.

9.2   PRELIMINARY TREATMENT OF DATA

For the purposes of crystal structure analysis, it is better to go in for the powder method and make use of the Debye-Scherrer (D.S.) camera technique or the diffractometer technique. Powder techniques using the Seemann-Bohlin camera or the pinhole camera are not able to cover a wide range of diffraction angles as possible in the D.S. camera or in the diffractometer. After getting the required diffraction pattern using the recommended methods, the values of $\sin^2\theta$ are calculated for each diffraction line and this set of $\sin^2\theta$ values are the raw data for the determination of the crystal structure. Thus, it should be ensured that all $\sin^2\theta$ values of this determined set are due only to the unknown substance and no extraneous lines are present. The possible sources of extraneous lines can be the following:

1. The diffraction of x-rays having wavelengths different from that of the principal component of radiation: Generally, for most diffraction purposes, β-filtered $K\alpha$ component of a strong characteristic radiation is used. But, as noted earlier, filtering is never perfect. A small intensity of $K\beta$ radiation commonly accompanies the filtered $K\alpha$ component of the characteristic spectrum. So, this $K\beta$ component of x-rays can give rise to the extraneous lines. If a particular set of planes (hkl) diffract the strong $K\alpha$ component at Bragg angle $\theta\alpha$, the same set of planes are capable of diffracting the weak $K\beta$ component at Bragg angle $\theta\beta$ according to the following equations of Bragg:

$$\lambda_{K\alpha} = 2d_{hkl} \cdot \sin\theta\alpha$$

and
$$\lambda_{K\beta} = 2d_{hkl} \cdot \sin\theta\beta$$

Squaring the equations and dividing one by the other,

$$\frac{\lambda_{K\alpha}^{2}}{\lambda_{K\beta}^{2}} = \frac{\sin^2\theta\alpha}{\sin^2\theta\beta}$$

Thus, if a diffraction line (generally weak in intensity) is suspected to be due to $K\beta$ component, from the above equation, it is possible to find the $\sin\theta\beta$ value based on $\lambda_{K\alpha}^{2}/\lambda_{K\beta}^{2}$ ratio and the $\sin^2\theta\alpha$ value of a strong nearby line of $K\alpha$. If the calculated $\sin\theta\beta$ value coincides with the practically observed value for the suspected weak line, then this line can be eliminated from the data as it is confirmed that it is due to $K\beta$ radiation. Sometimes, L-characteristic radiation of tungsten, which is generally present (particularly in old tubes due to reasons mentioned in section 3.1.2) as impurity on the target metal, may cause extraneous diffraction lines. These extraneous lines can also be eliminated in the same fashion as detailed for the elimination of $K\beta$ lines.

2. Diffraction from substances other than the specimen: Some impurities present in the specimen, and badly aligned slit material and specimen mount may all cause extraneous lines to appear on the diffraction pattern. Very careful specimen

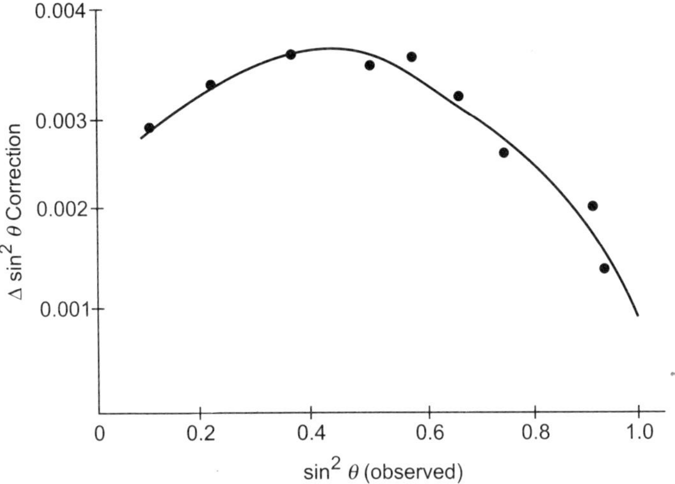

Fig. 9.1   Correction curve for $\sin^2\theta$.

preparation technique and proper precautions during experimentation can prevent the occurrence of extraneous lines on the diffraction pattern.

Some systematic errors always accompany $\sin\theta$ measurements. A simple method to eliminate these errors is to calibrate the D.S. camera or diffractometer with a standard substance of known lattice parameter. The difference between the observed and the calculated $\sin^2\theta$ values for the standard substance gives the error $\Delta\sin^2\theta$ and this error can be plotted as a function of the observed values of $\sin^2\theta$. Fig.9.1 illustrates such a plot schematically. This plot can be used as a calibration plot and all the observed $\sin^2\theta$ values for the specimen under investigation can be corrected by applying the appropriate correction $\Delta\sin^2\theta$ values obtained from the plot at the observed $\sin^2\theta$ values.

9.3   INDEXING THE DIFFRACTION PATTERNS OF CUBIC CRYSTALS

The plane spacing equation connecting the Miller indices (hkl) of a plane and its interplanar distance 'd' in a cubic crystal is given by

$$d^2 = a^2/(h^2 + k^2 + l^2) \tag{9.1}$$

where 'a' is the lattice parameter or the side of the cubic unit cell. Combining this equation with the Bragg's law,

$$\lambda^2 = \{4 \cdot a^2/(h^2 + k^2 + l^2)\} \cdot \sin^2\theta$$

or,
$$\sin^2\theta = \lambda^2 \cdot (h^2 + k^2 + l^2)/4a^2 \tag{9.2}$$

Since, $\lambda^2/4a^2$ of this equation 9.2 is a constant for a given diffraction pattern, it can be seen that $\sin^2\theta$ values must be proportional to $(h^2 + k^2 + l^2)$ values. Since, h, k and l are always integers, it is possible to find out the $(h^2 + k^2 + l^2)$ values in the increasing order by taking increasing integer values for h, k and l. It can be seen from table 4.2 that each type of cubic crystal structure, i.e., simple cubic, body centered cubic (BCC), face centered cubic (FCC) and diamond cubic (DC), has its own order of increasing $(h^2 + k^2 + l^2)$ values. This is shown in the table below:

Structure	$(h^2 + k^2 + l^2)$ *values of diffracting planes*
Simple cubic	1, 2, 3, 4, 5, 6, –, 8, 9, 10, 11, 12, 13, 14, –, 16, etc.
BCC	2, 4, 6, 8, 10, 12, 14, 16, 18, 20, etc.
FCC	3, 4, 8, 11, 12, 16, 19, 20, 24, 27, etc.
DC	3, 8, 11, 16, 19, 24, 27, 32, 35, etc.

As the possible $(h^2 + k^2 + l^2)$ values increase in a definite sequence in a given cubic crystal system, $\sin^2 \theta$ values also should increase proportionately as per equation 9.2. Thus, if the $\sin^2 \theta$ values, in the diffraction pattern of the unknown cubic crystal, increase in the ratios 1:2:3:4:5:6:8:9 and so on, then the specimen must belong to a simple cubic system. If the increase in $\sin^2 \theta$ values are in the ratios of 2:4:6:8:10:12:14 (or 1:2:3:4:5:6:7), then the system must be BCC. Similarly, the ratios of increasing $\sin^2 \theta$ values for the FCC and the DC structures are 3:4:8:11:12:16 etc. and 3:8:11:16:19 etc. respectively. It must be noted that certain values like 7, 15, 23 are not found in the simple cubic system or any other system as possible $(h^2 + k^2 + l^2)$ values for reflection. This is because, these values cannot be expressed as the sum of the squares of three integers.

Since, the positions (θ values or $\sin^2 \theta$ values) of the diffraction lines are connected with the $(h^2 + k^2 + l^2)$ values of a reflecting plane, the diffraction pattern is very much dependent on the sequence of possible $(h^2 + k^2 + l^2)$ values for a given cubic structure. Thus, the diffraction patterns of simple cubic and BCC structures would have diffraction lines almost equally spaced (since $(h^2 + k^2 + l^2)$ values are increasing in the fashion 1, 2, 3, 4, 5, 6. etc.) whereas, the FCC structure will have its diffraction pattern with two lines closely spaced, the next line a bit separated and this pattern repeating itself continuously for the remaining lines (since, for this structure, $(h^2 + k^2 + l^2)$ values increase in this fashion, i.e., 3 and 4 together and then 8 after a large gap and further 11 and 12 occurring together after a gap and so on). Thus, the diffraction patterns of the BCC and the FCC structures can easily be distinguished by close observation as noted above. But to distinguish a simple cubic pattern from a BCC pattern is not that easy since both have their lines almost equidistant from each other. If there are at least 7 lines in the pattern, one can definitely distinguish between them. If the pattern belongs to the BCC structure, the distance of separation between the 6th and the 7th lines would be almost similar to that between other lines in the pattern and if the pattern belongs to the simple cubic structure, it would be almost double the distance between other lines. This is because for the BCC structure the $(h^2 + k^2 + l^2)$ values increase in the same proportions even at the 7th line as seen above in the above table of $(h^2 + k^2 + l^2)$ values, and this is not so in the simple cubic where the $(h^2 + k^2 + l^2)$ values increase in the fashion 1, 2, 3, 4, 5, 6, 8, 9 and the 7th line has a value of 8 indicating a larger gap after the 6th line.

9.4 GRAPHICAL METHODS FOR INDEXING PATTERNS OF NON-CUBIC CRYSTALS

Indexing of non-cubic crystal patterns can be done either by graphical methods or by analytical methods. Graphical methods of indexing will be dealt with first.

9.4.1 Indexing of Tetragonal Crystals

The plane spacing equation for this system involves two lattice parameters 'a' and 'c'. The relevant equation is

$$1/d^2 = (h^2 + k^2)/a^2 + l^2/c^2$$

where 'd' and h, k, l have their usual meanings. This equation may be rewritten as

$$1/d^2 = 1/a^2 \cdot [h^2 + k^2 + l^2/(c/a)^2] \tag{9.3}$$

Taking logarithm to the base 10 on both sides,

$$2 \cdot \log d = 2 \cdot \log a - \log [h^2 + k^2 + l^2/(c/a)^2]$$

If this equation is written for two crystallographic planes $(h_1 \ k_1 \ l_1)$ and $(h_2 \ k_2 \ l_2)$ with spacings d_1 and d_2 respectively and subtracting one from the other,

$$2 \cdot \log d_1 - 2 \cdot \log d_2 = -\log [h_1^2 + k_1^2 + l_1^2/(c/a)^2]$$
$$+\log [h_2^2 + k_2^2 + l_2^2/(c/a)^2] \tag{9.4}$$

The equation 9.4 shows that the difference between $2 \cdot \log d$ values for any two planes depends only on the hkl values of the planes and c/a ratio of the system and not on the parameter 'a'. This fact is taken as the basis for devising the Hull-Davey chart which can be used for solving the powder patterns of tetragonal crystals.

The Hull-Davey chart is nothing but a plot of $\log [h^2 + k^2 + l^2/(c/a)^2]$ on x-axis versus c/a on the y-axis. The plot of $\log [h^2 + k^2 + l^2/(c/a)^2]$ on the x-axis extends for two ranges of log values from 0.1 to 10. Each (hkl) set of indices is plotted as a curve on this plot for varying c/a ratios. In other words, by assuming a c/a ratio, the value of $\log [h^2 + k^2 + l^2/(c/a)^2]$ can be calculated for a given set of hkl indices. Similarly, assuming different c/a ratios the whole set of $\log [h^2 + k^2 + l^2/(c/a)^2]$ are plotted as a curve for a set of hkl indices. Similarly, different curves, one for each set of hkl indices are presented in the Hull-Davey chart.

It can be observed that when l = 0, the curve is a straight line parallel to the c/a axis. So, curves corresponding to planes with indices (100), (110), (200), etc. will all be parallel to the c/a axis. Curves for planes with different indices but identical plane spacings like (100) and (010) would be represented by a single curve on the plot. The chart, in addition to the plot as mentioned above, contains a single range logarithmic 'd'(plane spacing) scale ranging from 0.1 to 1 nm and has the same dimension as the two range log $[h^2 + k^2 + l^2/(c/a)^2]$ scale. Further, this 'd' scale runs in a direction opposite to the log $[h^2 + k^2 + l^2/(c/a)^2]$ scale. This is done in order to satisfy the equation 9.3 which observes that the coefficient of log d is -2 times the coefficient of log $[h^2 + k^2 + l^2/(c/a)^2]$. Thus, for a given c/a ratio, the separation distance between the 'd' values on the chart for any two given planes would be exactly the same as the separation distance between the two corresponding curves on the chart at the same c/a ratio. A partial Hull-Davey chart is shown in Fig. 9.2.

The use of this Hull-Davey chart is done as detailed below. Initially, the plane spacing 'd' values are determined for each line on the pattern by Bragg's law where λ and θ are known. Suppose, the first seven lines of the diffraction pattern gives the 'd' values as 0.6, 0.4, 0.33, 0.3, 0.28, 0.255, and 0.24 nm. A paper strip is laid along the d scale of the chart and these values are marked as shown in Fig. 9.2. This paper strip is then placed on the chart and moved over the chart both vertically and horizontally until each of the marks made on the strip coincide with a line on the chart, taking care to see that the marked edge of the strip always remains parallel to the x-axis of the chart. Once this coincidence of all the marked 'd' values on the strip with the lines on the chart occurs (as shown in Fig. 9.2), the indices of each line can be simply read from the corresponding curves on the chart and the approximate c/a ratio can be read from the vertical position of the paper strip. In the example chosen, c/a value is 1.3 and the lines from the pattern can be indexed in order as (001), (100), (101), (110), (002), (111) and (102). After the lines have been indexed in this fashion, the 'd' values of the two highest angle lines can be used and equations of the form of 9.3 can be solved simultaneously to yield the values of 'a' and 'c' accurately. From these values accurate c/a ratio can be determined. The Fig. 9.2 represents only a few lines from the Hull-Davey chart. The complete Hull-Davey chart is presented in Fig. 9.3. It can be seen from this chart that there is too much crowding of high indices lines and it is difficult to index these lines by the method described above. Thus, generally low indices lines are indexed first and then the high indices lines are indexed on the basis of 'c' and 'a' values determined from low indices lines. $\sin^2 \theta$ values also can be used for indexing the lines by graphical method of Hull-Davey, instead of 'd' values. The equation 9.3 can be modified with Bragg's law for this purpose as shown below:

$$1/d^2 = 1/a^2 \cdot [h^2 + k^2 + l^2/(c/a)^2] \tag{9.3}$$

but, by Bragg's law $d^2 = \lambda^2/4 \cdot \sin^2 \theta$ substituting this value of d^2 in equation 9.3,

$$4 \cdot \sin^2 \theta/\lambda^2 = 1/a^2 \cdot [h^2 + k^2 + l^2/(c/a)^2]$$

or,
$$\sin^2 \theta = (\lambda^2/4a^2) \cdot [h^2 + k^2 + l^2/(c/a)^2].$$

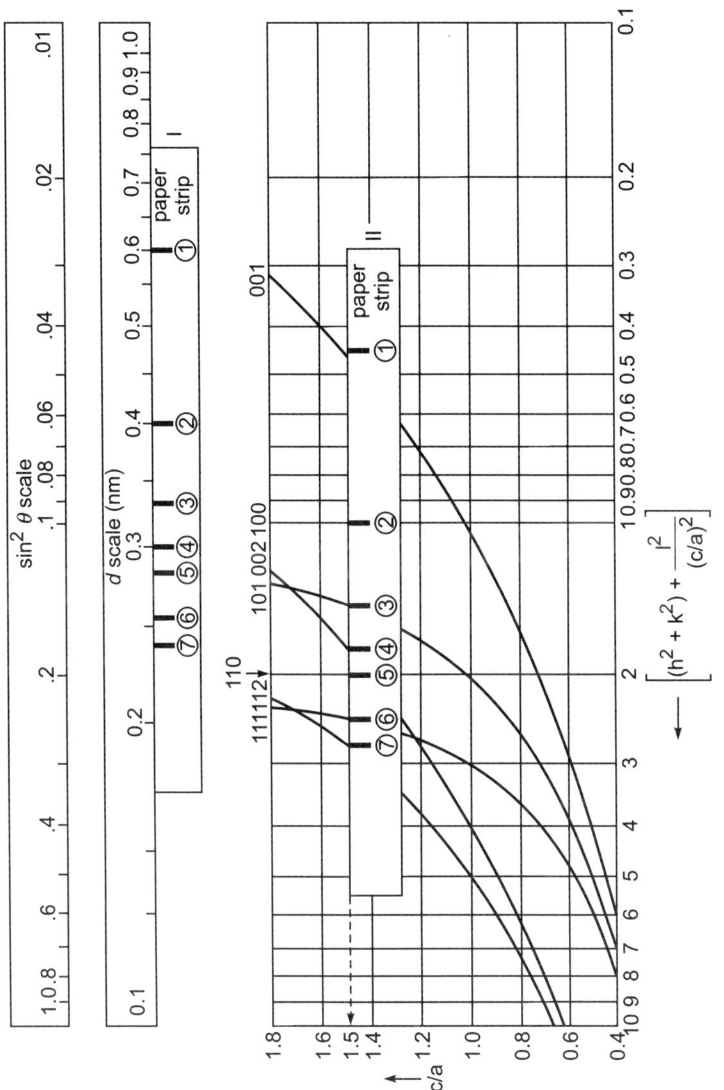

Fig. 9.2 Partial Hall-Davey chart for simple tetragonal lattices.

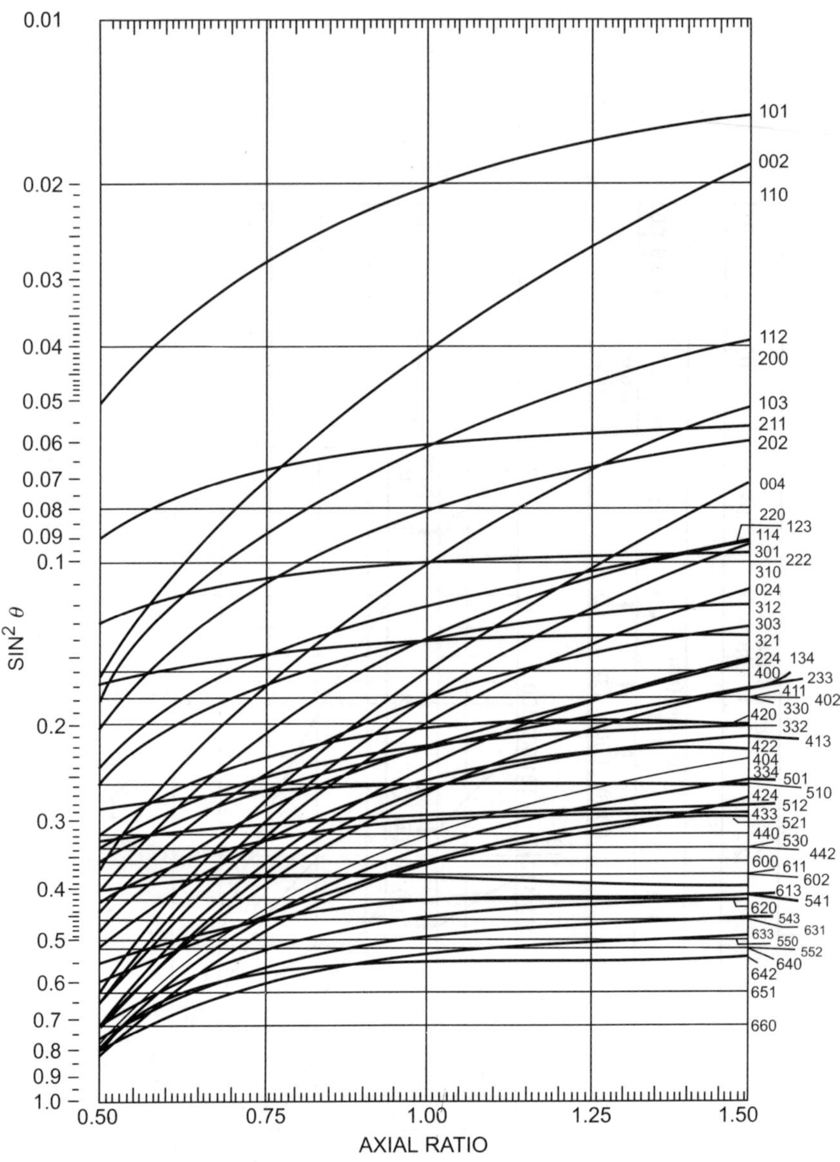

Fig. 9.3   Complete Hull-Davey chart for body-centered tetragonal lattices.

By taking logarithm on both sides,

$$\log \sin^2 \theta = \log (\lambda^2/4a^2) + \log [h^2 + k^2 + l^2/(c/a)^2].$$

By similar arguments as done in the case of 'd' values, one can use $\sin^2 \theta$ values for indexing. The $\sin^2 \theta$ scale can be chosen as having a two range logarithmic scale running from 0.01 to 1.0 in the same direction as that of $[h^2 + k^2 + l^2/(c/a)^2]$ scale. The $\sin^2 \theta$ scale is shown in Fig.9.2 and 9.3.

It should be noted that the tetragonal cell with a 'c/a' ratio of unity is nothing but a cubic cell and so the diffraction patterns of cubic crystals can also be indexed by the Hull-Davey chart with the marked paper strip held at the c/a ratio = 1. Anyway, it is not necessary to do this because cubic crystals can be indexed easily without the Hull-Davey (H.D.) chart. From Fig. 9.3, it can be seen that certain lines get split up into two or three lines if the c/a ratio deviates from a value of one on either side. This means that a given family of reflecting planes would give more lines in tetragonal systems than in cubic systems.

The Bunn chart devised by Bunn, is based on a different function of hkl and c/a than that used in the H.D. chart and can also be used to index tetragonal structures graphically. This chart also contains a log 'd' scale and the chart is used exactly in the same manner as the H.D. chart.

9.4.2 Indexing of Hexagonal Crystals

The powder diffraction patterns of hexagonal crystals can also be indexed by graphical methods similar to tetragonal crystals. The plane spacing equation for hexagonal crystal is used for the purpose, i.e.,

$$1/d^2 = (4/3) \cdot (h^2 + hk + k^2)/a^2 + l^2/c^2 \qquad (9.5)$$

If this equation is rewritten in logarithmic form,

$$2 \cdot \log d = 2 \cdot \log a - \log [(4/3) \cdot (h^2 + hk + k^2) + l^2/(c/a)^2].$$

As done earlier, taking differences between $2 \cdot \log d$ values for any two planes $(h_1 k_1 l_1)$ and $(h_2 k_2 l_2)$,

$$2 \cdot \log d_1 - 2 \cdot \log d_2 = -\log [(4/3) \cdot (h_1^2 + h_1 k_1 + k_1^2) + l_1^2/(c/a)^2]$$
$$+ \log [(4/3) \cdot (h_2^2 + h_2 k_2 + k_2^2) + l_2^2/(c/a)^2] \quad (9.6)$$

This equation 9.6 is in a form similar to the equation 9.4 for the tetragonal system. A Hull-Davey chart for the hexagonal system may be constructed by plotting c/a on the y-axis and $\log [(4/3) \cdot (h^2 + hk + k^2) + l^2/(c/a)^2]$ on the x-axis. This chart can be used in a fashion similar to the tetragonal H.D. chart. Alternatively, a Bunn chart for hexagonal system may also be constructed and used. Charts can also be constructed for specific cases like hcp structures by omitting all curves for which $(h + 2k)$ is an integral multiple of 3 and l is odd (see section 4.4.2).

9.4.3 Indexing of Rhombohedral Crystals

Rhombohedral crystals have two lattice parameters 'a' and 'α'. The rhombohedral crystals can easily be referred to hexagonal axes. The (hkl) indices of the rhombohedral lattice can be converted into (HKL) indices of hexagonal lattice by equations:

$$H = h - k, \; K = k - l, \; \text{and} \; L = h + k + l.$$

Thus, it follows that the powder pattern of a rhombohedral structure can be indexed on a hexagonal H.D. chart or Bunn chart. It can be seen from the above equations of conversion that $-H + K + L = 3k$, which means that if the lattice is rhombohedral, k is an integer and the sum $-H + K + L$ must be an integral multiple of 3. If this is not true then the lattice is assumed to be hexagonal. After the lattice is identified as rhombohedral by the above method, the lattice parameters 'a' and 'α' can be determined by the following three steps:

1. The lattice parameters aH and cH of the equivalent hexagonal structure can be determined in the usual manner, as done for hexagonal crystals.

2. The rhombohedral parameters 'a' and 'α' are given by

$$a_R = (1/3) \cdot [a_H^2 + c_H^2]^{1/2}, \quad \sin \alpha/2 = 3/[2 \cdot \{3 + (c_H/a_H)^2\}^{1/2}].$$

3. For converting the hexagonal indices (HKL) into rhombohedral indices (hkl), the following equations may be used:

$$h = [2H + K + L]/3, \; k = [-H + K + L]/3$$

and
$$l = [-H - 2K + L]/3.$$

It must be pointed out here that whenever the H.D. or the Bunn chart is used, care must be taken to see that every mark on the paper strip coincides with a curve on the chart except for the identified extraneous lines. But, it is not necessary that the paper strip should contain a mark for every line on the chart since some lines may be too weak to be observed.

9.5 ANALYTICAL METHODS OF INDEXING NON-CUBIC PATTERNS

Substances having orthorhombic, monoclinic and triclinic crystal structures yield powder patterns which are almost impossible to be indexed by graphical methods. Analytical methods are available which can be used for indexing tetragonal, hexagonal, rhombohedral and orthorhombic crystals. It can be observed that tetragonal, hexagonal and rhombohedral systems have two independent lattice parameters each. But orthorhombic crystals have three lattice parameters (a, b and c). As the number of independent lattice parameters increase, the indexing becomes more and more difficult. Monoclinic and triclinic crystals, which have 4 (a, b, c and β) and 6 (a, b, c, α, β, λ) lattice parameters respectively, can only be indexed if the structure is known. If the structures are unknown, the indexing of monoclinic and the triclinic powder patterns becomes complicated. But, these structures as well as any other may be determined by going in for the rotating crystal diffraction technique or one of its variations which is beyond the scope of this book.

9.5.1 Indexing of Tetragonal Crystals

The plane spacing equation for tetragonal system is

$$1/d^2 = (h^2 + k^2)/a^2 + l^2/c^2.$$

Substituting $d = \lambda/2 \cdot \sin \theta$ (from Bragg's law) in this equation,

$$(4 \cdot \sin^2 \theta)/\lambda^2 = (h^2 + k^2)/a^2 + l^2/c^2$$

or,
$$\sin^2 \theta = A \cdot (h^2 + k^2) + B.l^2 \tag{9.7}$$

where A and B are constants and are equal to $\lambda^2/4a^2$ and $\lambda^2/4c^2$, respectively. It can be seen from this equation that if such lines are chosen whose indices are (hk0), then for these lines the equation reduces to

$$\sin^2 \theta = A \cdot (h^2 + k^2).$$

Thus, by inserting increasing integral values of h and k, it can be found that the permissible values of $(h^2 + k^2)$ are 1, 2, 4, 5, 8, 13, and so on. Therefore, (hk0) lines must have their $\sin^2 \theta$ values increasing in the ratio of these integers 1, 2, 4, 5, 8, 13, and so on. Thus, it is possible to identify those lines having these specific ratios. Once these lines are detected, it is a simple matter to determine the constant A since in these identified series, the $\sin^2 \theta$ value of the first line must be equal to A and that of the second line 2A and that of the third line 4A and so on.

Once the (hk0) lines are identified and separated, for detecting the other lines, the general equation 9.7 could be modified and used as

$$\sin^2 \theta - A \cdot (h^2 + k^2) = B \cdot l^2.$$

Here, the differences rendered by the left hand side (LHS) of the above equation could be determined and these must be equal to the right hand side (RHS). Or, the LHS must be increasing in the order $B \cdot 1^2, B \cdot 2^2, B \cdot 3^2, B \cdot 4^2$ and so on since 'l' can take values 1, 2, 3, 4 and so on. In other words, the lines other than (hk0) must have their $\{\sin^2 \theta - A \cdot (h^2 + k^2)\}$ values in the proportion of 1, 4, 9, 16 (i.e., $1^2, 2^2, 3^2, 4^2$ respectively) Thus, it is possible to identify the lines which can give these specific ratios. So, the value of B can be determined.

From A and B determined as detailed above, the lattice parameters 'a' and 'c' can be obtained using relations $A = \lambda^2/4a^2$ and $B = \lambda^2/4c^2$, where λ is the wavelength used for obtaining the diffraction patterns.

9.5.2 Indexing Hexagonal and Rhombohedral Crystals

For hexagonal crystals the plane spacing equation is

$$1/d^2 = (4/3) \cdot (h^2 + hk + k^2)/a^2 + l^2/c^2 \tag{9.5}$$

Modifying this with Bragg's law,

$$\sin^2 \theta = (\lambda^2/4) \cdot [(4/3a^2) \cdot (h^2 + hk + k^2) + l^2/c^2]$$

or, rewriting,

$$\sin^2 \theta = C \cdot (h^2 + hk + k^2) + D.l^2 \qquad (9.8)$$

where, $C = \lambda^2/3a^2$ and $D = \lambda^2/4c^2$ are constants for a given crystal structure. As done in section 9.5.1, by selecting those (hk0) lines for which the index l is zero, the equation 9.8 reduces to $\sin^2 \theta = C \cdot (h^2 + hk + k^2)$. For these lines $\sin^2 \theta$ values must increase in the ratios of increasing $(h^2 + hk + k^2)$ values. By putting increasing integral values for h and k, it can be found that the permissible $(h^2 + hk + k^2)$ values are 1, 3, 4, 7, 9 and so on. Selecting those $\sin^2 \theta$ values from the pattern which are increasing in the ratio of 1:3:4:7:9, the 'C' value can be determined as discussed in section 9.5.1 for the determination of the A value. Once the 'C' value is determined, the equation 9.8 can be used in the form

$$\sin^2 \theta - C \cdot (h^2 + hk + k^2) = D.l^2$$

and the procedure discussed in section 9.5.1 can be similarly adapted which establishes the 'D' value. The lattice parameters 'a' and 'c' can be determined through the relations $C = \lambda^2/3a^2$ and $D = \lambda^2/4c^2$.

Rhombohedral crystals also can be indexed in the same manner as hexagonal crystals and after the equivalent hexagonal parameters are determined, they can be converted into rhombohedral parameters as discussed in section 9.4.3.

9.5.3 Indexing of Orthorhombic Crystals

The plane spacing equation for orthorhombic crystals is

$$1/d^2 = h^2/a^2 + k^2/b^2 + l^2/c^2 \qquad (9.9)$$

where a, b and c are lattice parameters of orthorhombic crystal unit cell. Using Bragg law, this equation can be modified as

$$\sin^2 \theta = (\lambda^2/4a^2) \cdot h^2 + (\lambda^2/4b^2) \cdot k^2 + (\lambda^2/4c^2) \cdot l^2$$

or
$$\sin^2 \theta = E \cdot h^2 + F \cdot k^2 + G \cdot l^2$$

where $E = \lambda^2/4a^2$, $F = \lambda^2/4b^2$ and $G = \lambda^2/4c^2$.

The three unknowns E, F and G are to be determined. The method of indexing here, involves identifying the pairs of lines having identical indices h and k but differing only in the third index l. For example, if any two lines with indices $(h_1\ k_1\ 0)$ and $(h_1\ k_1\ 1)$ are present in the pattern then their $\sin^2 \theta$ values are (as per the above equation)

$$\sin^2 \theta_1 = E \cdot h_1 + F \cdot k_1 + G \cdot 0$$
$$\sin^2 \theta_2 = E \cdot h_1 + F \cdot k_1 + G \cdot 1^2.$$

So, the difference

$$\sin^2 \theta_2 - \sin^2 \theta_1 = G \cdot 1 - G \cdot 0 = G.$$

Thus, if the index l for any pair of lines differ from each other by 1, 2, 3, etc., the $\sin^2 \theta$ values differ by G, 4G, 9G, and so on, respectively provided h and k indices for these pairs are the same. With this information, it is possible to identify such pairs and the G value can be determined. Once the G value is found, the constants E and F can be determined by setting up simultaneous equations as noted in earlier sections and solving them.

In the case of indexing of orthorhombic crystals, one might get into problems since it is possible that some of the lines of interest (One of the pair needed for solving) may be missing from the diffraction pattern due to its low intensity because of certain factors discussed in section 4.4. But, with a little patience, it is possible to apply the analytical technique successfully to orthorhombic patterns and many such structures are indexed by this method.

Crystal structures with more than three independent parameters, like monoclinic (4 lattice parameters) and triclinic (6 parameters), cannot be easily indexed from their powder patterns. But, as said earlier, they can be indexed by the rotating crystal method or one of its variations which makes use of single crystals of the substance of interest.

9.6 DETERMINATION OF THE NUMBER OF ATOMS PER UNIT CELL

From sections 9.3, 9.4 and 9.5, it is clear that the lattice parameters of individual substance, could be determined after indexing the lines, which represent the shape and size of the unit cell of the crystal structure involved. It is now necessary to determine the number of atoms present in the unit cell of the crystal structure concerned. The volume of the unit cell can be calculated by knowing the shape and size of the unit cell using equations listed under Table 2.2. If the density of the substance is known, or determined by an experiment practically, then the weight of the unit cell is obtained by multiplying the volume of the unit cell by density. The weight of the unit cell corresponds to the weight of the total number of atoms (of the same kind or of different kinds) present in the unit cell. Suppose, it is known that the unit cell contains atoms of only one kind, with atomic weight 'w', then the weight of a single atom of this substance is equal to w/N where N is the Avogadro number, 6.02×10^{23}. The number of atoms in the unit cell is obtained by dividing the weight of the unit cell, determined earlier, by the weight of the single atom. If the unknown substance is a chemical compound, then the unit cell would contain 'molecules' of the compound and correspondingly the calculation is to be made based on molecular weight and the number of molecules in the unit cell may be determined. The number of molecules or atoms per cell determined in this fashion is generally an integer except in the case of defect structures like FeO.

9.7   DETERMINATION OF ATOM POSITIONS

From the discussion in the previous sections, it is clear that the shape and the size of the unit cell as well as the number of atoms in it can be determined by methods discussed therein. The positions of the atoms in the unit cell are still unknown and have to be determined in order to complete the structure analysis. The intensities of the diffraction lines are to be looked into for this purpose. It is seen earlier in section 4.4.2 that the structure factor is connected with the atom positions. Recalling equation 4.10, i.e.,

$$F = F_{hkl} = \sum_{1}^{n} f_n \cdot e^{\,2\Pi i(hun+kvn+lwn)} \tag{4.10}$$

It can be seen that the fractional indices of atomic positions (un, vn, wn) control the structure factor. Since, the intensity of a given line, its multiplicity factor 'p', and its position θ are known, it is possible to determine the relative amplitude of the structure factor, |F|, by the equation 4.12, i.e.,

$$I = |F|^2 \cdot p \cdot \frac{1 + \cos^2 2\theta_B}{\sin^2 \theta_B \cdot \cos \theta_B} \tag{4.14}$$

But, | F | as calculated from equation 4.14 only tells the relative amplitude of the structure factor and not the phase angle of the reflection. And it is the phase angle which we need for the determination of the fractional indices of atomic positions, (un, vn, wn). So, it is not possible to use equation 4.10 to determine the atom positions directly.

Thus, the only alternative is to determine the atomic positions by trial and error. A most probable set of atomic positions is assumed in a given case and the structure factor is calculated on the basis of equation 4.10. This structure factor is then substituted into equation 4.14 and the intensities of the lines calculated. These calculated intensities of the lines are compared with the observed intensities on the diffraction pattern of the substance concerned. If the intensities match well, then the assumed positions of the atoms are correct positions in the unit cell of the substance. If not, another assumption is made as to the positions of the atoms and the process repeated until a proper match is obtained between the calculated and the observed intensities of the diffraction lines. The question now is how to assume certain atomic positions in the unit cell of an unknown substance. One way is to derive the knowledge from the already accumulated structural details of several substances. Substances can be grouped into specific classes with similar atomic positions in unit cells. Classification may also be done on the basis of the type of bonding existing in the substance. The unknown substance may be identified to be belonging to one of the known classes. The atomic positions of this particular identified group may be assumed for the trial. Some of the simple structures may be solved in this fashion. There are certain powerful methods like the theory of space groups and the Fourier series which can be used while solving complicated structures. The theory of space groups relates crystal symmetry to the possible atomic arrangements. Thus, it is possible to get a certain definite number of atomic position sets for a given crystal symmetry. All these atomic position sets could be used for the trial and error of intensity calculations. With the Fourier series method, it is possible to map out the actual electron density throughout the unit cell by the intensity of the diffraction pattern lines. As the electron density is highest at the location where the atom is present, the electron density map reveals the possible atom positions in the unit cell.

9.8 PRECISION IN THE DETERMINATION OF LATTICE PARAMETERS

For many purposes, the determination of the lattice parameter with a high precision is required. For example, as the concentration of the solute varies in a solid solution, its lattice parameter changes and this can be taken as a basis for the determination of the solute concentration and this requires a high degree of accuracy in lattice parameter measurements. Another example would be the determination of the coefficients of thermal expansion by accurate measurements of lattice parameter at different temperatures. Here, the determination of the precise parameter will be discussed particularly with reference to cubic crystals which may also be applied with slight modifications to other crystal structures.

Basically, Bragg's law is used for the determination of plane spacing 'd' of a given set of lattice planes (hkl) reflecting a given wavelength λ at a particular θ value and then the parameter 'a' is calculated by the plane spacing equations. It should be noted here, that, in Bragg's law, θ does not occur directly but occurs as a sine function. The error in θ measurement introduces an error in $\sin \theta$ which in turn causes an error in the 'd' value determined and consequently gives erroneous 'a'. The variation of $\sin \theta$ with θ is shown in Fig.9.4. This figure shows how the error in $\sin \theta$, i.e., $\Delta \sin \theta$, is different at high angles and at low angles considering a constant error in θ, i.e., $\Delta\theta$, during measurements. It is seen that the error in $\sin \theta$ is least as θ approaches 90° for a constant error $\Delta\theta$ during measurements. This is fortunate, since if the calculations of lattice parameter is made on the basis of high angle lines, the error in the parameter

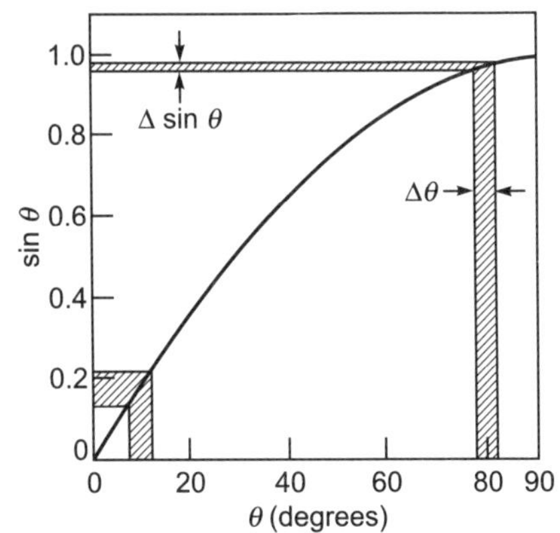

Fig. 9.4 The variation of $\sin \theta$ with θ. The error in $\sin \theta$ caused by a given error in θ decreases as θ increases ($\Delta\theta$ exaggerated).

value determined would approach zero even when there is an error in θ measurement. In other words, a very accurate determination of lattice parameter is possible making use of high angle lines of powder diffraction patterns. This idea can also be obtained by differentiating the Bragg's law with respect to θ, i.e., $\lambda = 2d \cdot \sin \theta$, or,

$$\lambda/2 = d \cdot \sin \theta$$

differentiating

$$0 = d \cdot \cos \theta \cdot d\theta + \sin \theta \cdot dd \text{ (since, } \lambda \text{ is a constant)}$$

or, writing in terms of small differences $\Delta\theta$ and Δd,

$$\Delta d/d = -(\cos \theta/\sin \theta) \cdot \Delta\theta = -\cot \theta \cdot \Delta\theta.$$

For the cubic system,

$$a = d \cdot (h^2 + k^2 + l^2)^{1/2}.$$

So,
$$\Delta a/a = \Delta d/d = -\cot \theta \cdot \Delta\theta \tag{9.10}$$

As θ approaches $90°$, $\cot \theta$ approaches zero and $\Delta a/a$ (fractional error in 'a') also approaches zero. So, if the lattice parameter 'a' values calculated from different diffraction lines are plotted against the θ values (or $\sin \theta$ values) of these lines it must be possible to extrapolate and find the accurate value of 'a' at $\theta = 90°$ (or at $\sin \theta = 1$). But this curve of 'a' versus θ (or versus $\sin \theta$) is not a linear one and extrapolation of a non-linear curve does not yield an accurate value. So, it is necessary that a function of θ must be found which has a linear relationship with the lattice parameter so that an accurate extrapolation can be done. The kind of precision in lattice parameter determination, that can be obtained by using high angle lines for calculation is around + 0.001 nm. With better experimental techniques and the use of a proper extrapolation function, this accuracy can be increased to + 0.0001 nm. The maximum accuracy that can be obtained with considerable effort, both experimental and computational, is expected to be + 0.00001 nm.

9.9   SOURCES OF ERROR IN DEBYE-SCHERRER CAMERAS

The main sources of experimental error in Bragg angle determination are

1. film shrinkage
2. incorrect camera radius
3. off-centering of specimen
4. absorption in specimen

As seen in the previous section, back-reflection region (high θ region) is the most suitable one for precise parameter measurements. Fig. 9.5 shows the geometry of diffraction in a Debye-Scherrer (D.S.) camera. In this figure 2θ is the diffraction angle and let an angle 2φ be supplementary to it, i.e., $(\pi - 2\theta) = 2\varphi$ or $\varphi = 90 - \theta$. Let s' be the distance between a pair of back reflection lines which is also equal to the arc length defined by the angle 4φ as shown in Fig. 9.5.

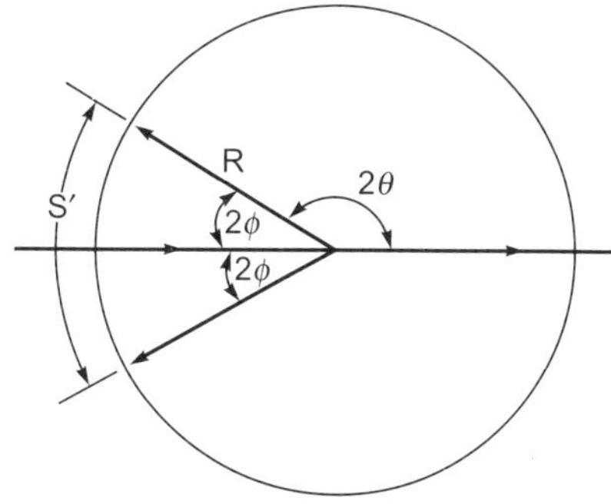

Fig. 9.5

By geometry, $\varphi = s'/4R$.   (9.11)

where R is the radius of the camera.

The film shrinkage occurs during the development and drying of the exposed x-ray film after the diffraction experiment. Let $\Delta s'$ be the shrinkage which occurs in the x-ray film (during processing) in a distance s' defined earlier. Let the camera radius also have an error ΔR either due to design or wear out in usage. The equation 9.11 can be written in logarithmic form as noted below.

$$\ln \varphi = \ln s' - \ln 4 - \ln R$$

Differentiating the above equation,

$$\Delta\varphi/\varphi = (\Delta s'/s') - (\Delta R/R).$$

Thus, the error in φ due to shrinkage and radius errors is given by

$$(\Delta\varphi)_{S'R} = [(\Delta s'/s') - (\Delta R/R)] \cdot \varphi \qquad (9.12)$$

The shrinkage error and the error due to incorrect camera radius may be eliminated using unsymmetrical or Straumani's film loading technique as discussed in section 7.6.3.

If the specimen is not placed exactly along the axis of the cylindrical camera, an off-center error is introduced. The off-centering of the specimen maybe in the direction of the incident beam (Δx) or perpendicular to it (Δy). When the error is caused due to the off-centering of the specimen in the direction Δy, the error can be almost neglected since on either side of the incident beam the errors are opposite in sense i.e., AC – BD as illustrated in Fig. 9.6 (b). The error Δx causes the error (AC + BD) in s' measurement as shown in Fig. 9.6 (a).

i.e., $\qquad\qquad\qquad \Delta s' = AC + BD = 2\,ON = 2 \cdot \Delta x \cdot \sin 2\varphi.$

This error $\Delta s'$ causes an error $\Delta\varphi$ in the measurement of φ. The equation above can be rewritten as

$$\Delta\varphi/\varphi = \Delta s'/s' = (2 \cdot \Delta x \cdot \sin 2\varphi)/(4R\varphi)$$

or, $\qquad\qquad\qquad \Delta\varphi_C = (\Delta x/R) \cdot \sin \varphi \cdot \cos \varphi \qquad\qquad (9.13)$

An error in φ can also be caused due to the specimen itself, particularly, if it has a high absorption coefficient. This effect is often the largest single cause of error in the parameter measurements. And it is very difficult to calculate this effect with

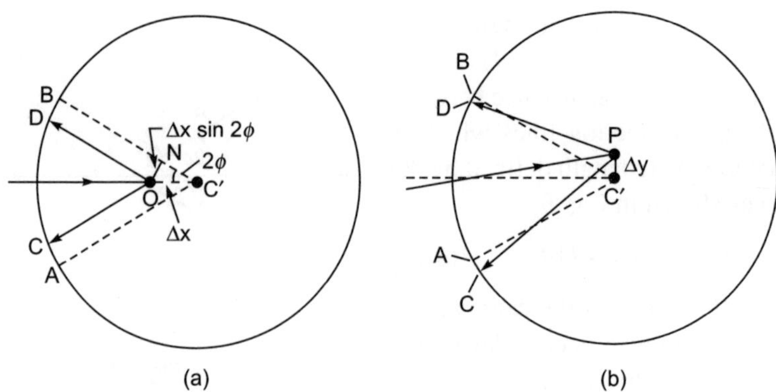

Fig. 9.6 Effect of specimen displacement on line positions

accuracy. Precise parameter measurements are generally made using back reflection lines. For a highly absorbing specimen, the lines of back reflection originate mainly from the specimen surface facing the collimator as shown in Fig. 9.7. So, to a first approximation, this is equivalent to an off-centering error and it can be presumed to be included in equation 9.13.

So, the overall error in φ, due to film shrinkage, radius error, centering error and absorption can now be written combining equations 9.12 and 9.13 as,

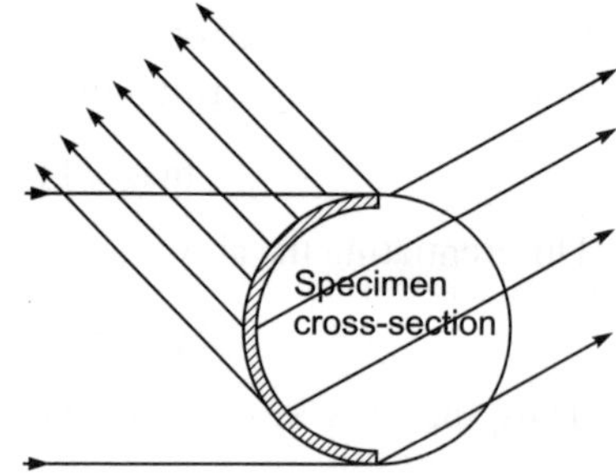

Fig. 9.7 Back reflection from a highly absorbing specimen

$$(\Delta\varphi)_{Total} = [(\Delta s'/s') - (\Delta R/R)] \cdot \varphi + (\Delta x/R) \cdot \sin\varphi \cdot \cos\varphi \qquad (9.14)$$

Putting $\varphi = 90° - \theta$, $\Delta\varphi = -\Delta\theta$, $\sin\varphi = \cos\theta$ and $\cos\varphi = \sin\theta$, the equation 9.10 can be modified as

$$\Delta d/d = -(\cos\theta/\sin\theta) \cdot \Delta\theta = (\sin\varphi/\cos\varphi) \cdot \Delta\varphi$$

Now, putting the value of $\Delta\varphi$ from equation 9.14,

$$\Delta d/d = (\sin\varphi/\cos\varphi) \cdot \{[(\Delta s'/s') - (\Delta R/R)] \cdot \varphi + (\Delta x/R) \cdot \sin\varphi \cdot \cos\varphi\}$$

In the back reflection region, φ is small and may be replaced by $\sin\varphi \cdot \cos\varphi$ since $\sin\varphi = \varphi$ and $\cos\varphi = 1$ for small values of φ. So the above equation can be rewritten as

$$\Delta d/d = [(\Delta s'/s') - (\Delta R/R) + (\Delta x/R)] \cdot \sin^2\varphi.$$

In this equation, the terms in the bracket are constant for any given film so that

$$\Delta d/d = K \cdot \sin^2\varphi = K \cdot \cos^2\theta,$$

where K is a constant. Further, for a cubic system,

$$\Delta d/d = \Delta a/a = K \cdot \cos^2\theta \qquad (9.15)$$

Thus, for a cubic substance, if the computed 'a' value for each line is plotted versus $\cos^2 \theta$, a straight line should result and the precise parameter 'a_o' can be found by extrapolating this line to $\cos^2 \theta = 0$. It should also be noted that the above extrapolation holds only for small values of φ or large values of θ (>60).

If the various sources of error are analyzed more rigorously than are being done here, it can be shown that the following relation holds quite accurately to even lower values of θ:

$$\Delta d/d = K \cdot [(\cos^2 \theta/\sin \theta) + (\cos^2 \theta/\theta)] \qquad (9.16)$$

Here, the bracketed value approaches zero as θ approaches 90°.

For tetragonal and hexagonal crystals, there are two lattice parameters for each crystal, namely 'a' and 'c'. It is not possible to determine both of them by a single diffraction line. The procedure, then, is to separate the lines into two groups: one containing lines belonging to reflections from (hk0) planes and another containing lines belonging to reflections from (00l) planes. The former group can be used to calculate 'a' values and the latter group to calculate 'c' values. Two separate extrapolations can be made (one for each parameter) to determine precise parameters 'a_o' and 'c_o'.

In general, for crystal structures having more parameters, methods similar to the above suiting the specific structure may be employed to determine precise parameters. In all extrapolations, it is better to go in for the best fit line employing the method of least squares.

9.10 SOURCES OF ERROR IN FOCUSING CAMERAS

A camera of focusing type is preferred for work requiring high precision, since the position of a diffraction line on the film is twice as sensitive to small changes in plane spacing with this camera as it is with a D.S. camera of identical diameter. Sources of error in focusing cameras are

1. film shrinkage
2. incorrect camera radius
3. displacement of the specimen from the camera circumference
4. absorption in specimen

By detailed analysis, it has been found that the error produces fractional errors in 'd' values which are very closely proportional to $\varphi \cdot \tan \varphi$ where φ is equal to $(90^\circ - \theta)$. This function $\varphi \cdot \tan \varphi$ can be used for extrapolation.

9.11 SOURCES OF ERROR IN PINHOLE CAMERAS

Here, the sources of error are due to film shrinkage, incorrect specimen to film distance and absorption in specimen. It may be shown that the fractional error in 'd' values is proportional to $\sin 4\varphi \cdot \tan \varphi$ or to the equivalent expression $\cos^2 \theta \cdot (2 \cos^2 \theta - 1)$. This function can therefore be used for extrapolation.

9.12 SOURCES OF ERROR IN DIFFRACTOMETERS

The chief sources of error in diffractometers are due to
1. misalignment of the instrument
2. use of a flat specimen instead of a specimen curved to conform to the focusing circle
3. absorption in the specimen
4. displacement of the specimen from the diffractometer axis
5. vertical divergence of the incident beam

It is found that in diffractometers, $\Delta d/d$ is approximately proportional to $\cos^2 \theta$ and so, a fairly accurate value of lattice parameter can be obtained by extrapolation against $\cos^2 \theta$.

SOLVED EXAMPLE

Example 9.1   The powder pattern of an unknown cubic metal showed initial eight lines with 2θ values of 38.2, 44.4, 64.6, 77.4, 81.6, 98.0, 110.6, 115.0 degrees. The pattern was obtained using Cu Kα radiation in an x-ray diffractometer. Index the lines, determine the lattice parameter and identify the metal.

Solution: The following table shows the systematic calculations

Line	2θ	θ	$sin^2 \theta$	$h^2 + k^2 + l^2$	hkl	a, nm
1	38.0	19.0	0.1060	3	111	0.4101
2	44.2	22.1	0.1415	4	200	0.4098
3	64.4	32.2	0.2840	8	220	0.4092
4	77.4	38.7	0.3916	11	311	0.4086
5	81.6	40.8	0.4272	12	222	0.4086
6	98.0	49.0	0.5696	16	400	0.4086
7	110.6	55.3	0.6765	19	331	0.4086
8	115.0	57.5	0.7121	20	420	0.4086

The calculations till column 4 are simple. The $sin^2 \theta$ values in column 4 are to be examined to see whether they increase in certain ratios sequentially. As discussed in section 9.3, the $(h^2 + k^2 + l^2)$ values increase in definite order for each type of cubic structures. The $sin^2 \theta$ values (which are proportional to the $h^2 + k^2 + l^2$ values) in column 4 in the table are in the ratios of 3:4:8:11:12:16:19 and so on indicating the structure to be FCC and the $h^2 + k^2 + l^2$ values can filled in column 5. The (hkl) values can be easily deduced from $h^2 + k^2 + l^2$ values. The exercise till this step is termed as indexing. The lattice parameter 'a' can be calculated using equation 9.2, i.e.,

$$sin^2 \theta = \lambda^2 \cdot (h^2 + k^2 + l^2)/4a^2, \text{ or rewritten in another form,}$$
$$a^2 = \lambda^2 \cdot (h^2 + k^2 + l^2)/4sin^2 \theta$$

It can be seen that the calculated 'a' value of 0.4086 nm coincides well with the standard 'a' value 0.40856 nm of FCC silver.

EXERCISES

9.1 A powder pattern of a cubic metal obtained by x-ray diffractometer run shows eight lines with 2θ values in degrees as follows: 40.5, 58.5, 73.3, 87.1, 100.8, 115.1, 131.4, 154.0. The radiation used is Cu $K\alpha$ (wavelength = 0.15418 nm). Index these lines, determine the lattice parameter and identify the metal.

[Ans: a = 0.31648 nm, BCC tungsten]

9.2 Debye-Scherrer (camera dia = 57.3 mm) pattern of a material obtained using Co $K\alpha$ radiation (wavelength = 0.17902 nm) contains 5 pairs of lines. The distance in mm between pairs of lines are 56.6, 77.4, 99.9, 124.2, 162.0.

 (i) Index the lines and calculate the lattice parameter.

 (ii) If Cr $K\alpha$ (wavelength = 0.22909 nm) would have been used as the radiation for diffraction, how many pairs of lines would have appeared on the pattern?

[Ans: BCC iron, a = 0.2866 nm, 3 pairs]

9.3 A powder pattern of a cubic metal obtained by x-ray diffractometer run between 2θ values of 10 and 100 degrees shows eight lines with 2θ values in degrees as follows: 39.2 (m), 43.6 (s), 45.4 (w), 50.6 (m), 66.0 (w), 74.2 (m), 79.2 (w), and 89.8 (s). The letters 's', 'm', and 'w' indicate the rough estimates of intensities of lines as strong, medium and weak respectively. The radiation used is unfiltered Cu K radiation containing Cu $K\alpha$ (wavelength = 0.15418 nm) and Cu $K\beta$ (wavelength = 0.13922 nm). Index these lines after identifying and separating the lines due to $K\alpha$ and $K\beta$ radiations. Determine the lattice parameter and identify the metal.

[Ans: Copper]

9.4 Predict the line positions 2θ in degrees and their relative integrated intensities in a diffractometer powder pattern of gray tin (diamond structure with a = 0.647 nm) using Cu $K\alpha$ radiation.

[Ans: 23.8, 39.4, 46.6, 56.9, 62.6, 71.4, 76.5, 84.7]

9.5 (i) Construct a Hull-Davey chart using semi-log graph sheet and accompanying 'd' scale in nm units for HCP lattices. Cover a range 0.5 to 2.0 for c/a ratio and plot the curves 00.2, 10.0, 10.1, 10.2 and 11.0.

 (ii) Use the chart and index the lines of a HCP metal showing 'd' values in nm units of 0.2556, 0.2368, 0.2254, 0.1724 and 0.1489. Determine the lattice parameters 'a' and 'c'.

[Ans: α titanium with a = 0.2950 nm and c = 0.4683 nm]

9.6 The powder pattern of a cubic substance using Cu Kα1 radiation (wavelength = 0.154051 nm) showed the following data for five lines.

(i) Determine the precise lattice parameter using graphical extrapolation of 'a' versus $\cos^2 \theta$.

(ii) Determine the lattice parameter using the method of least squares.

$h^2 + k^2 + l^2$	2θ in degrees
16	98.2
19	110.8
20	115.3
24	135.4
27	157.9

[Ans: FCC gold with 'a' = 0.40783 nm]

10

C H A P T E R

SINGLE CRYSTAL ORIENTATION

10.1 GENERAL

The properties of polycrystalline materials are basically determined by the properties of single crystals which make up the polycrystalline substance. It is therefore necessary to study the properties of single crystals. It is well-known that single crystals are mechanically anisotropic, i.e., their mechanical properties are not the same in different crystallographic directions. So, it is necessary to know the crystallographic orientation of a given single crystal, so that its mechanical properties could be studied with reference to the crystallographic orientation. There are three x-ray diffraction methods for the identification of the crystallographic orientation of single crystals, viz., the back reflection Laue method, the transmission Laue method and the diffractometer method.

10.2 THE BACK REFLECTION LAUE METHOD

As discussed in section 4.3.1, the back reflection Laue technique utilizes heterochromatic or white x-radiation and the diffraction spots are recorded on a photographic film placed at the incident beam side of the specimen as shown in Fig. 4.6 b). The diffracted beams under such condition form an array of spots which are seen to lie on certain imaginary curves, generally hyperbolas. The spots lying on any one hyperbola are diffractions from planes belonging to one zone. Laue reflections from planes belonging to a zone lie on the surface of an imaginary cone whose axis is the zone axis. Fig. 10.1 shows the situation for the back reflection Laue technique where the curved surface of the imaginary diffraction cone also contains the transmitted beam and the angle of inclination φ of the zone axis to the transmitted beam is equal to the semi-apex angle of the diffraction cone. It can be noted that in the back reflection technique, the angle φ exceeds $45°$ and the intersection of the diffracted beams with the film gives spots lying on imaginary hyperbolas.

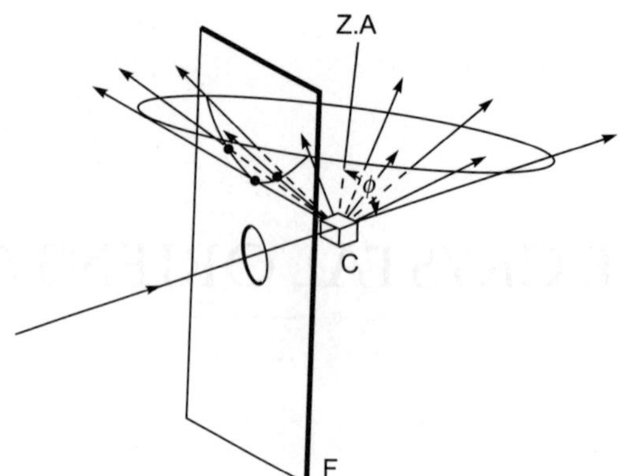

Fig. 10.1   Intersection of a conical array of diffracted beams with a film placed in the back-reflection position. C = crystal, F = film, Z.A. = zone axis (Ref: B.D. Cullity, Elements of x-ray Diffraction).

The positions of the diffraction spots on the film depend on the orientation of the crystal relative to the incident beam. As the aim is to determine the orientation of the crystal specimen, it is important and necessary to orient the specimen relative to the film in some known way during the diffraction experiment. If the specimen is in the form of a wire or a rod, it can be best mounted with its axis vertical and parallel to one edge of the rectangular photographic film and a fiducial mark on the surface placed nearest to the film. Sheet or plate specimens can be conveniently mounted with their planes parallel to the sheet film and one noted edge is made parallel to a noted edge of the photographic film. If the specimen is of irregular shape, a few fiducial marks (as many as needed) can fix its orientation relative to the film.

Though it is possible to measure the Bragg angle θ corresponding to any given diffraction spot, it is not possible to identify the plane causing the diffraction spot in the Laue technique because the wavelength being diffracted is not known. Each plane depending on its 'd' value chooses a particular wavelength from the continuous spectrum available (since, the Laue technique makes use of white x-radiation) so as to satisfy the Bragg's law and diffracts that particular radiation. Even though, it is not possible to identify the plane causing the spot, it is possible to locate the position of the normal to the plane causing the spot by the geometry of the experimental setup. Once the directions of these plane normals are detected, they can all be plotted on a stereographic projection, angles between these normals can be measured and can be compared with the standard interplanar angles for the crystal system studied, so that the planes causing the diffraction spots can be identified.

The first step is therefore, to identify the normal to the plane causing a given diffraction spot and then plotting this normal position as a pole on a stereographic projection. Fig. 10.2 shows the geometry of the Laue spot 'S' and the plane normal CN causing this spot as viewed from the crystal side in a back reflection

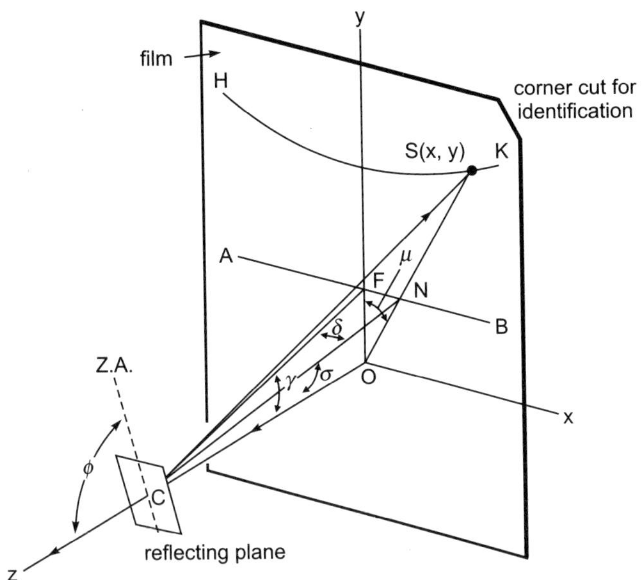

Fig. 10.2   Location of back-reflection Laue spot. Note that $\gamma = 90° - \phi$ (from Cullity).

Laue experiment. Incident beam direction OC can be taken as the 'z' axis and the coordinate axes 'x' and 'y' can be imagined to lie on the plane of the film such that the zone axis of the reflecting plane lies on the 'yOz' plane. The angle between the zone axis and the transmitted beam, φ, is more than 45°, so that the imaginary cone, on whose surface the diffracted beams lie, cuts the film plane in a hyperbola KH which is symmetrical with respect to 'y' axis. The diffraction spot 'S' with coordinates (x, y) is caused by the reflecting plane 'c' of the crystal shown whose normal cuts the film at N. It should be noted that the lines CN, CO and CK are coplanar as they are the normal of the reflecting plane, the incident beam and the reflected beam respectively. The positions of the planes belonging to the zone shown, can be visualized by rotating the reflecting plane of Fig. 10.2 about its zone axis through 360°. As this plane is rotated, the diffraction spots may be formed on specific locations on the hyperbola HK (depending on the availability of a crystallographic plane in those orientations) provided these positions satisfy the Bragg's law. At the same time, as the plane is rotated, the plane normal cuts the photographic film plane in a line AB. It should be noted that the positions of the plane normals belonging to a zone would all lie in a plane perpendicular to the zone axis itself. This plane containing the plane normals cuts the the film plane in a straight line AB (since, the intersection of two planes is a straight line). Now, the position of the plane normal N could be identified with the help of the vertical shift angle γ and the horizontal shift angle δ. It is thus necessary to find a relation between the (x, y) coordinates of the spot 'S' and the angular coordinates γ and δ of the corresponding reflecting plane normal N. It can be noted from the Fig. 10.2 that

$$x = OS \cdot \text{Sin } \mu, \quad y = OS \cdot \text{Cos } \mu \quad \text{and} \quad OS = OC \cdot \tan 2\sigma,$$

or,
$$x = OC \cdot \tan 2\sigma \cdot \sin \mu \quad \text{and} \quad y = OC \cdot \tan 2\sigma \cdot \cos \mu,$$

where OC is the specimen to film distance.

$$\text{Further, } \tan \mu = \frac{FN}{ON} = \frac{(CF \cdot \tan \delta)}{(CF \cdot \sin \mu)} = \frac{\tan \delta}{\sin \mu}.$$

and
$$\tan \sigma = \frac{ON}{OC} = \left(\frac{FN}{\sin \mu} \right) \left(\frac{1}{CF \cdot \cos \gamma} \right)$$

$$= \left(\frac{CF \cdot \tan \delta}{\sin \mu} \right) \cdot \left(\frac{1}{CF \cdot \cos \gamma} \right)$$

$$= \frac{\tan \delta}{(\sin \mu \cdot \cos \gamma)}$$

With these equations, it is possible to determine the angular coordinates γ and δ for any given diffraction spot with cartesian coordinates (x, y) on the diffraction film. Using these equations Greninger devised a chart as shown in Fig. 10.3. The chart shown is devised for specimen to film distance of 3 cm (the commonly used distance) and graduated at 2° intervals. The Greninger chart should be drawn on a transparent sheet, so that γ and δ values corresponding to a diffraction spot can be directly read by placing the chart on the Laue pattern. In use, the Greninger chart is to be placed over the film with its center coinciding the film center and the edges of the film and the chart are parallel. For the purpose of proper identification and orientation, it is necessary to cut the top right corner of the film before the experiment as shown in Fig. 10.2. Since, the plane normal CN intersects the film on the side closer to the x-ray source, the projection must be viewed (and read) from the x-ray source side. So, while reading the γ and δ angles by the Greninger chart, the cut corner must be on the left hand top corner as shown in Fig. 10.4 a). After reading the γ and δ values from the Greninger chart, the same can be plotted on a stereographic projection as shown in Fig. 10.4 b). While plotting the pole on a tracing sheet, the underlying Wulff net must be kept such that the longitudes run from left to right and not from top to bottom as done usually. The reason is that all the poles belonging to a zone lie on a single longitude (which is a great circle) and it is known by the geometry of the Laue experiments that all planes belonging to a zone cause diffraction spots which lie on a single curve and the normal to these planes will have the same vertical shift γ value. Only with the Wulff net orientation in which the longitudes run from left to right, the poles having the same γ values can be plotted on a single longitude. It can be noted that the spots on the bottom half of the film can be read by keeping the chart upside down and while plotting, the projection also correspondingly can be turned upside down and the poles plotted.

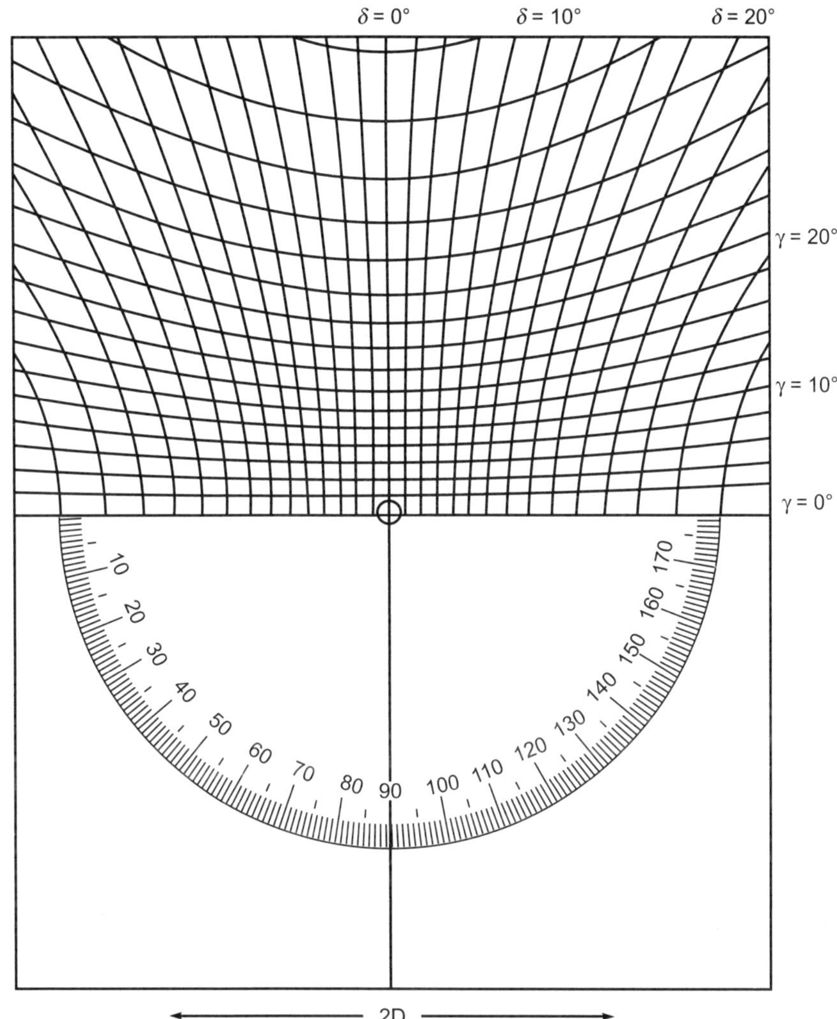

Fig. 10.3   Greninger chart for the solution of back-reflection Laue patterns, reproduced in the correct size for a specimen-to-film distance D of 3 cm (Ref: B.D. Cullity, Elements of x-ray Diffraction).

The next step after obtaining the stereographic projection is to index the poles. With the aid of the Wulff net it is possible to draw zone circles connecting the poles. It is well-known that the intersections of zone circles are generally poles of low indices like {100}, {110} and {111}. The axes of the zones are also poles of low indices. The angles between the zone intersection poles and zone axes poles are measured and compared with the standard interplanar angles of the crystal system to which the single crystal under investigation belongs. It is then possible to identify the low indices poles on the stereographic projection. Once this is done, the orientation determination is complete. With respect to the external orientation

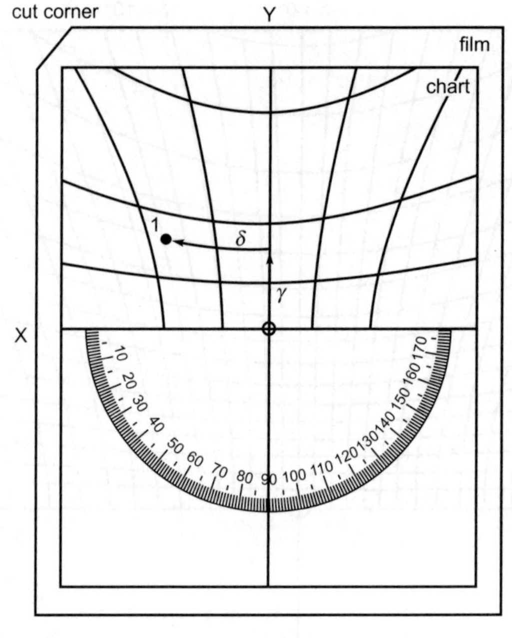

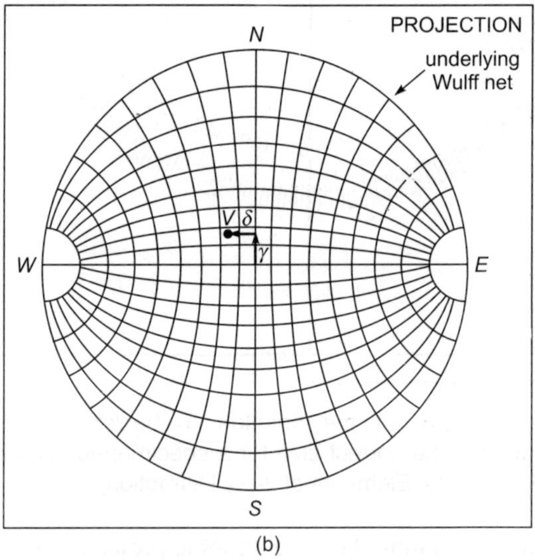

(b)

Fig. 10.4   Use of the Greninger chart to plot the pole of a reflecting plane on a stereographic projection. Pole 1′ in (b) is the pole of the plane causing diffraction spot 1 in (a).

of the single crystal fixed at the time of the experiment, the crystallographic orientation is shown by the stereographic projection.

Another way of manipulating the Greninger chart is as shown in Fig. 10.5 for reading the vertical shift of plane normals corresponding to diffraction spots lying

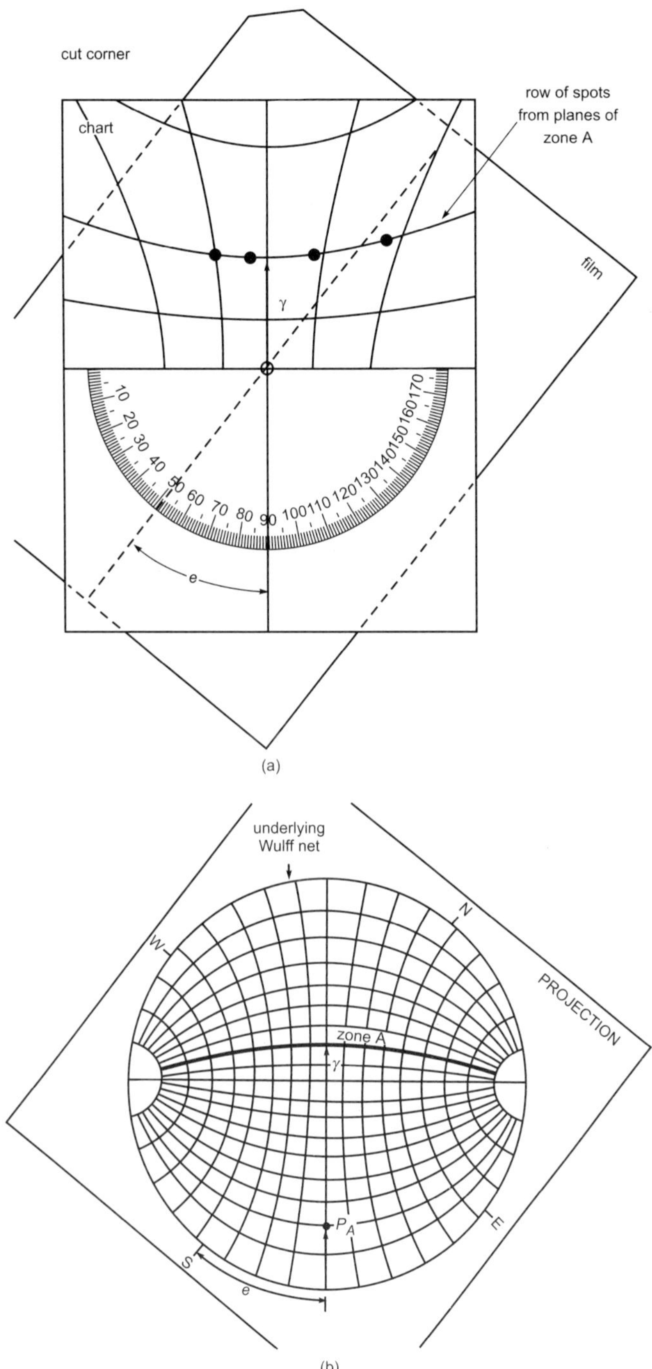

Fig. 10.5   Use of the Greninger chart to plot the axis of a zone of planes on the stereographic projection. P_A is the axis of zone *A* (Ref: B.D. Cullity, Elements of x-ray Diffraction).

on a given hyperbola. Here, keeping the centers of the film and chart coincident, the film is rotated such that diffraction spots from a given zone of planes fall on any one hyperbola on the chart as shown. The vertical shift γ of all these spots is read and plotted on the projection. While plotting this zone circle on a stereographic projection, care must be taken to rotate the projection about the center of the underlying Wulff net by the same angle of rotation, ε, as used during the rotation of the film about the center of the Greninger chart while reading γ value. This method is quite convenient as the zone circles are directly obtained on the projection without resorting to plotting of the individual poles.

10.3   TRANSMISSION LAUE METHOD

Generally, the back reflection Laue method is convenient to determine the orientations of single crystals since the intensity of spots are higher in this method than those in the transmission Laue method. This is because of the absorption of the diffracted beam during its transmission through the specimen in the latter method. But, given a thin specimen of sufficiently low absorption, a transmission Laue pattern can be obtained and used to determine the orientation of the single crystal.

Fig. 10.6 shows the transmission Laue diffraction geometry wherein the diffraction spots seem to lie on the surface of an imaginary cone. The transmitted beam forms a part of the surface of this imaginary cone which has a semi-apex angle φ which is also the angle between the transmitted beam and the zone axis of the zone concerned. This imaginary cone cuts the photographic film in the form of a complete ellipse for sufficiently small values of φ. For larger values of φ, an incomplete ellipse is recorded because of the restriction of the finite size of the film to record the large resulting ellipse. Thus, the spots in transmission Laue pattern lie on imaginary ellipses.

The angular relationships in the transmission Laue method are shown in Fig. 10.7. A reference sphere is centered at the position of the crystal C and is

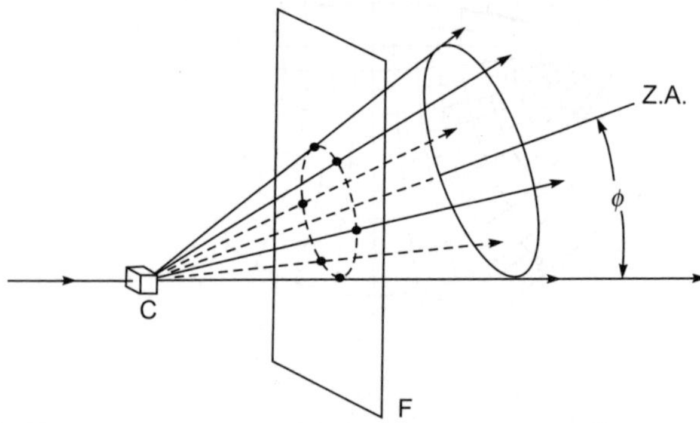

Fig. 10.6 Intersection of a conical array of diffracted beams with a film placed in the transmission position. C = crystal. F = film, Z. A. = zone axis (Ref: B.D. Cullity, Elements of x-ray Diffraction).

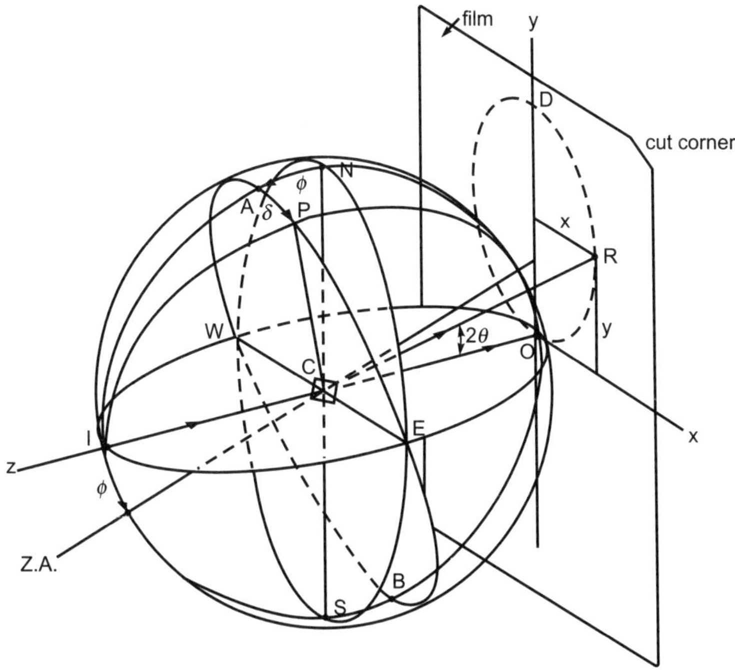

Fig. 10.7 Relation between plane normal orientation and diffraction spot position in the transmission Laue method (Ref.: B.D. Cullity, Elements of X-ray Diffraction).

tangential to the film which is cut at the right hand top corner. The x, y, and z axes are as defined in section 10.2. It can be seen from Fig. 10.7 that the diffraction spot R (x, y) is caused by the reflecting plane whose normal cuts the reference sphere at pole P, whose vertical shift angle is φ and the horizontal shift angle is δ. As discussed in section 10.2, here too, it is possible to establish relationships between the diffraction spot coordinates (x, y) and the angles φ, δ so that the reflecting plane normal can be plotted on a stereographic projection. Leonhardt devised a chart for the purpose of conversion of the diffraction spot coordinates to the angular coordinates φ and δ of reflecting plane normal. The chart is as shown in Fig. 10.8 which is drawn for 10 intervals. This chart is used in the same way as Greninger charts are used. Here, the projection plane is tangent to the reference sphere at the point 'I' of Fig. 10.7 and the projection is made from the point 'O'. This means that the film must be read from the side facing the crystal, i.e., with the cut corner of the film at the upper right as shown in Fig. 10.9 (a). The Fig. 10.9 (b) shows the plotting of the reflecting plane normal on a stereographic projection. Alternately, the entire zone circle may be directly plotted analogous to the procedure discussed under back reflection Laue method. This is shown in Fig. 10.10 (a) and (b). The remaining procedure after obtaining the zone circle or pole positions is again similar to that described under section 10.2 for the determination of the orientation of the single crystal.

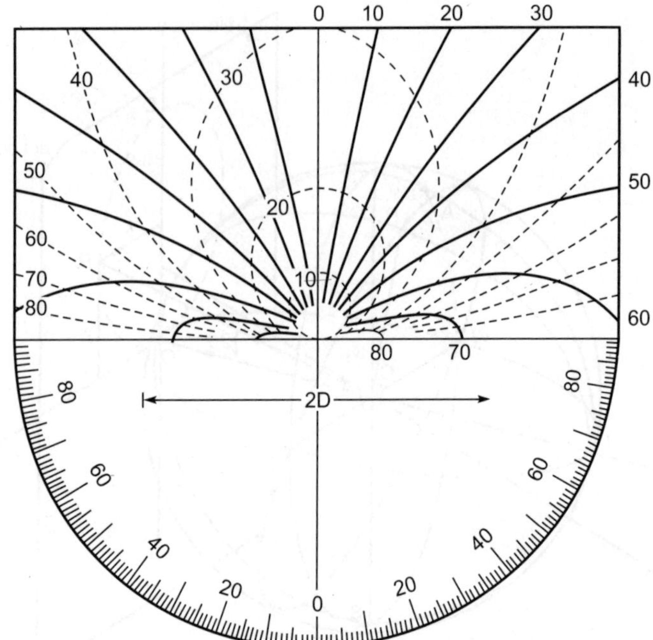

Fig. 10.8 Leonhardt chart for the solution of transmission Laue patterns, reproduced in the correct size for a specimen-to-film distance of 3 cm. The dashed lines are lines of constant ϕ and the solid lines are lines of constant δ (Ref: B.D. Cullity, Elements of x-ray Diffraction).

10.4   THE DIFFRACTOMETER METHOD

This method is convenient and simple compared to Laue methods of determination of orientation of single crystals. The diffractometer method utilizes a monochromatic beam and requires the rotation of the single crystal specimen until a reflection is observed for a fixed diffraction angle 2θ. Once a reflection is observed by the counter of the diffractometer, the plane which satisfies the Bragg law is directly identified.

The method involves the use of a three circle goniometer for holding the single crystal specimen. The goniometer has three possible axes of rotation as shown in Fig. 10.11 where one axis coincides with the diffractometer axis, the second (AA′) lies in the plane of the incident beam 'I' and the diffracted beam 'D' and tangent to the specimen surface and the third (BB′) is normal to the specimen surface.

For the purpose of the experiment, it is necessary to select a crystallographic plane in the crystal for which the intensity of reflection is high. For example, in face centered cubic crystals, the strongest intensity reflection can be obtained from the {111} planes. The interplanar distance of this plane being known, it is

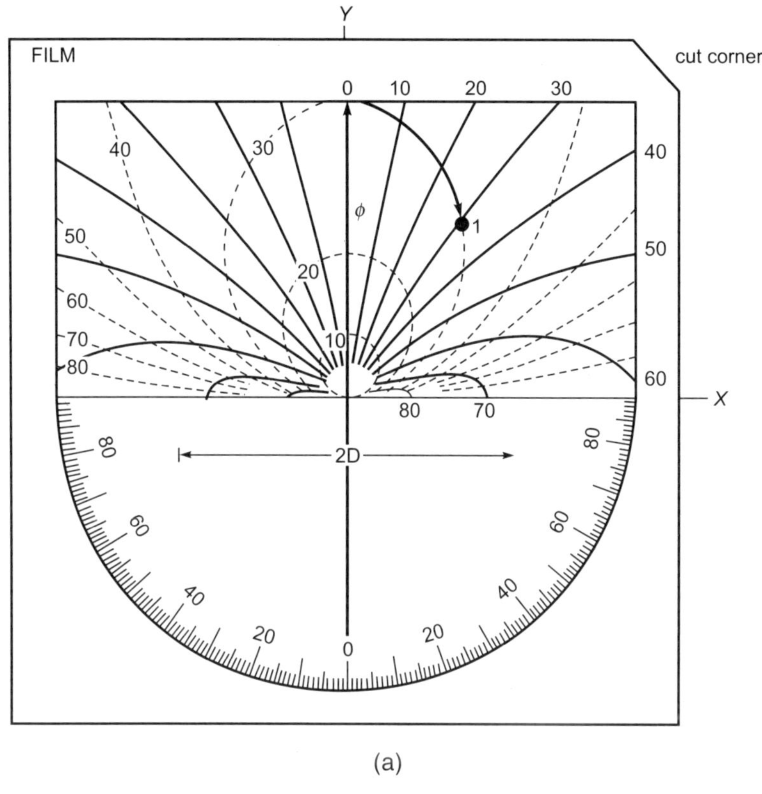

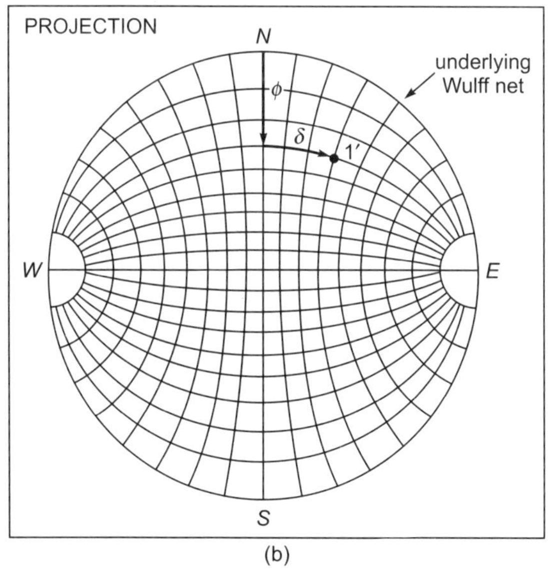

Fig. 10.9 Use of the Leonhardt chart to plot the pole of a plane on a stereographic projection. Pole 1' in (b) is the pole of the plane causing diffraction spot 1 in (a).

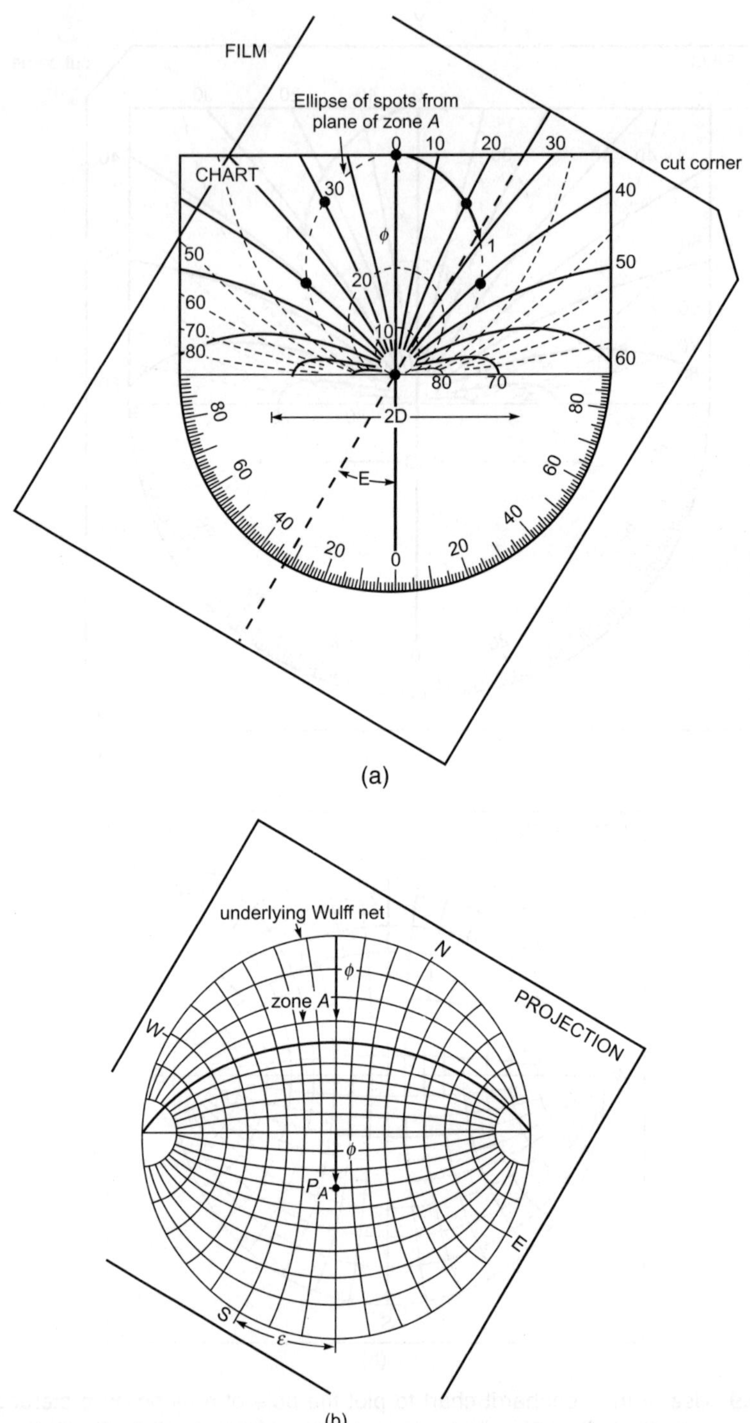

Fig. 10.10 Use of the Leonhardt chart to plot the axis of a zone of planes on the projection. P_A is the axis of zone A.

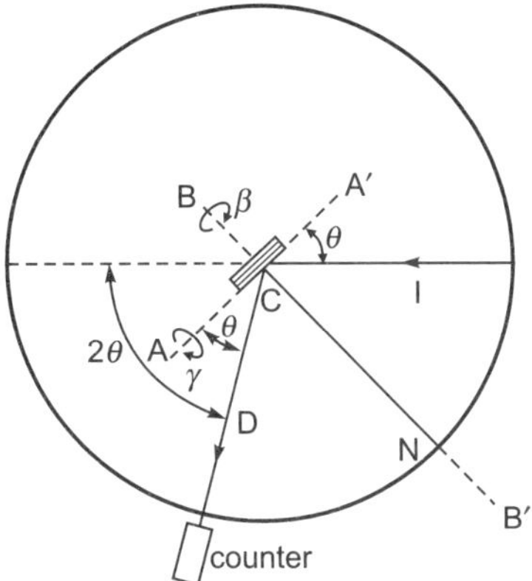

Fig. 10.11 Crystal rotation axes for the diffractometer method of determining orientation (Ref.: B.D. Cullity, Elements of X-ray Diffraction).

possible to calculate the θ value using the Bragg's law (along with the knowledge of the monochromatic wavelength used in the diffractometer). The counter of the diffractometer is then fixed at the diffraction angle 2θ so as to receive the diffraction, as shown in Fig. 10.11, if it occurs. The specimen holder along with the specimen is then rotated about the diffractometer axis till its surface makes an angle θ with respect to the incident beam and the axis AA' coincides with the specimen surface. The specimen holder is then fixed in this position and it is not disturbed further till the end of the experiment. The specimen crystal can then be rotated about the axis BB' until a line drawn on the specimen or an edge of the specimen is parallel to the diffractometer axis. This is the initial position of the crystal and the stereographic projection is done based on this initial position.

After fixing the initial position of the crystal as described above, it can be slowly rotated about the axes AA' (through a certain angle γ) and BB' (through a certain angle β) until a strong reflection is indicated by the counter. Once a diffracted beam is located in this fashion, it means that a chosen strong reflecting plane from the selected family (selected by the chosen diffraction angle 2θ) has come to the reflecting position. This means that the reflecting plane at the current position makes equal angles with the incident and the diffracted beams and the normal to this plane bisects the angle between the incident and the diffracted beams (i.e., the direction CN shown in Fig. 10.11). The pole can now be plotted directly on the stereographic projection where the projection plane is parallel to the initial surface position of the crystal. This means that, if a strong reflection is observed in this initial position itself, the reflecting plane normal and the direction CN are coincident and the pole of the reflecting plane would be at the center of the stereographic projection. Now, let a strong reflection occur at certain angles of rotations γ and β in the sense shown in Fig. 10.11. In Fig. 10.12, the rotations β and γ, with reference to the crystal rotation shown in Fig. 10.11, are done to locate

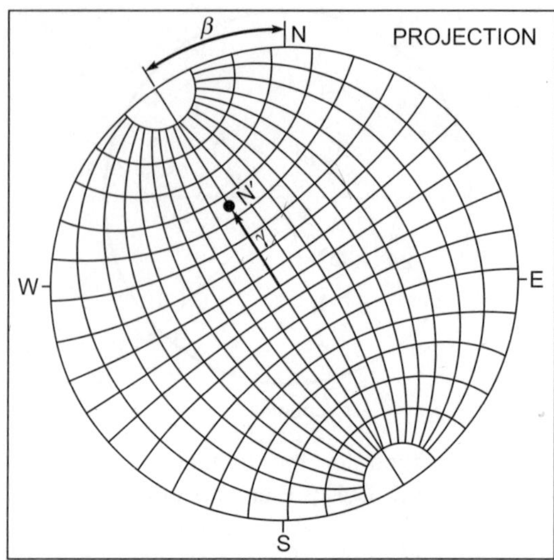

Fig. 10.12 Plotting method used when determining crystal orientation with the diffractometer. (The directions of the rotations shown here correspond to the directions of the arrows in Fig. 10.11) (Ref: B.D. Cullity, Elements of x-ray Diffraction).

the reflecting plane normal position at N' in the stereographic projection. In other words, N' is the position of the normal to the reflecting plane which comes into position of diffraction by rotating the γ axis downward about AA' axis and the β angle clockwise about the BB' axis. Once the orientation of one set of a family of strong reflecting planes is determined, the determination of the orientations of the other planes in this family is not difficult, since, the interplanar angles within the family are generally well-known. It is sufficient to determine the orientations of two planes in a family so as to fix the orientation of a single crystal. But, the orientation of the third plane in the family can be determined as a confirmatory procedure.

It should be noted that there is no focusing action during diffraction here, since the incident beam is a parallel x-ray beam and the crystal is an undeformed one.

The diffractometer method is faster and easier than Laue methods. With an experienced operator, the diffractometer method can lead to more accurate determination of orientation provided extremely narrow slits are used to reduce the beam divergence. On the other hand, the Laue patterns, in addition to fixing the orientation, can also give indications of the crystal distortion, if any, by the shape of the spots. If the Laue pattern contains clear, well-defined and equiaxed spots, the crystal is generally undistorted and perfect. If the crystal is distorted, the spots would be of non-uniform intensity and elongated along the imaginary diffraction curves (ellipses or hyperbolas) as shown in Fig. 10.13 a) and b).

Thus, the diffractometer method in general can be utilized for routine determination of orientations of a large number of crystals. The Laue methods need be resorted to only when it is necessary determine the crystal perfection in addition to orientation studies.

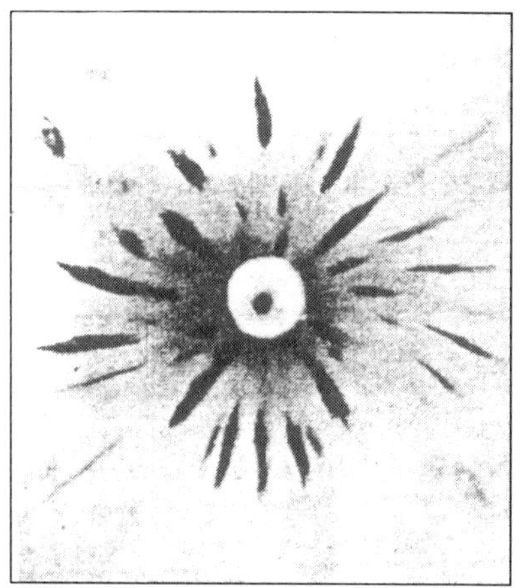

(a) Transmission

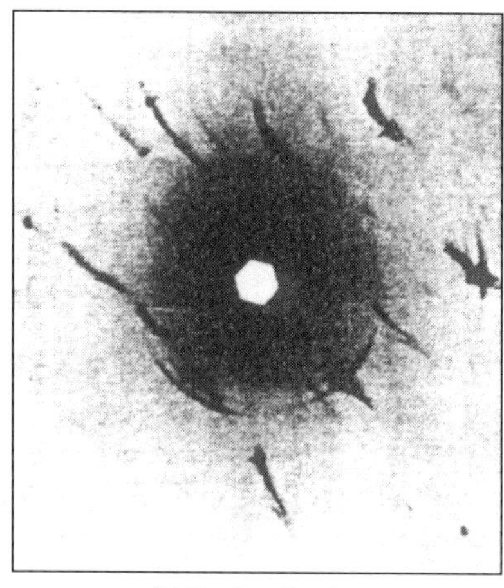

(b) Back reflection

Fig. 10.13 Laue photographs of a deformed aluminium crystal. Specimen-to-film distance 3 cm, tungsten radiation, 30 kv (Ref: B.D. Cullity, Elements of x-ray Diffraction).

10.5 APPLICATIONS

Basically the orientation determination is required in studies of crystal anisotropy. But, there are other uses of orientation identification. The process of twinning can be studied in detail by knowing the relative orientation with respect to each other of twinned and untwinned parts of a given crystal. Further, in these studies, the Miller indices of the twinning plane can also be determined. The relative orientation determination of a precipitate and the matrix in which it resides can be quite useful in the study of precipitation and solid state reactions. The determination of habit planes are quite useful in studying the kinetics of the process. These are some of the additional applications of the orientation determination.

EXERCISES

10.1 An aluminium single crystal diffraction pattern when viewed from the x-ray source shows the following 'x' and 'y' coordinates (measured in mm from the center of the pattern) for back-reflection Laue spots with a crystal to film distance of 30 mm. Plot these spots on a graph sheet. With the help of a Greninger chart, determine the orientation of the crystal. Write down the coordinates of the cube poles {100} on the projection in terms of latitudes and longitudes.

x	y	x	y
+6.6	+2.3	−11.2	+31.5
+11.4	+17.8	−28.0	+45.7
+31.8	+45.7	−30.7	+10.2
+33.5	+10.2	−43.2	+30.2
+3.3	−40.9	−19.3	−35.8
+7.1	−30.7	−20.1	−24.1
+13.0	−17.5	−23.4	−6.6
+18.8	−7.9		

11

CHAPTER

NATURE OF POLYCRYSTALLINE AGGREGATES

11.1 GENERAL

The properties of polycrystalline aggregates basically depend on the properties of single crystals of which they constitute. If the aggregate contains crystals of more than one phase then the properties of the aggregate depends on the properties and volume fractions of each of the phases. The following are some of the other factors which also contribute to the properties of polycrystalline aggregates:

(a) The size of the crystals.

(b) The perfection of the crystals.

(c) The orientations of the crystals with respect to the external shape of the bulk aggregate.

X-ray diffraction methods are available to determine qualitatively and in most of the cases quantitatively each of the above factors.

11.2 SIZE OF THE CRYSTALS

The grain size of a polycrystalline aggregate has a considerable effect on its properties and the effect on the strength property is illustrated by the well-known Hall-Petch equation, i.e., $\sigma_o = \sigma_i + k/d^{1/2}$, in which σ_o is the yield strength, σ_i is the inherent frictional resistance of the material to dislocation movement, d is the grain size and k is a constant. The determination of the grain size is therefore very important and necessary in order to assess the properties of the aggregate and to control them. The grain sizes generally encountered in metals and alloys range from about 1 μm to $10^3 \mu$m. When the size of the crystals is less than about 1 μm, the crystals are termed as particles instead of grains.

11.2.1 Grain Size Determination

The most accurate method of measuring grain size is by microscopic examination. The grain size is quantified by the ASTM grain size index using the equation $n = 2^{N-1}$, where n is the number of grains per square inch at 100X and N is the ASTM index.

The x-ray method of assessing the grain size is qualitative or may be semi-quantitative at best. But, in addition to giving a semi-quantitative information, the x-ray method can also yield information on the perfection and orientation of the crystals. For this purpose, a back reflection or a transmission pinhole photograph of the specimen with filtered radiation is obtained. If the back reflection method is to be used, the surface of the specimen must be etched to remove any disturbed layer (Beilby layer) which might be present and might mask the results and lead to incorrect conclusions. This is especially important in the case of the back reflection method because here most of the diffracted radiation originates from a thin surface layer only.

The schematic back reflection pinhole patterns with varying grain sizes are illustrated in Fig. 11.1. A coarse grain sized material shows a pattern of the type shown in Fig. 11.1 a. Here, only a few crystals or grains take part in diffraction because of their large sizes and the limited cross-section of the incident beam. This pattern is like superimposed Laue patterns from a few crystals due to the low

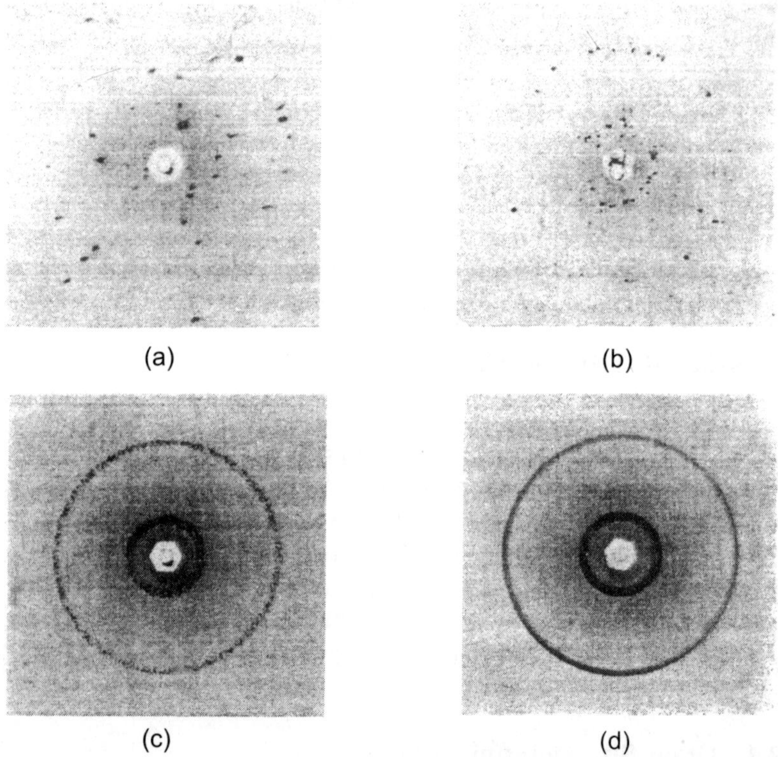

(a)   (b)

(c)   (d)

Fig. 11.1   Back-reflection pinhole patterns of recrystallized aluminum specimens; grain size decreases in the order (a), (b), (c), (d). Filtered copper radiation (Ref: B.D. Cullity, Elements of x-ray Diffraction).

intensity white radiation inherently present in the filtered beam. As the grain sizes become finer, the number of Laue spots increase (because of the increasing number of crystals interacting with the incident beam) and those which lie on potential Debye rings are of high intensity since they are formed by highly intense characteristic radiation. Thus, the potential Debye rings begin to show up as in Fig. 11.1 b. When the grain sizes further decrease, the number of Laue spots increase so much that they merge into an evenly darkened background intensity whereas, the spots which arise from the characteristic radiation clearly show up the Debye rings due to their high intensity as in Fig. 11.1 c. If the grain size further decreases, continuous Debye rings are obtained as shown in Fig. 11.1 d due to the increased number of spots arising from the increased number of grains merging together. The pinhole pattern shows a continuous Debye ring at a grain size of approximately 1mm. Below this size till 0.1 μm, the pinhole pattern does not show any detectable change.

It is possible to prepare standard specimens of various ASTM grain size indices and obtain standard pinhole patterns under identical diffraction conditions and use them as a standard pattern chart. This chart can be used for the purpose of comparing and determining the grain size quantitatively by the pinhole diffraction method. Proper care should be taken to see that the experimental conditions used for the standard patterns and those used for the unknown are identical.

11.2.2   Particle Size Determination

The term particle size is used for crystal sizes below about 0.1 μm in a polycrystalline material. The crystals in the size ranges below 0.1 μm (100 nm) cause broadening of diffraction lines in powder patterns. It has been already seen in section 4.4.4 that the expression governing the line broadening is

$$B = 0.9\lambda/(\cos\ \theta_B \cdot t) \qquad (4.13)$$

where B is the broadening measured in radians at half the maximum intensity of a given line, λ is the wavelength used, θ_B is the Bragg angle of diffraction concerned and t is the particle size.

It is to be noted that all the diffraction lines have a normal measurable width or breadth even when the particle size just exceeds 0.1 μm due to causes such as the divergence of the incident beam, width of the x-ray source and the size of the sample. The line width B referred to in the equation 4.13 is the additional width due the particle size effect alone. Thus, B in equation 4.13 is zero when the particle size is just above 0.1 μm. As the particle size t decreases below 0.1 μm, the width B increases as per the equation noted.

So, the width B_m of any powder diffraction line is due to the normal width referred to in the previous paragraph (standardized by a material with a grain size just above 0.1μm) plus the additional width B_m due to fine particle size. One of the simplest methods to determine this additional width is proposed by Warren. Here, the unknown powder sample is mixed with a standard powder sample which has a particle size just greater than 0.1 μm and also which produces a diffraction line close to the line of the unknown which is going to be used for the determination of the particle size. A diffraction pattern of the mixture is then obtained either using a Debye-Scherrer camera or a diffractometer. The pattern would contain diffraction lines from the standard substance in addition to the broad lines from the unknown powder sample due to fine particle size effect. If B_m is the width of a line from the unknown sample and B_S is the normal width of a line close to it from the standard substance, then the width B due to small particle size effect is given by :

$$B^2 = B_m{}^2 - B_S{}^2.$$

This equation follows from the assumption that the shape of the intensity versus 2θ curve for a given diffraction line follows a normal distribution. The value of B obtained is then substituted in the equation 4.11 to calculate the value of t, the particle size of the unknown specimen. The use of back reflection lines are better for particle size measurement, since, at higher θ values, cosine θ values are smaller thereby increasing the width B as per equation 4.13 for a given particle size thus, ensuring more accurate measurement of the line width. The method described is used commonly in measuring the particle sizes of carbon blacks, catalysts and industrial dusts.

The line broadening may also occur due to the effect of cold work which distorts the crystals and introduces non-uniform strain. The next section deals with this aspect of line broadening. Thus, cold worked polycrystalline aggregate (containing particles instead of grains) would show line broadening due not only to small particle size effect but also to non-uniform strain present. Complete freedom from non-uniform strain cannot be ensured even with annealing after cold working. But, the powder samples of brittle materials are almost free from non-uniform strain and their particle size can be measured by the method discussed above. The particle sizes of powder samples obtained by filing or grinding of ductile materials can also be determined by the above method provided they are properly annealed under vacuum to remove, as far as possible, the non-uniform strain arising due to cold work (filing or grinding).

11.3   PERFECTION OF CRYSTALS

Crystalline matter can contain a number of varieties of imperfections each of which affects the properties of a polycrystalline aggregate. Under this section, the focus is on the type of imperfection caused by the strain present in the material as a result of the cold work it has received. The extent of cold work received by a crystalline material is shown up in most of the cases as non-uniform strain. The individual crystals in a polycrystalline specimen are not free to deform under the action of working stresses due to constraint from the neighbouring crystals during most of the commercial forming processes such as forging, rolling, extrusion and drawing. This causes the crystals to deform in a non-uniform manner (like twisting or bending).

How does the the strain present in the material affect the x-ray diffraction phenomenon? Fig. 11.2 illustrates these effects. The intensity versus 2θ of the diffraction line corresponding to a given plane (hk1) with interplanar spacing d_0 is shown schematically in Fig. 11.2 (a). The curve shows a sharp peak at a specific 2θ value, namely $2\theta_0$ which satisfies the Bragg's law. Fig. 11.2 (b) shows a similar curve for a material whose (hk1) planes have increased interplanar distance as a result of uniform strain in the material. Since, the wavelength λ used is identical in both cases, the 2θ value decreases as d value increases to d_1 from the original value of d_0. The diffraction now occurs at a different, smaller, value of 2θ, i.e., $2\theta_1$. Fig. 11.2 (c) shows the situation in which the material being diffracted is non-uniformly strained, in the case shown schematically, actually bent.

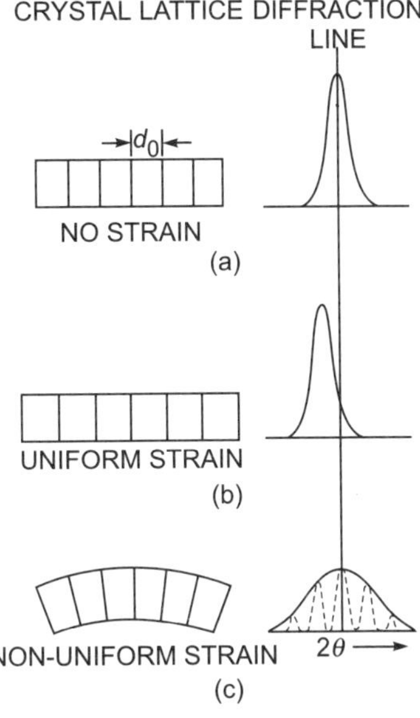

Fig. 11.2   Effect of lattice strain on Debye-line width and position

It can be observed that the top portion of the crystal is in an expanded state in which the interplanar distance is increased to d_1 and the bottom part is in a contracted state in which the interplanar distance is decreased to d_2. Further, the interplanar distance varies continuously from a small value d_2 to a large value d_1 lying on either side of the original value d_0. The diffraction, likewise, takes place continuously from an angle $2\theta_1$ to $2\theta_2$ peaking at $2\theta_0$ corresponding to the original interplanar distance value, d_0, as shown in Fig. 11.2 (c). This leads to line broadening. It is possible to find a relation between the line broadening produced and the extent of non-uniform strain by the differentiation of Bragg's law,

i.e.,   line broadening $b = \Delta 2\theta = -2 \cdot (\Delta d/d) \cdot \tan \theta$ (11.1)

where b is the broadening due to a fractional variation in plane spacing ($\Delta d/d$). With this equation it is possible to find the fractional variation in plane spacing using the value of the line broadening b due to the non-uniform strain in a material. However, as noted in Fig. 11.2 (c), this strain includes both the tensile and the compressive parts and if it is assumed that both of these are equal, then the tensile and the compressive strains individually are equal to half the strain calculated using the equation 11.1. Using this value of the strain, the maximum stress present may also be determined by Hooke's law.

It is well known that the diffraction lines of any metal or alloy become broader when the concerned metal or alloy is cold worked. Under cold working, the grains of the material may get fragmented and at the same time some non-uniform strain may be introduced. Both of these changes have their own share in causing line broadening. It is generally difficult to study the effect of one factor on line broadening by isolating the other. The effects of both these factors together on line broadening can be studied by comparing the diffraction pattern of cold worked metal with that of the same metal under annealed condition. The separation of the $K\alpha$ doublet ($K\alpha_1$ and $K\alpha_2$), particularly in the back reflection region of the diffraction pattern, can furnish a good internal standard in the sense that as the cold work increases, this separation decreases because each component, $K\alpha_1$ or $K\alpha_2$, individually broaden without changing their original diffraction angles. The diffraction pattern in the back reflection region of a highly cold worked material therefore shows an unresolved single broad line as a result of the overlapping of the broadened lines of $K\alpha_1$ and $K\alpha_2$ components. An unresolved $K\alpha$ doublet can therefore be taken as an evidence of cold working. The $K\alpha$ doublet will be clearly resolved into two separate lines in the diffraction pattern if the pattern is obtained after annealing the cold worked material. This is because, on annealing the internal stresses are relieved which reduces the strain and puts the material back into almost the original condition due to the phenomenon of recrystallization. The effect of cold working and annealing can be studied by photographic methods like Debye-Scherrer or pinhole techniques or by x-ray diffractometer.

11.4 ORIENTATION OF CRYSTALS

Crystallographic orientation of the individual crystals or grains is an important factor influencing the properties of polycrystalline materials. The very fact that a given material is a polycrystalline one is an indication that the individual crystals have their own independent crystallographic orientation and the regions of mismatch between the individual grains form the grain boundaries. If the orientations of all the crystals are identical, the entire material would have been a single crystal with no necessity for grain boundaries. The orientations of crystals are generally determined with respect to a set of a reference plane and a direction. If the polycrystalline material is in the form of a sheet, then the reference plane and direction may be conveniently taken as the plane of the sheet (rolling plane) and the rolling direction. If the material is in the form a wire or a rod, then the reference axis can be the wire axis and the reference plane can be any selected plane containing the wire axis. Preferred orientation (texture) is said to exist in a polycrystalline material, when any particular set of crystallographic directions/ planes in most of its crystals show a common orientation. If not, the orientation is said to be random.

Texture is more a rule than exception. In general, it is uncommon to find a polycrystalline material without a texture. A sheet metal has a sheet texture, a wire has a fibre texture or, in general, a plastically deformed material has a deformation texture. Annealing of a deformed metal produces annealing or recrystallization texture. Castings, coatings, electrodeposits, evaporation deposited coatings, etc. have their own characteristic textures. Texture is not confined to metallic materials alone. There are textures in rocks, natural and artificial fibres and sheets. By controlling these textures, it is possible to induce desired properties in any given direction in materials. Grain oriented silicon steels for transformers and texture controlled rolling of nuclear fuel claddings are two specific examples for this type of property control. Further, a sheet material used for deep drawing is supposed to exhibit the same plastic properties in all directions within the sheet plane (planar isotropy) in order to prevent a defect called earing. At the same time, a high normal anisotropy (a ratio of width strain to thickness strain) is desirable, which decreases the reduction in thickness during deep drawing, preventing the premature failure of the sheet. The anisotropy of physical and mechanical properties may play a role in all classes of materials, viz., metals, polymers and ceramics. Another wide field of application of texture is the study of processes occurring in crystalline solids. Many processes like plastic deformation, phase transformation and recrystallization are orientation dependent. Thus, the texture can be used as a sensitive probe to follow solid state processes in materials.

It is true that the forces responsible for causing deformation of grains also cause preferred orientations termed as deformation textures, viz. rolling textures, drawing textures, etc. But, it is to be recognized that the shape of the grains of a material as seen under a microscope has nothing to do with the crystallographic orientation of the grains. Even materials with equiaxed grain structure obtained by recrystallization after deformation show preferred orientation. Only diffraction studies can give evidence regarding the presence or absence of preferred orientation.

Textures can be best expressed graphically by means of pole figures. This method of expressing the texture was first used by Wever, a German metallurgist, in 1924. A pole figure is a stereographic projection of a given set of poles hkl in most of the grains in a polycrystalline material with respect to a given convenient reference plane and direction. The reference plane can be rolling plane in a sheet, a plane containing the axis in a wire, or simply a plane containing two reference lines marked on the external surface of the concerned material.

Assume that a sheet material has only five grains present in it and it is intended to plot the (100) pole figure of the sheet. That is, the orientation of the 100 poles in each of the five grains is to be plotted on a stereographic projection with the projection plane parallel to the plane of the sheet. Each grain has three 100 poles (namely, 100, 010 and 001) and 15 poles are present totally in the five grains of the polycrystalline material sheet. If the grains are oriented at random, then the projection will have poles distributed uniformly on the projection as shown in Fig. 11.3 a. But, if there is a preference in orientation, the poles may cluster around certain points on the projection. One such type of clustering of poles is shown in Fig. 11.3 b. This type of clustering is present in materials which are said to have cube texture. Here, 100 poles are clustered near the rolling direction (RD) and the

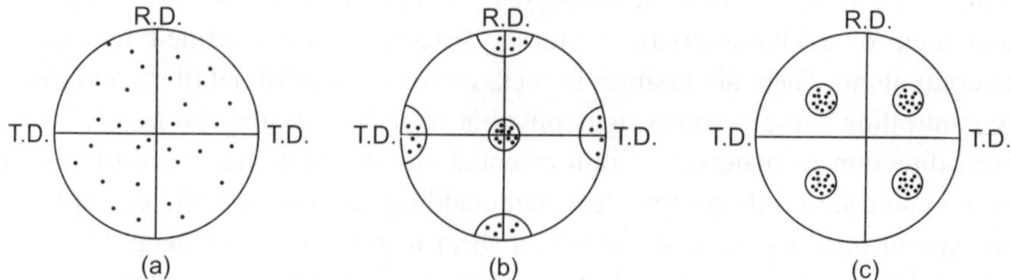

Fig. 11.3   (100) pole figures for sheet material, illustrating (a) random orientation and (b) preferred orientation. R.D. (rolling direction) and T.D. (transverse direction) are reference directions in the plane of the sheet (Ref: B.D. Cullity, Elements of x-ray Diffraction).

transverse direction (TD), which indicates that a set of {100} planes in grains are parallel or nearly parallel to the rolling plane and another set are perpendicular to it. It can also be said that a set of <001> directions in these grains are parallel to the rolling direction. This type of texture is referred to as (100) [001] texture, which is common among recrystallized face centred cubic metals and alloys. This texture, if plotted using 111 poles instead of 100, would result in a pole figure as shown in Fig. 11.3 c. This is obvious, considering the fact that (111) plane is at an equiangular position between the cube planes (100), (010) and (001) {Refer, Table 2}. Thus, it can be understood that the appearance of the pole figure for a given texture depends on the indices of the poles chosen to plot it. It is always convenient to choose simpler poles and poles of planes which can offer high intensity reflections like 100 or 111 in cubic materials and 0001 in hexagonal materials for plotting pole figures. The choice of poles for pole figures also depends on the aspects of the texture to be studied.

The pole figures discussed so far are called *direct* pole figures. In contrast, an *inverse* pole figure is one which represents the density distribution of an important direction in the polycrystalline sample (for example, the axis of a wire sample) on a stereographic projection of the crystal lattice in some standard orientation like (001) projection. Thus, in a direct pole figure the basic circle of the projection might represent the normal to the wire axis, whereas in the inverse pole figure it would represent the (001) plane. However, direct pole figures are most commonly used to represent textures in materials.

In the above paragraph, only five grains were considered for analyzing the orientation. But, any practical material would contain a large number of grains and the individual determination of orientations of even a small representative number of them would be very difficult if not impossible. So, x-ray methods are used in which the diffraction effects from a large number of grains in the path of incident beam are averaged automatically. The (hk1) pole figure of the material can then be constructed by analyzing the distribution of intensity at various locations of the corresponding Debye ring. There are two methods of getting the diffraction data required for the texture analysis, namely, the photographic method and the diffractometer method. The photographic method is qualitative and has almost become obsolete whereas the diffractometer method is more accurate and quantitative. The diffractometer method of texture determination and analysis can be easily computerised.

11.4.1 Determination of Fibre Texture—Photographic Method

Cold drawn wires or rods have a texture in which a certain crystallographic direction [uvw] in most of the grains is parallel or nearly parallel to the axis of the wire or rod. A similar texture is found in natural and artificial fibres and hence this texture is called fibre texture and the axis of the fibre (or wire or rod as the case may be) is called the fibre axis. Materials having a fibre texture have their planes and directions in rotational symmetry about the fibre axis. Thus, a fibre axis is to be expected in cases where the forces in the manufacturing process of the material have a rotational symmetry about an axis, for example, in drawing, swaging or extrusion. Sometimes, fibre texture is also found in sheet formed by simple compression, in columnar crystals of castings and in coatings formed by electro-deposition, hot-dipping and evaporation. Fibre textures vary in perfection, i.e., in the extent of scatter of the crystallographic axis [uvw] in different grains about the fibre axis. Sometimes, a single material may have double fibre texture. A cold drawn aluminium wire has a single [111] fibre texture whereas cold drawn copper has a double, i.e., [111] + [100] texture. There are two sets of grains in drawn copper wire with one set having [111] fibre axis and the other set [100].

Fibre texture determination generally means the determination of the Miller indices [uvw] of the fibre axis. The basic principle in the determination of the fibre axis can be best understood by considering an ideal case of a wire of a cubic material having a perfect [100] fibre texture, i.e., all the grains in the material have their [100] crystallographic axis exactly parallel to the wire axis. Fig. 11.4 shows the wire specimen at 'C' with its axis along 'NS', normal to the incident beam 'IC'. The line 'CP' is normal to a set of (111) planes. Diffraction from these (111) planes can only occur when they are inclined to the incident beam at the Bragg angle θ, and this requires that the pole of the plane (111) lie somewhere on the circle 'PUQV' drawn in the figure, since then the angle between the plane normal and the incident beam will always be $90° - \theta$. This circle 'PUQV' is therefore called the reflection circle. If the grains of the wire had completely random orientation, then (111) poles would lie at all positions on the reflection circle and the (111) reflection would consist of the complete Debye ring as indicated by dotted lines on the film in the figure. But, if the wire has a perfect [100] fibre texture, then the diffraction pattern produced by this wire is identical to that obtained from a single crystal (of the same material) rotated about its [100] axis. During this rotation of the single crystal, the (111) pole is confined to a small circle 'PAQB' all points of which make a constant angle $\rho = 54.7°$ with the [100] direction 'N' shown in Fig. 11.4. Under such a condition, diffraction can occur only when the (111) pole lies at the intersections of the reflection circle and the circle 'PAQB'. These

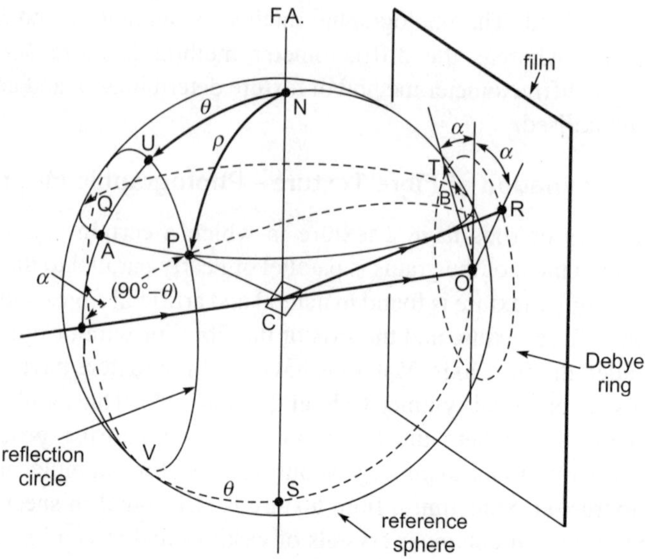

Fig. 11.4   Geometry of reflection from material having a fiber texture. F.A. = fiber axis (Ref: B.D. Cullity, Elements of x-ray Diffraction).

intersections are located at 'P' and 'Q' and the corresponding diffraction spots at 'R' and 'T', at an azimuthal angle α from a vertical line through the centre of the film. Two other spots due to the intersection of the reflection circle with the bottom counterpart of the circle 'PAQB', not shown, are located on the lower half of the film. If the texture is not perfect, each of these spots will broaden peripherally into an arc whose length is a function of the degree of scatter in the texture. By solving the spherical triangle 'IPN' the following general relation between ρ, θ and α can be found:

$$\cos \rho = \cos \theta \times \cos \alpha \qquad (11.2)$$

These angles are shown stereographically in Fig. 11.5, projected on a plane normal to the incident beam. Fig. 11.5 (a) shows the (111) pole figure of the material having a perfect [100] fibre texture which consists of two arcs which are the paths traced out by {111} poles during rotation of a single crystal about [100] (a situation identical to the stationary polycrystalline material having a perfect [100] texture). In Fig. 11.5 (b), this pole figure has been superimposed on a projection of the reflection circle in order to find the locations of the reflecting plane normals. Radii drawn through these points (P, Q, P′ and Q′) then enable the angle α to be measured and the appearance of the diffraction pattern to be predicted. In practice, the angle α on the film is measured and ρ is obtained from equation 11.2 since, θ can be calculated using equation 7.7 for a given indexed (hkl) plane on the pattern. The values of angle ρ are determined for a number of different hkl reflections in the manner noted above and these set of hkl and ρ

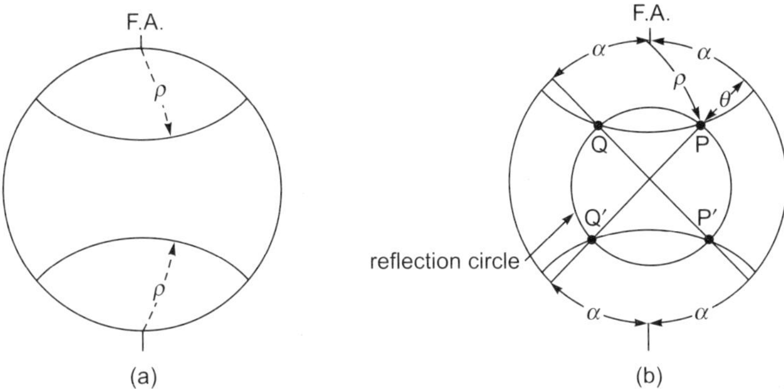

Fig. 11.5   Perfect [100] fiber texture: (a) (111) pole figure; (b) location of reflecting plane normals.

values enable one to determine the fibre axis by comparison with the standard interplanar angles in cubic crystals. For example, if the ρ values found by pinhole diffraction experiments are 70° and 55° for (111) and (200) reflections respectively then the fibre axis has got to be [111] since, the angle between any two planes of the family of (111) is 70.5° and that between (111) and (100) is 54.7° and these values agree with the values noted above within the experimental error.

Another interesting aspect of fibre texture is to be carefully noted. In materials having fibre textures, the individual grains have a common crystallographic direction parallel to the fibre axis but they can have any rotational position about that axis. It follows that the diffraction pattern of such materials will have continuous Debye rings if the incident x-ray beam is *parallel* to the fibre axis. However, the relative intensities of these rings will not be the same as those calculated for a specimen containing randomly oriented grains. Thus, continuous Debye rings are not, by themselves, evidence for a lack of preferred orientation.

11.4.2   Determination of Sheet Texture: Photographic Method

Metallic sheets either in the rolled condition or in the recrystallized condition have textures with lower symmetry than the textures of wires or fibres. The rotational symmetry of fibre texture is not present in sheet textures. Thus, the determination of wire texture is simpler in the sense that it suffices to determine the fibre axis and the extent of spread of Debye ring arc of a given hk1 reflection by just one pinhole diffraction photograph. But, in the case of sheet texture determination, it is necessary to obtain a number of pinhole patterns at different angles to the sheet normal and synthesize these information into a pole figure.

The photographic method of determination the pole figure of sheets is similar to the one discussed for the determination of wire textures. A transmission pinhole technique is used and the diffraction experiment is conducted making use of a characteristic radiation. The specimen sheet must be thinned down by etching to a thickness of a few hundreds of a micrometer and mounted initially perpendicular to the incident beam with the rolling direction vertical. The diffraction pattern resembles that of a drawn wire with a fibre texture. It contains Debye rings of non-uniform intensity and the pattern is symmetrical about the vertical line coinciding with the rolling direction of the sheet. If the sheet is now rotated by, say, 10° about the rolling direction and another pattern made, this pattern will differ from the first because the texture of the sheet does not have a rotational symmetry about the rolling direction. Further, the azimuthal positions of the high intensity regions of the Debye rings on the second pattern would be different from those of the first. Fig. 11.6 illustrates this effect for cold-rolled aluminium.

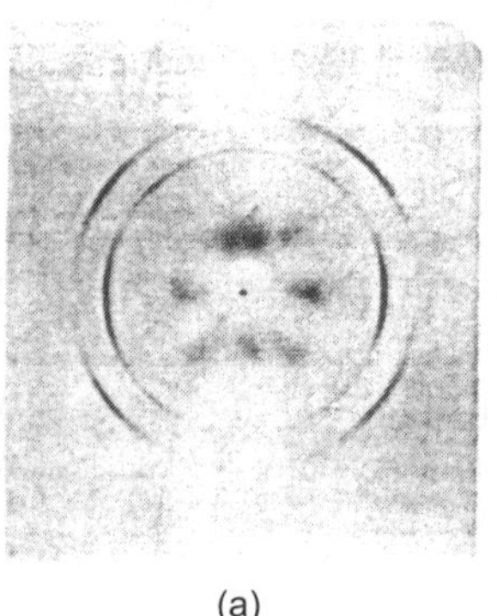 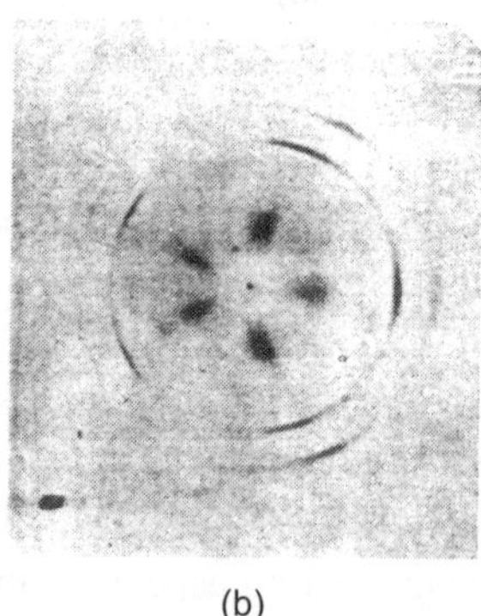

(a)   (b)

Fig. 11.6 Transmission pinhole patterns of cold-rolled aluminium sheet, rolling direction vertical: (a) sheet normal parallel to incident beam; (b) sheet normal at 30° to incident beam (the specimen has been rotated clockwise about the rolling direction, as in fig. 11.7). Filtered copper radiation (Ref: B.D. Cullity, Elements of x-ray Diffraction).

Hence, the necessity for getting a number of photographs at various rotational positions about the rolling direction and put the information together in a single pole figure.

The experimental arrangement is shown in Fig. 11.7 and it also defines the angle β which is the angle between the sheet normal and the incident beam. The intensity of the diffracted rays in any given Debye cone is decreased by absorption in the specimen by an amount which depends on the angle β. It can be seen from the figure that the diffracted beams on the left undergo more absorption due to the longer travel through the specimen than those on

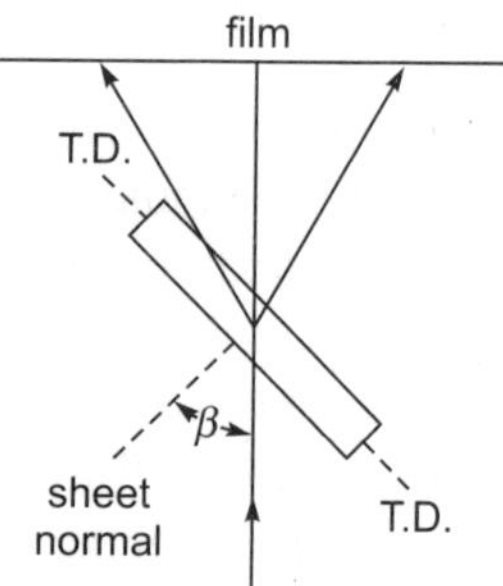

Fig. 11.7 Section through sheet specimen and incident beam (specimen thickness exaggerated). Rolling direction normal to plane of drawing. T.D. = transverse direction

the right. So, it is advisable to make measurements only on the right side of the film particularly when β is large. The normal practice is to obtain pinhole patterns at about 10° intervals from $\beta = 0$ to $\beta = 80^\circ$ and to measure the intensity distribution around a particular Debye ring on each pattern. The procedure for plotting the pole figure from measurements on the patterns is described below for an idealized case like the one shown in Fig. 11.8 where the intensity of the Debye ring is constant over certain angular ranges and zero between them. The range of blackening of the Debye arcs is plotted stereographically as a range of reflecting pole positions along the re-

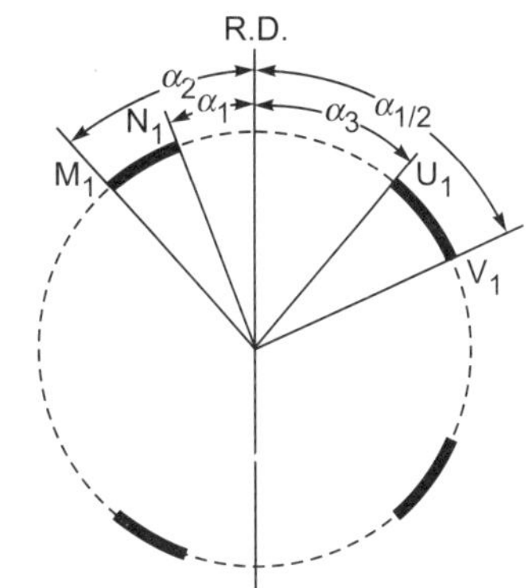

Fig. 11.8   Measurement of azimuthal position of high-intensity arcs on a Debye ring. $\beta = 40^\circ$, R.D. = rolling direction (Ref: B.D. Cullity, Elements of x-ray Diffraction).

flection circle, the azimuthal angle α on the film being equal to the azimuthal angle α on the projection.

When $\beta = 0$, the reflection circle is concentric with the basic circle of the stereographic projection and θ degrees inside it, as shown in Fig. 11.9, which is drawn for $\theta = 10^\circ$. when the specimen is rotated, for example by an angle $\beta = 40^\circ$, in the sense shown in Fig. 11.7, the new position on the reflection circle is found by rotating two or three points on the 0° reflection circle by 40° to the right along latitude lines and drawing circle arcs, centred on the equator or its extension, through these points. This new position of the reflection circle is indicated by the arcs ABCDA in Fig. 11.9. A part of this reflection circle, namely, CDA lies in back hemisphere. The arcs in Fig. 11.8 are first plotted on the 0° reflection circle as though the projection plane were still perpendicular to the incident beam, and then rotated to the right along latitude circles on to the 40° reflection circle. Thus, arc M_1N_1 in Fig. 11.8 becomes M_2N_2 and then, finally, M_3N_3 in Fig. 11.9. Similarly, Debye arc U_1V_1 is plotted as U_3V_3, lying on the back hemisphere.

The texture of the sheet is normally such that two planes of symmetry exist, one normal to the rolling direction (RD) and another normal to the transverse direction (TD). For this reason, arc M_3N_3 may be reflected in the latter plane to give the arc M_4N_4, thus, helping to fill out the pole figure. These symmetry elements are also the justification for plotting the arc U_3V_3 as though it were situated on the front hemisphere, since reflection in the centre of the projection (to bring it to the front hemisphere) and successive reflections in the two symmetry planes will bring it to this position anyway. If the diffraction patterns indicate that these symmetry planes are not present, then these short-cuts in plotting may not be used.

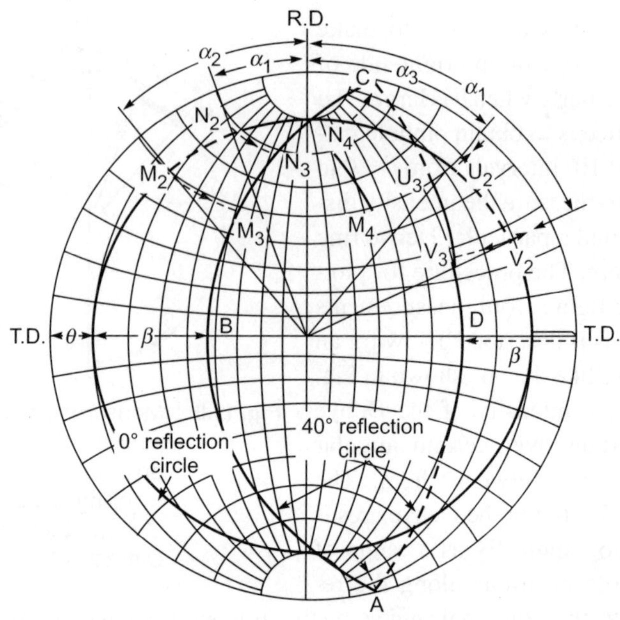

Fig. 11.9 Method of plotting reflecting pole positions for non-zero values of β. Drawn for $\theta = 10°$ and $\beta = 40°$ (Ref: B.D. Cullity, Elements of x-ray Diffraction).

By successive changes in β, the reflection circle can be made to move across the projection and to disclose the positions of reflecting poles. The result of such an exercise may look like what is represented in Fig. 11.10 involving measurements at $\beta = 0$, $20°$, $40°$, $60°$ and $80°$ (with RD vertical) and $\beta = 5°$ (with TD vertical, for exploring regions near the N and S poles of the projection). The arcs in Fig. 11.9 are plotted here with the same symbols, and the arcs E_1F_1 and E_2F_2 lie on the $5°$ reflection circle with the transverse direction vertical. The complete set of arcs defines areas of high pole density and by reflecting these areas in the symmetry planes mentioned above, the complete pole figure can be obtained as shown in Fig. 11.11.

In practice, the variation of intensity around a Debye ring is not abrupt as in Fig. 11.8 but gradual. This can be taken into account by plotting ranges in which the intensity is substantially constant, and generally four such ranges are used, namely, zero, weak, medium and strong. The result is a pole figure in which areas distinguished by different kinds of cross-hatching, represents various degrees of pole density from zero to a maximum. Figure 11.12 is a photographically determined pole figure showing the primary recrystallized texture of 70 – 30 brass which has been cold rolled and annealed.

The texture of a sheet is often described in terms of an 'ideal orientation' i.e., the orientation of a single crystal whose poles would lie in the high density regions of the pole figure. For example, in Fig. 11.12 the solid triangular symbols mark the positions of the {111} poles of a single crystal which has its (113) plane parallel to

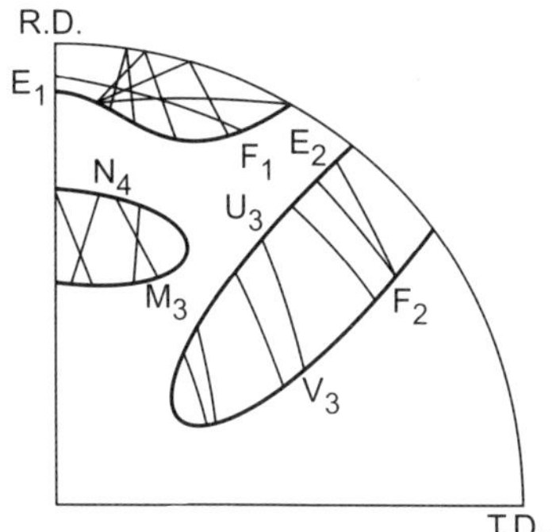

Fig. 11.10   Plotting a pole figure (Ref: B.D. Cullity, Elements of x-ray Diffraction).

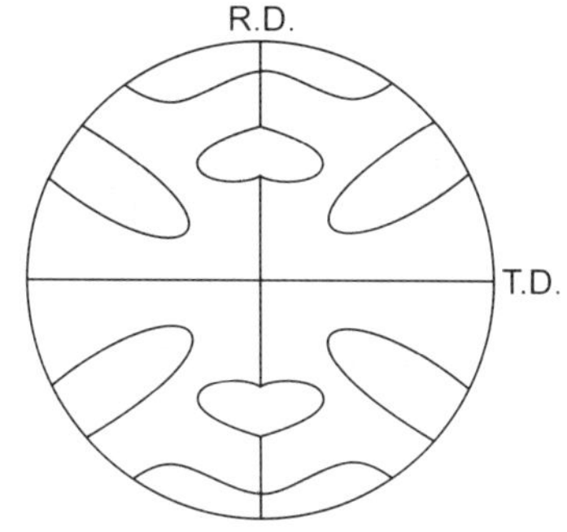

Fig. 11.11   Hypothetical pole figure derived from fig. 11.10

the plane of the sheet and [211] direction in this plane parallel to the rolling direction. This orientation when reflected in the two symmetry planes normal to the rolling and transverse directions will approximately account for all the high density regions on the pole figure. Accordingly this texture has been called a (113) $[\overline{2}\,\overline{1}\,1]$ texture. The actual pole figure, however, is a far better description of the texture than any statement of an ideal orientation, since the latter is

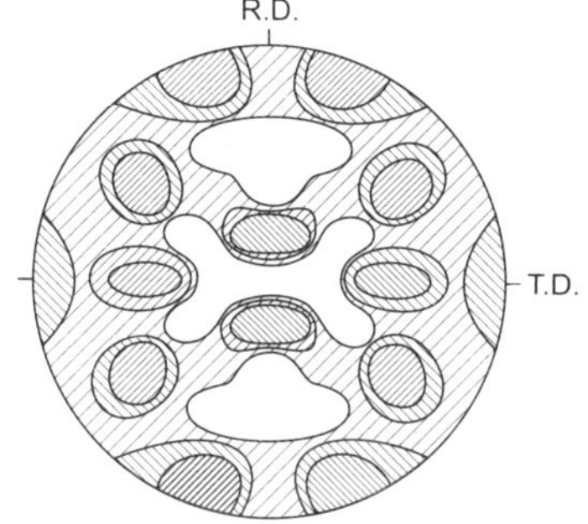

Fig. 11.12   (111) pole figure of recrystallized 70-30 brass, determined by the photographic method. (R. M. Brick, Trans. A.I.M.E. 137, 193, 1940).

frequently not very exact and gives no information about the degree of scatter of the actual texture about the ideal orientation. The inaccuracies of photographically determined pole figures are due to two factors:

1. Intensity measurements made on the film are usually only visual estimates and
2. No allowance is made for the change in the absorption factor with changes in β and α. This variation makes it difficult to relate intensities observed on one film at a particular value of β to those observed on another film at a different value of β even when the exposure time is varied for different films in an attempt to allow for changes in absorption.

11.4.3   Determination of Sheet Texture: Diffractometer Method

This method is capable of high precision because the intensity of the diffracted rays is measured quantitatively with a counter and either the intensity measurements are corrected for absorption or the x-ray optics are so designed that the absorption is constant and no correction is required. Further, this method can be computerized so as to directly give the pole figure. Two different methods, namely, the transmission method and the reflection method, are used generally to cover the entire pole figure.

The principal features of the transmission method (developed by Decker, Asp and Harker) are illustrated in to determine an (hk1) pole figure, the counter is fixed in position at the correct angle 2θ to receive the hk1 reflection with respect to the incident beam. The sheet specimen, in a special holder, is fixed initially with the rolling direction vertical and coincident with the diffractometer (vertical-axis type) axis, and with the plane of the specimen bisecting the angle between the incident and diffracted beams. The specimen holder allows rotation of the specimen about the diffractometer axis and about a horizontal axis normal to the specimen surface. Although it is impossible to move the counter around the Debye ring for exploring the variation of diffracted intensity around this ring, it is possible to accomplish the same by keeping the counter fixed and and rotating the specimen in its own plane. This rotation combined with the other rotation about the diffractometer axis, moves the pole of the (hk1) reflecting plane over the surface of the pole figure, which is plotted on a projection plane parallel to the specimen plane as in the photographic method. At each position of the specimen, the measured intensity of the diffracted beam, after correction for absorption, gives a figure which is proportional to the pole density at the corresponding point on the pole figure.

The method of plotting data is indicated in (Fig. 11.13 b). The angle α measures the amount of rotation about the diffractometer axis. The angle α is zero when the sheet bisects the angle between the incident and diffracted beams. The positive direction of α is conventionally taken as anticlockwise. The angle δ measures the amount by which the transverse direction is rotated about the sheet normal out of the horizontal plane and is zero when the transverse direction is horizontal. The reflecting plane normal bisects the angle between the incident and diffracted beams, and remains fixed in position whatever be the orientation of the specimen. While plotting the pole of the reflecting plane on the pole figure, it can be noted that the pole coincides initially, when both α and δ are zero, with the left transverse direction. A rotation of the specimen by δ degrees in its own plane then moves the pole of the reflecting plane δ degrees around the circumference of the pole figure, and a rotation of - α degrees about the diffractometer axis then moves it α degrees from the circumference along a radius. Thus, to plot the pole figure by this method, it is convenient to use the polar net as shown in (Fig 11.13 b). To explore the pole figure, it is convenient to make intensity readings at intervals of 5° or 10° of - α for a fixed value of δ: the pole figure is thus mapped out along series of radii. By this procedure the entire pole figure can be determined except for a region at the centre extending from about $\alpha = -50^\circ$ to $\alpha = -90^\circ$; in this region not only does the absorption correction become inaccurate but also the frame of the specimen holder obstructs the diffracted x-ray beam.

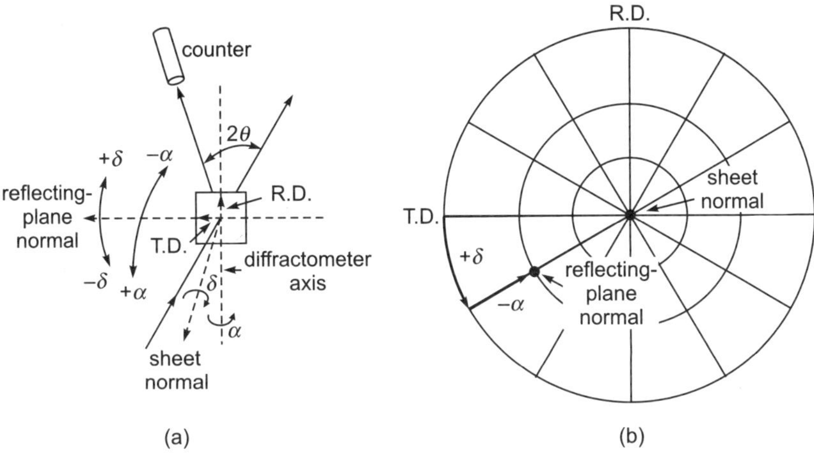

Fig. 11.13   Angular relationships in the transmission pole-figure method (a) in space and (b) on the stereographic projection. (On the projection, the position of the reflecting plane normal is shown for $\delta = 30°$ and $\alpha = -30°$) (Ref.: B.D. Cullity, Elements of X-ray Diffraction).

An absorption correction is necessary in this method because variations in α cause variations in both the volume of diffracting material and the path length of the x-rays within the specimen. Variations in δ have no effect. The correction factor 'R', which is the ratio of the integrated intensity of diffraction at a particular angle α ($I_D\alpha$) to the integrated intensity at an a angle of zero (I_{DO}), is given by

$$R = I_D\alpha/I_{DO} = \frac{\cos\theta \cdot [e^{-\mu t/\cos(\theta+\alpha)} - e^{-\mu t/\cos(\theta+\alpha)}]}{\mu t \cdot e^{-\mu t/\cos\theta}[\cos(\theta-\alpha)/\cos(\theta+\alpha) - 1]} \qquad (11.3)$$

where t = thickness of sheet specimen and θ = Bragg angle. The variation of the correction factor 'R' with a for clockwise rotation from zero position is shown in Fig. 11.14 for typical values involved in the 111 reflection for aluminium with copper K_α radiation, namely $\mu t = 1$ and $\theta = 19.25°$. The value of μt used in equation 11.3 must be obtained by direct measurement, since it is not sufficiently accurate to use a tabulated value of μ together with the measured thickness t of the specimen. To determine μt, the intensity of a strong diffracted beam from any convenient material may be measured when the sheet material is interposed in its path and again measured when the sheet is not interposed. The value of μt may

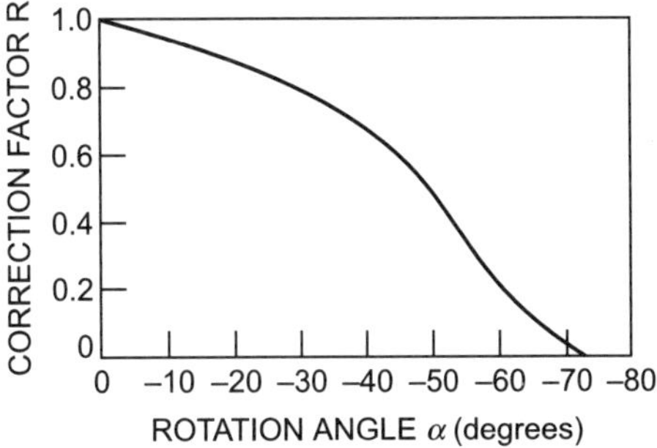

Fig. 11.14   Variation of the correction factor R with α for clockwise rotation from the zero position. $\mu t = 1.0$, $\theta = 19.25°$ (Ref.: B.D. Cullity, Elements of X-ray Diffraction).

then be obtained from the absorption equation, $I_t = I_o \cdot e^{-\mu t}$, where I_o and I_t are the intensities of the beams incident on and transmitted by the sheet specimen respectively.

The central part of the pole figure, i.e., in the α range of $-50°$ to $-90°$, cannot be covered by the transmission method as described earlier. This region can be conveniently explored by the reflection method developed by Schultz, a method in which the measured diffracted beam issues from that side of the sheet on which the primary beam is incident. It requires a special holder which allows rotation of the specimen in its own plane about an axis normal to its surface and about a horizontal axis; these axes are shown as BB′ and AA′ respectively in Fig. 11.15. The horizontal axis AA′ lies in the specimen surface and is initially adjusted, by rotation about the diffractometer axis, to make equal angles with the incident and diffracted beams. After this is done no further rotation about the diffractometer axis is made. Since, the axis AA′ remains in a fixed position during the other rotations of the specimen, the irradiated surface of the specimen is always tangent to a focusing circle passing through the x-ray source and counter slits. A divergent beam may therefore be used since the diffracted beam will converge to a focus at the counter slits.

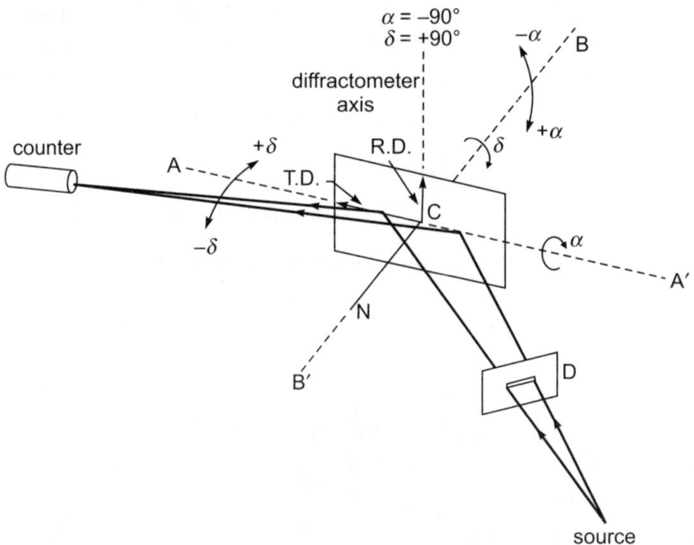

Fig. 11.15   Reflection method for pole-figure determination

When the specimen is rotated about the axis AA′, the axis BB′ normal to the specimen surface rotates in a vertical plane, but CN, the reflecting plane normal, remains fixed in a horizontal position normal to AA′. The rotation angles α and δ are defined in Fig. 11.15. The angle α is zero when the sheet is horizontal and has a value of $-90°$ when the sheet is in the vertical position shown in the figure. In this position of the specimen, the reflecting plane normal is at the centre of the projection. The angle δ measures the amount by which the rolling direction is rotated away from the left end of the axis AA′ and has a value of $+90°$ for the position illustrated. With these conventions the reflecting plane normal, for any angular positions α and δ, may be plotted on the pole figure in the same way as in the transmission method.

The great merit of the reflection method is that no absorption correction is required for values of α between $-90°$ and about $-40°$. To achieve a minimum error due to absorption, the reflecting surface of the specimen must be adjusted to accurately coincide with the axis AA′ for all values of α and δ.

It is evident that the transmission and reflection methods complement one another in their coverage of the pole figure. The usual practice is to use the transmission method to cover the range of α from 0 to $-50°$ and the reflection method from $-40°$ to $-90°$. This produces an overlap of $10°$ which is useful in checking the accuracy of one method against the other, and necessary in order to find a normalizing factor for one set of readings which will make them agree with the other set in the region of overlap.

When this is done, the numbers which are proportional to pole density can then be plotted on the pole figure at each point at which a measurement was made. Contour lines are then drawn at selected levels connecting points of the same pole density, and the result is pole figure such as that shown in Fig. 11.16, which represents the deformation texture of 70–30 brass cold-rolled to a reduction in thickness of 95 percent. The numbers attached to each contour line give the pole density in arbitrary units.

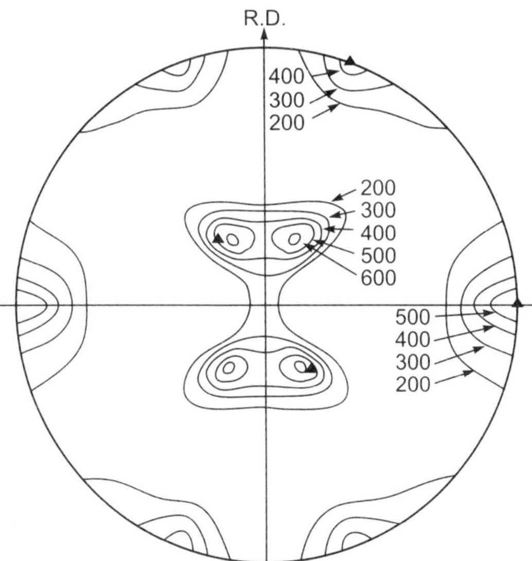

Fig. 11.16   (111) pole figure of cold-rolled 70-30 brass, determined by the diffractometer method. (H. Hu, P. R. Sperry, and P. A. Beck, *Trans.* A.I.M.E. 194, 76, 1952).

In Fig. 11.16, the solid triangular symbols representing the ideal orientation (110) $\left[1\bar{1}2\right]$ lie approximately in the high density regions of the pole figure. But here again, the pole figure itself must be regarded as a far better description of the texture than any bare statement of an ideal orientation.

Two sources of error in the diffractometer method are described below:

1. When an $(h_1k_1l_1)$ pole figure is being determined, the counter is set at the approximate angle 2θ to receive $K\alpha$ radiation reflected from the $(h_1k_1l_1)$ planes. But at some specific position of the specimen, there may be another set of planes, $(h_2k_2l_2)$, so oriented that they can reflect a component of the continuous spectrum at the same angle 2θ. If the $(h_2k_2l_2)$ planes have a high reflecting power, this reflection may be so strong that it may be taken for an $(h_1k_1l_1)$ reflection of the $K\alpha$ wavelength. The only way of eliminating this possibility is to use a monochromatic radiation obtained through balanced filters.

2. The crystal structure of the material investigated may be such that a set of planes, $(h_3k_3l_3)$, has very nearly the same spacing as the $(h_1k_1l_1)$ planes. The $K\alpha$ reflections of these two sets will therefore occur at very nearly the same angle 2θ. Thus, if the counter is set to receive the $h_1k_1l_1$ reflection, then there is a possibility that some of the $h_3k_3l_3$ reflection may also be received, especially in the transmission method for which a wide receiving slit is used. The best way to avoid this difficulty is to select another reflection, $h_4k_4l_4$, well separated from its neighbours, and construct an $h_4k_4l_4$ pole figure instead of an $h_1k_1l_1$.

Another important point about the pole figure determination is the necessity for integrating devices when the grain size of the specimen is large, as in recrystallized metals and alloys. With such specimens, the incident x-ray beam will not strike enough grains to give a good statistical average of the orientations present. This is true of both the methods, that is, photographic and the diffractometer. With coarse grained specimens, it is therefore necessary to use some kind of integrating device, which will move the specimen back and forth, or in a spiral, in its own plane and so expose a large number of grains to the incident beam.

11.5   ORIENTATION DISTRIBUTION FUNCTION (ODF)

Pole figure interpretation can be done on a quantitative fashion by the recently introduced orientation distribution function (ODF). This function has to be calculated from several pole figures. It is obtained by neglecting the shape and the position of the grains or crystallites in the sample and taking into account only the orientation of the crystallographic axes with respect to the macroscopic sample axes. This orientation may be symbolized by 'g'. Thus, each crystallite can be characterized by its orientation 'g_i' and its volume 'v_i'. But, this description is still too complicated. For simplification, crystals having the same or very similar orientations are grouped together. If Δv (g) is the volume of all the grains having orientations in the range from g to g + Δg and 'v' is the volume of the whole sample, then the orientation distribution function is defined by

$$\Delta v \ (g)/v = f \ (g) \cdot dg \tag{11.4}$$

The function f (g) is the orientation distribution function (ODF). By an appropriate choice the definition of the orientation element 'dg' it is possible to normalize the function f (g) in such a way that it is unity in the case of random orientation distribution. In the case of non-random distribution, the function f (g) is expressed in multiples of random density sometimes abbreviated as 'mrd-units'.

EXERCISES

11.1 (i) Determine the line breadth in degrees of 2θ due to particle size effect alone of powder pattern lines of particles of size 100, 80, 60, 40 and 20 nm. Assume that the Bragg angle of diffraction is 50 degrees and the wavelength of radiation used is 0.1542 nm. (Note that the breadth 'B' is in radians in the Scherrer formula).

[Ans: 0.12, 0.15, 0.20, 0.30, 0.60]

(ii) In the above problem, calculate the breadths for Bragg angles of 20, 45 and 80 degrees for particles of 20 nm.

[Ans: 0.41, 0.55, 2.23]

11.2 An x-ray diffraction pattern of a cold worked polycrystalline metallic material showed a broadening of 1.45 ($\Delta 2\theta$) over and above the normal line breadth of the recrystallized material at a 2θ value of $140°$. If this broadening is assumed to be due to residual stresses of value close to yield strength, determine the yield strength of the material. Assume the modulus of elasticity as 205 GPa. (Hint: See section 11.3)

[Ans: 472 MPa]

12
C H A P T E R

MEASUREMENT OF STRESS

12.1 GENERAL

Stresses present in a structural material may be classified into two types, namely, applied stress and residual stress. The applied stress is a stress applied on the material by some external means. It is possible for a material to have certain inherent residual stress as a result of thermal or mechanical treatment during its manufacture. Generally, during metal working, the surface of the workpiece undergoes more deformation than the inner bulk. So, the surface tends to force the inner bulk to adjust to the former's increased dimension which results in the development of tensile stresses in the bulk and compressive stresses on the surface. The residual stress can also develop in a material during heat treatment because of differential expansion and/or contraction due to temperature variations in different localities of a bulk specimen. It should be noted that the total tensile residual force must balance the total compressive residual force in a bulk specimen. Further, if the tensile stress localities and compressive stress localities are mechanically separated from each other (by cutting, drilling, etc.), then residual stresses would vanish. Thus, the residual stresses exist mainly because there is a restraint on the change in dimension of one part or locality of a specimen with respect to another part or locality. It should also be noted that the maximum value of the residual stress in any locality of a material component cannot exceed the yield strength of the material because if it does, then that locality plastically deforms and the stress decreases to the level of the yield strength.

The elastic stresses cause a change in interplanar distance 'd' and the extent of change (Δd) is governed by the modulus of elasticity, E, of the material since E = stress/strain and strain is $\Delta d/d$ in this case. As the interplanar distance d is changed, the Bragg angle 'θ', at which reflection occurs from these sets of planes, also changes provided the wavelength of the incident beam does not alter. It can be seen from the Bragg's law that as the d value increases the θ value decreases for a given constant λ value. Thus, by observing the change in θ or 2θ values of reflections, it is possible to deduce the new interplanar spacing and the corresponding strain $\Delta d/d$ and also consequently the stress associated. It is to be noted that only elastic strains cause changes in θ value. In section 11.3 it has been shown that plastic deformation or cold work causes broadening of diffraction line because of the occurrence of microscopic non-uniform strains during the process of working.

There are certain limitations of x-ray diffraction methods of stress measurements. The penetration of x-rays into most of the metals is only a few microns and consequently the x-ray method is capable of measuring stresses at and close to the surface and not at a point too deep into the bulk. Further, difficulties are encountered when examining cold worked materials since the diffracted rays are broad and weak in intensities. Measuring precisions of the order of $\pm$ 20 MPa can be easily achieved with annealed steel. But with cold worked and hardened steel this precision is very difficult to achieve, if not impossible.

12.2   MEASUREMENT OF UNIAXIAL STRESS

In Fig. 12.1, a bar of length 'L_o' is subjected to a uniaxial tensile stress σ_x in the x axis which also happens to be the axis of the bar. Let us assume the stress σ_x causes only elastic strain in the material. Then the elastic strain ε_x, in x direction caused by the stress σ_x, is given by

$$\varepsilon_x = (L_f - L_o)/ L_o$$

where L_f is the length of the bar after the introduction of the stress σ_x and since the material is in the elastic region

$$\sigma_x = E \cdot \varepsilon_x \tag{12.1}$$

where E is the Young's modulus.

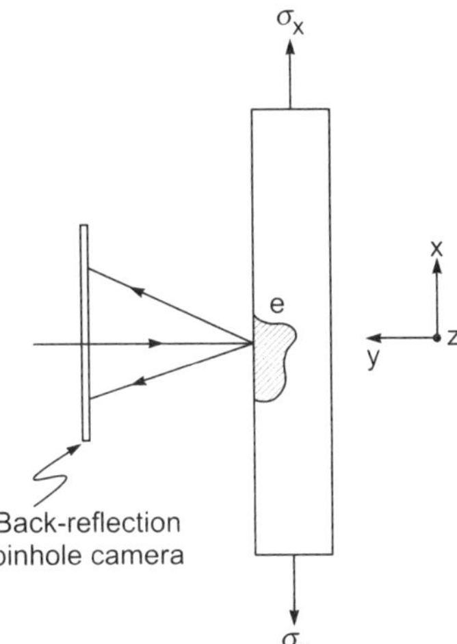

Fig. 12.1   Arrangement for stress measurement in a bar subjected to tensile stress (schematic)

From Fig. 12.2 it can be seen that when the material is under stress the interplanar distance d increases for those planes whose normals are parallel or nearly so to the axis of the specimen (the stress axis) and it decreases for those whose normals are perpendicular or nearly so to the specimen axis. It is possible to obtain a back reflection pinhole pattern (arrangement of Fig. 12.1) and determine the changed d values of those planes whose normals are almost perpendicular (i.e., y axis) to

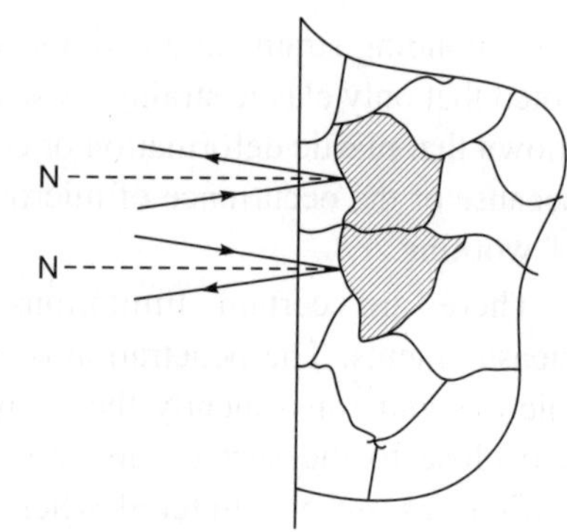

Fig. 12.2   Region 'C' of Fig. 12.1 magnified view (schematic)

the tensile axis. If d_y is the changed d value in the y direction and d_o is the normal d value of the same plane in the absence of stress, then the strain in the y direction is given by

$$\varepsilon_y = (d_y - d_o)/d_o \qquad (12.2)$$

But, Poisson's ratio, $v = -\varepsilon_y/\varepsilon_x$

Substituting for ε_y from equation 12.2

$$\varepsilon_x = -\varepsilon_y / v = -1/v \cdot (d_y - d_o)/d_o \qquad (12.3)$$

Rewriting the equation 12.1 using 12.3

$$\sigma_x = -E/v \cdot (d_y - d_o)/d_o \qquad (12.4)$$

Thus, if the constants E and v for the material of the bar is known, the stress σ_x can be calculated from the equation 12.4. It should be noted that the equation 12.2 gives the strain in the y direction accurately only if the reflecting plane normal is exactly in line with the incident beam, so that it is possible to determine d_y (i.e., interplanar distance of hk1 planes in y direction). But, in practice, this direction can only be approached and the reflecting plane normal can be close to y direction as shown in Fig. 12.2. So, the real d_y value is slightly less than the value actually measured by this technique.

A convenient way of determining the altered d value with good accuracy would be to make use of a standard specimen. The powder of the standard specimen could be smeared on the surface of the specimen bar which is being stressed and then the pinhole pattern obtained. The pattern would then contain Debye rings from the standard along with the Debye ring corresponding to the selected hk1 plane of the specimen. It is now possible to calculate Bragg angle θ for the diffraction from the hk1 plane of the specimen by the measurement of the radius of the Debye ring of the specimen and that of the standard by using equation of the type 5.2 without the necessity of knowing the specimen to film distance (Bragg angle for the Debye ring of the standard should be known). Once the Bragg angle

θ is calculated, it is a simple matter to calculate the d value for the hkl plane of the specimen using Bragg's law.

For this method it is necessary to know the d_o value of the hkl plane of the specimen under no-stress condition for using the equation 12.4. If the interest is to determine the residual stress present in the material, then the unstressed condition of the material (for measuring the d_o value for use in the equation 12.4) can be achieved only by cutting the material in a proper manner so as to reduce the residual stress to zero. As discussed earlier it is important under such circumstances to carefully study the stress situation and completely separate the tensile and the compressive stressed regions of the specimen by cutting parallel to the stress directions. In practice, where the stress pattern is not simple, it is easier said than done. Usually, the x-ray diffraction method is capable of measuring surface or near surface stresses. And only two-dimensional stresses generally encountered on the surface can be determined, since the stress perpendicular to the free surface is zero. This somewhat simplifies the situation and affords the determination of residual stresses on the surface or close to the surface.

12.3 MEASUREMENT OF BIAXIAL STRESS

Any multiaxial state of stress on a structural member can always be resolved into any three mutually perpendicular set of axes x, y, z. The stress system can then be defined by three normal stresses σ_x, σ_y and σ_z acting along the three directions x, y and z respectively and the three shear stresses τ_{xy}, τ_{yz} and τ_{zx} acting on the xy, yz and zx planes respectively. In other words, a minimum of these six stresses are necessary to define a multiaxial stress system. But, it is possible to select a particular set of three mutually perpendicular axes such that all the shear stresses vanish. Such, a set of axes are termed principal axes or directions 1, 2 and 3 and the normal stresses acting in these three directions are termed principal stresses σ_1, σ_2 and σ_3. By definition, σ_1 is the highest of the three stresses, σ_2 is the intermediate and σ_3 is the lowest. Recalling from the previous section that no stress can exist normal to a free surface the state of stress on a surface is only biaxial or planar. It is possible to find the principal stresses σ_1 and σ_2 in the biaxial stress state as in a triaxial one. Though the stress σ_3 normal to the surface does not exist (i.e., σ_3 is zero), the strain ε_3 in this direction will exist due to the stresses σ_1 and σ_2 on the surface plane. The stress σ_1 would cause a lateral strain of $-v \cdot \sigma_1/E$ and the stress σ_2 would, in addition cause a lateral strain of $-v \cdot \sigma_2/E$ and the total principal strain normal to the surface, ε_3 is given by

$$\varepsilon_3 = -v \cdot (\sigma_1 + \sigma_2)/E \qquad (12.5)$$

The value of ε_3, the strain perpendicular to the free surface, can be determined by x-ray diffraction as noted in the previous section by making use of the equation 12.2 i.e.,

$$\varepsilon_3 = (d_y - d_o)/d_o = -v \cdot (\sigma_1 + \sigma_2)/E \qquad (12.6)$$

Thus, it is possible to determine the sum of the principal stresses σ_1 and σ_2 lying in the surface plane of any specimen.

It is generally required, in many instances, to determine a stress σ_x acting in some specified direction x on the surface as shown in Fig. 12.3. In such a case two back reflection pinhole patterns may be obtained, one with the incident beam normal to the surface (position 1 shown in Fig. 12.3) and the other with the incident beam inclined at angle ψ (along OA) with the surface normal (position 2 in Fig. 12.3). It should be noted that the direction OA lies in the plane containing the surface normal and σ_x. If ε_3 is the strain measured normal to the surface and ε_ψ is the strain measured at an angle ψ to the surface, then by the theory of elasticity it can be shown that

$$\sigma_x = (\varepsilon_\psi - \varepsilon_3) \cdot E/(1 + v)\sin^2 \psi \qquad (12.7)$$

But, $\varepsilon_3 = (d_y - d_o)/d_o$ from equation 12.2 and, if d_ψ is the plane spacing in ψ direction,

$$\varepsilon_\psi = (d_\psi - d_o)/d_o$$

Therefore,

$$\varepsilon_\psi - \varepsilon_3 = [(d_\psi - d_o)/d_o] - [(d_y - d_o)/d_o] = (d_\psi - d_y)/d_o$$

Here if d_o is substituted by d_y, there will be no necessity for the measurement of d_o (i.e., d value in the unstressed state which needs cutting of the specimen). Further, this substitution does not cause much of an error since d_o is approximately equal to d_y.

Then the equation 12.7 can be rewritten as

$$\sigma_x = (d_\psi - d_y)/d_y \cdot E/(1 + v)\sin^2 \psi \qquad (12.8)$$

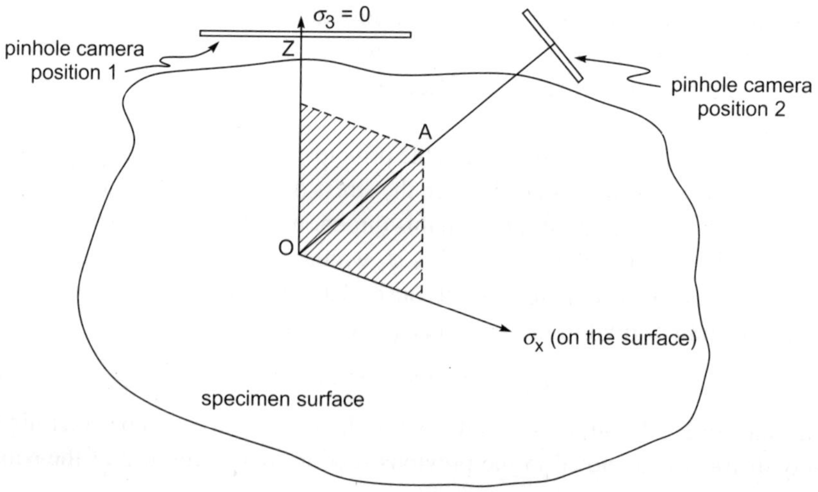

Fig. 12.3   Stress on a free surface and camera positions for measurement of stress

Thus, the stress in any direction x on the surface can be determined using equation 12.8. The knowledge of d_y is not necessary for the use of this equation.

Here too, a standard specimen powder could be smeared over the surface for obtaining a precise measurement of the interplanar spacings, as explained in section 12.2. The Debye rings of the standard substance and the specimen would be concentric in the photograph taken at normal incidence exposure, but it will not be so in the inclined incidence exposure. Fig. 12.4 shows the Debye ring schematically from an inclined incidence exposure. It shows that the Debye ring of the specimen is no longer circular. It should be noted that the stress normal to the specimen surface is zero and the stress in a direction lying at an angle φ increases as φ approaches 90° and attains a maximum value at 90° i.e., along the surface. Hence, reflecting hk1 planes will have an average plane spacing if their normals are along the sheet normal and the plane spacing d keeps on increasing as the normals approach the specimen surface because of the increasing stress. As the d values of the planes increase, θ values decrease and consequently tan $(180 - 2\theta)$ values and the radii of the Debye rings increase [since, tan $(180 - 2\theta) = R/D$ eq. 5.2]. This is the reason for the Debye ring of the specimen being non-circular. This is as shown in Fig. 12.4 where the upper part of the ring is formed by sets of reflecting planes with smaller d values and the lower part is formed by planes of greater d values.

For convenience, the equation 12.8 can be modified slightly. The Debye ring radius, R, is related to specimen-to-film distance, D, by the relation

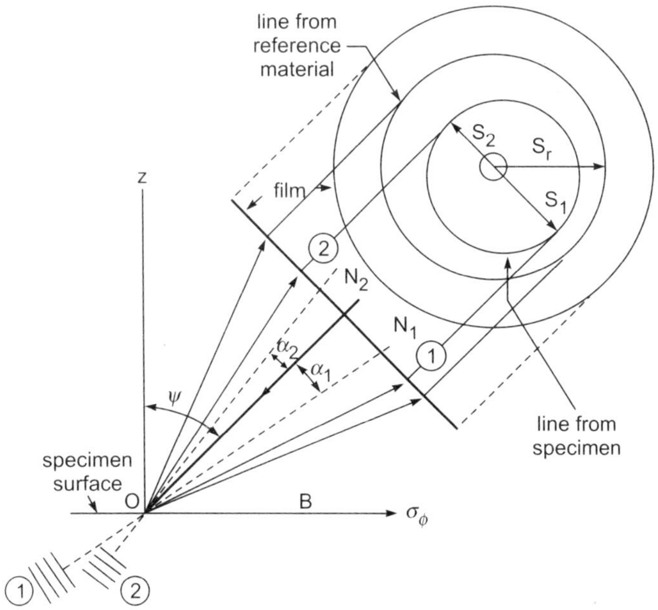

Fig. 12.4   Back-reflection method at inclined incidence
(Ref: B.D. Cullity, Elements of x-ray Diffraction).

$$R/D = \tan(180 - 2\theta) = -D \cdot \tan 2\theta$$

Or,
$$\Delta R = -2D \sec^2 2\theta \cdot \Delta\theta \qquad (12.9)$$

By differentiating Bragg's law,

$$\Delta d/d = -\cot\theta \cdot \Delta\theta \text{ or, } \Delta\theta = -\tan\theta \cdot \Delta d/d$$

Combining this with equation 12.9,

$$\Delta R = 2D \sec^2 2\theta \cdot \tan\theta \cdot \Delta d/d.$$

In this equation $\Delta d/d$ can be written as $(d_\psi - d_y)/d_y$ and ΔR can be replaced by $(R_\psi - R_y)$ where R_ψ is the radius S_1 in Fig. 12.4 (i.e., the Debye ring in the inclined incidence photograph. The radius S_1 rather than S_2 is taken because the radius at S_1 side of the ring is more sensitive to strain.)

Rewriting the above equation,

$$R_\psi - R_y = 2D \sec^2 2\theta \cdot \tan\theta \cdot (d_\psi - d_y)/d_y$$

or
$$(d_\psi - d_y)/d_y = (R_\psi - R_y)/(2D \sec^2 2\theta \cdot \tan\theta)$$

Substituting this value of $(d_\psi - d_y)/d_y$ in equation 12.8,

$$\sigma_x = (R_\psi - R_y) \cdot E/[2D \sec^2 2\theta \cdot \tan\theta (1+v)\sin^2\psi]$$

or
$$\sigma_x = K \cdot (R_\psi - R_y) \qquad (12.10)$$

where K is a constant for a given experimental set-up with a fixed specimen-to-film distance, a specified Bragg angle θ (i.e., a particular reflecting plane hk1), a fixed inclined incidence angle ψ and a given specimen material. This constant can be evaluated once for all and can be used for the specimen under different stress levels. This equation 12.10 is a most convenient one for practical purposes.

12.4   EXPERIMENTAL METHODS

The two-exposure (one at normal incidence and the other at inclined incidence) method is generally used to measure the stress along any given direction on the surface of the specimen. The experimental techniques which can be used to determine the stress are the pinhole camera technique and the diffractometer technique.

12.4.1   The Pinhole Camera Method

Certain modifications are required in the pinhole cameras to suit the requirements of the stress measurement. Firstly, the specimens in which the stresses are to be determined are generally quite large and so it is necessary to take the camera to the specimen rather than vice versa. Secondly, as a very high precision is needed in the measurement of Debye ring radii, the Debye ring should be smooth and continuous which can be ascertained by rotating or oscillating the film about the incident beam axis.

Thus, it can be noted from the above, that a portable x-ray tube, with a pinhole camera rigidly attached, is essential (see Fig.12.5). Further, it is necessary to have an arrangement for the rotation or oscillation of the camera to obtain a continuous and smooth Debye ring. The circular photographic film can have a metallic cover of the type shown in Fig. 12.5 (b), so that both the normal and inclined incidence exposures can be obtained on a single film by just rotating the cover after one exposure by 90° about the camera axis. Since, the metallic cover has two diametrically opposite openings, one exposure will give two diametrically opposite parts of each Debye ring. The other exposure will produce Debye ring parts at 90° to the first as shown in Fig. 12.5 (c).

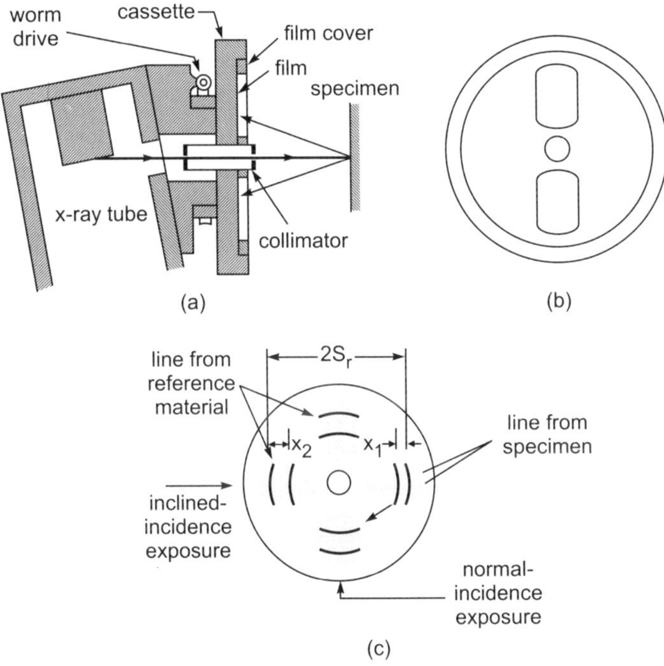

Fig. 12.5 Pinhole camera for stress measurement (schematic): (a) section through incident beam; (b) front view of cassette; (c) appearance of exposed film. (Ref: B.D. Cullity, Elements of x-ray Diffraction).

The x-ray beam used can be a divergent type or collimated type. The collimated type of beam and its use is shown in Fig. 12.5 (a) schematically. This can be utilized for pinpointing the location of stress measurement on the specimen surface so that the stress existing at a given location could be determined. A divergent beam can also be used and it is possible to obtain a focusing condition under this beam as shown in Fig. 12.6. Two slits A and B in the Fig. 12.6 are so arranged that divergent beams are allowed to hit the specimen and the film can be so located as to register a focused Debye ring. It should be noted that the focusing action for any particular specimen-to-film distance is possible for only one

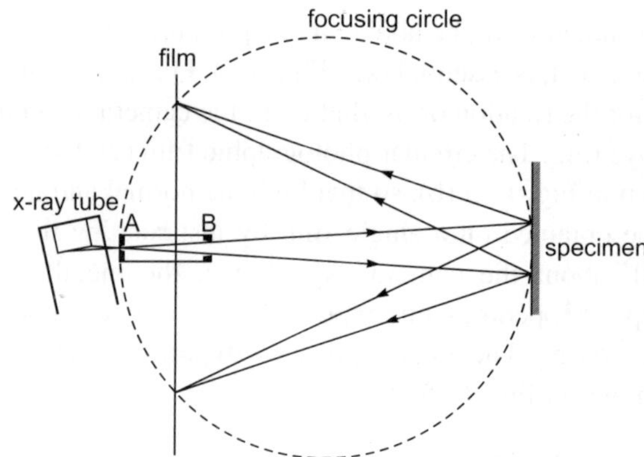

Fig. 12.6   Back-reflection pinhole camera used under semifocusing conditions (Ref: B.D. Cullity, Elements of x-ray Diffraction).

particular Debye ring. With this type of arrangement it is possible to obtain a sharper ring with reduced exposure time. But, since the beam is divergent and covers a large area on the specimen surface compared to the collimated beam the measured stress is indicative of the average stress on the entire surface area under diffraction. In this type of arrangement it is not possible to measure the stress at a point which can be done only by using the collimated beam technique.

A lot of care must be exercised in preparation of the specimen for stress measurement. The surface of the specimen may be ground and then etched deeply to remove the deformed surface layer. Then the surface may be lightly polished with 3/0 or 4/0 emery papers and lightly etched again. All this procedure is done so that the surface is smooth since a rough surface generally has stress variations at elevated and depressed locations.

12.4.2   The Diffractometer Method

A x-ray diffractometer may also be used for the measurement of the stress. The specimen preparation technique is similar to the one discussed in the previous section. In this method, the specimen holder should be such that it allows the independent rotation of the specimen about the diffractometer axis. Fig. 12.7 illustrates schematically the experimental set-up required. Fig. 12.7 (a) shows the specimen equally inclined to the incident and diffracted beams. Here, ψ is zero and the specimen normal N_s coincides with the plane normal N_p. Divergent radiation from the source S is diffracted to a focus at F on the diffractometer circle. With this arrangement the strain normal to the surface is measured. In Fig. 12.7 (b) the specimen has been rotated through an angle ψ about the diffractometer axis for the measurement in the inclined incidence position of the specimen. Note that the reflecting plane normal N_p and the specimen normal N_s are separated by the angle ψ in the plane of the diffractometer circle. Since, the focusing circle is always

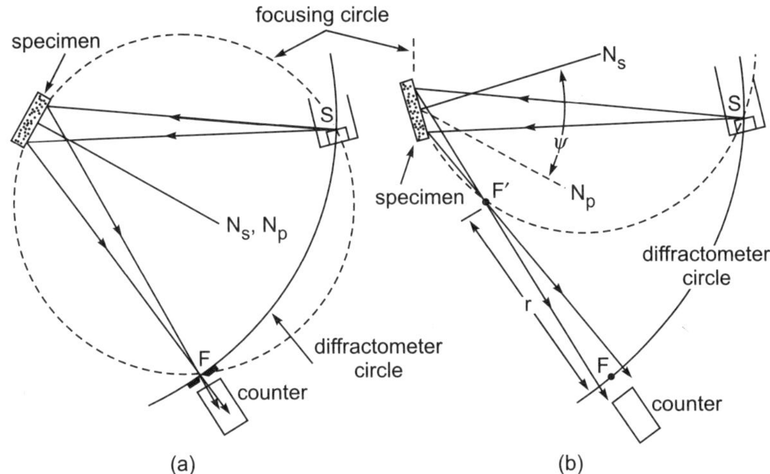

(a)   (b)

Fig. 12.7   Use of a diffractometer for stress measurement; (a) $\psi = 0$; (b) $\psi = \psi$.

tangent to the specimen surface, rotation of the specimen changes the focusing circle both in position and radius. The diffracted rays now come to a focus at F' located at a distance r from F. If R is the radius of the diffractometer circle, it may be shown that

$$r/R = 1 - \{\cos\,[\psi + (90° - \theta)]\}/\{\cos\,[\psi - (90° - \theta)]\}$$

Thus, when ψ is not zero, the focal point of the diffracted beam lies between F (the usual position of the counter receiving slit) and the specimen. At F the intensity of the beam received by the counter will be low. If a wide slit is used, to increase the intensity received, the resolution will suffer. The proper procedure in such cases is to put a narrow slit at F' and move the counter to a position just behind it. Alternately, a narrow slit can be used at F' and a wide one at F in front of the counter. But in practice neither of the above procedures are used and instead a compromise is struck between the intensity and resolution by placing a narrow slit at some point between F and F' where satisfactory results are indicated experimentally. The slit is then left in this position for both the measurements.

As the diffractometer measures directly 2θ of the diffracted beam it is convenient to write the stress equation in terms of 2θ rather than plane spacings. The equation 12.10 can be modified to give

$$\sigma_x = C \cdot (2\theta_n - 2\theta_i) \tag{12.11}$$

where $2\theta_n$ is the observed value of the diffraction angle in the normal incidence measurement ($\psi = 0$) and $2\theta_i$ its value in the inclined incidence measurement ($\psi = \psi$).

The quantity measured in the diffractometer is $\Delta 2\theta$ or $(2\theta_n - 2\theta_i)$, the shift in the diffraction line due to stress as angle ψ is changed. Certain geometrical factors introduce small errors which cause a change in 2θ even for a stress-free specimen, when ψ is changed from zero to certain angle. It is therefore necessary to determine this change experimentally and apply it as a correction $(\Delta 2\theta)_0$ to all $\Delta 2\theta$ values measured on stressed specimens.

12.5   SUPERIMPOSED MACROSTRESS AND MICROSTRESS

A specimen may contain both a uniform macrostress and a non-uniform microstress. If the diffraction pattern of such a specimen is taken, it will be found to contain diffraction lines which are shifted as well as broadened. When a polycrystalline metallic component is deformed plastically (due to mechanical or thermal treatment) during its manufacture, the lattice planes usually get distorted in such a fashion that the lattice spacing of any particular hk1 planes varies not only from one grain to another but also from one part of a grain to another. So, as a result, there exists a non-uniform microstrain. In addition, as a result of a residual elastic stress, which is inevitable, a uniform macro-level stress also exists in such a component. It can be recalled from section 11.2 that a uniform macrostress causes a shift in the Bragg angle for a given hk1 plane and a non-uniform strain causes line broadening. Thus, a plastically deformed polycrystalline metal would show both shifting and broadening of the diffraction lines. This effect is quite commonly seen in hardened steels. Here, the non-uniform microstress is caused by the austenite to martensite transformation due to change in specific volumes of these phases and superimposed on this is a uniform residual macrostress due to quenching, prior plastic deformation, welding, etc.

It is the stress which causes the diffraction line shift that is generally determined by x-ray diffraction method. If the lines are sharp, it is a simple matter to measure this shift in the diffraction line. The measurement of the line shift becomes difficult if the lines are broadened (a breadth of the order of 5 to 10° of 2θ is quite common in hardened steels) for the reason discussed above. In such cases, the profile of the diffraction line is determined and the centre of it located for the purpose precise determination of the line position.

There are two methods for fixing the position of the broadened lines. The first one uses the method of extrapolation of the linear portions on either side of the peak of the broadened line as shown in Fig. 12.8 (a). This method is useful whenever the profile contains recognizable linear portions. The intersection of the extrapolated linear portions is taken as the centre of the line for the purposes of 2θ determination. The second method makes use of the fact that the profile near the peak of a broad line has the shape of a parabola with a vertical axis, as shown in Fig. 12.8 (b). The general equation of a parabola whose axis is parallel to the y axis is given by

$$y = ax^2 + bx + c. \tag{12.12}$$

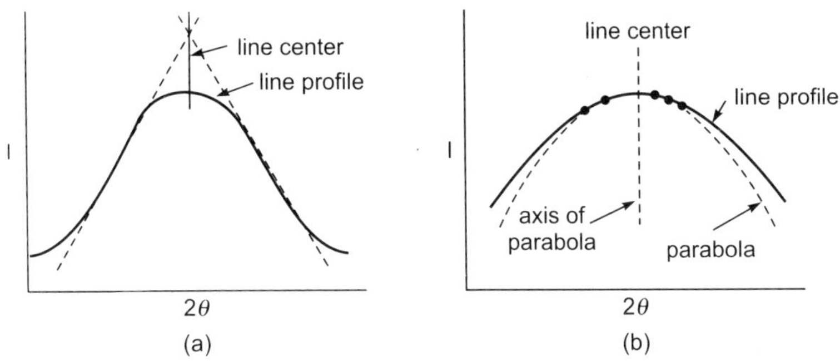

Fig. 12.8   Methods of locating the centers of broad diffraction lines.

The maximum on this curve occurs when the first differentiation of this equation becomes zero, i.e., when

$$dy/dx = 2ax + b = 0 \text{ or when } x = -b/2a \qquad (12.13)$$

If several pairs of $2\theta(x)$ and corresponding intensity values (y) are substituted into the equation 12.12, it is possible to determine the constants a and b by the method of least squares. The equation 12.13 then gives the exact value of x at which the peak occurs. By this method precision of the order of $0.02°$ is possible.

It is necessary to calibrate the stress factor K or C used in equations 12.10 and 12.11. These factors depend on the material constants E and v. It is well known that both E and v vary with crystallographic directions. It is therefore necessary to determine the stress factor K experimentally in the required crystallographic directions by subjecting the materials to known stresses. The values obtained by such measurements differ from the values calculated directly from the mechanically measured elastic constants by as much as 40%.

12.6   APPLICATIONS

The fact that the stress at a localized spot (as small as the diameter of the incident beam) of the specimen can be determined is the greatest advantage of the x-ray diffraction method. Therefore, the x-ray method is preferred for highly localized stress measurement over methods involving electrical (resistance strain gauges) or mechanical gauges because the latter methods measure average stresses over a considerable length (of the order of a few mm or cm) of the specimen. Further, the x-ray method always measures only the elastic portion of the strain even when the specimen has undergone or is undergoing plastic deformation unlike the mechanical or electrical methods which measure the total (elastic + plastic) strain. This is because of the fact that the lattice spacings are altered not by plastic deformation or plastic strain but only by elastic deformation or elastic strain. The x-ray method therefore reveals the residual elastic stress actually present at the time of the measurement. However, there is a destructive method called mechanical relaxation which can also be used to determine the residual stresses. In this method the specimen dimension in a stressed condition is measured and then the localities of the specimen component subjected to opposite stresses are separated by cutting or grinding and once again the dimensions of the cut or ground piece is determined. The difference in dimensions in the stress direction divided by the original dimension indicates the residual strain. There are many variations of this mechanical method of residual stress determination and all are destructive and are dependent on the removal of a part of the stressed metal. The x-ray method on the other hand is completely non-destructive. It can therefore be concluded that the x-ray method is most suitable for the non-destructive measurement of residual stress especially when the stress varies from point to point on the specimen surface. The rapid variation of the residual stress is quite frequently encountered in welded structures and the residual stress measurement in welded structures is one of the important applications of the x-ray method. The electrical or mechanical methods are decidedly superior for the measurement of applied stress since they are faster, more accurate and less expensive compared to the x-ray method.

13

CHAPTER

ORDER-DISORDER TRANSFORMATION

13.1 GENERAL

Most of the substitutional solid solutions are random or disordered at all temperatures. But, some of them prefer to have their solute atoms arranged in an orderly fashion at lower temperatures. At elevated temperatures, these types of solid solutions exist as random solutions wherein the solute (assume B) atoms are distributed in near statistical randomness in the crystal lattice of solvent (assume A) atoms. As these solid solutions are cooled below certain critical temperatures (T_c), both A and B atoms segregate more or less completely into certain designated atomic sites, so that the resulting arrangement can be described as a lattice of A atoms interpenetrating a lattice of B atoms. The segregation of atoms to particular lattice sites take place with little or no deformation of the lattice, creating an ordered solid solution or superlattice.

The formation of superlattices, normally described as long-range ordering, takes place at relatively low temperatures and usually at compositions expressed by a simple formula such as AB or A_3B or at compositions near these. At all temperatures above T_c the usual randomness persists; when the temperature is lowered through T_c order sets in and increases as the temperature drops approaching perfection only at low temperatures. Nevertheless, over certain ranges of composition and at temperatures above T_c, the structures may at times be neither perfectly random nor perfectly ordered. Even in such a disordered state a certain correlation may be present between nearest, second nearest and more distant neighbours, which may be described as short-range order. The presence of both long-range and short-range order can be measured experimentally by their effect on the scattering of x-rays, electrons and neutrons. The structures of ordered alloys usually produce diffraction patterns that have additional Bragg reflections called the superlattice lines which are not present in patterns of the disordered alloys. Bain in 1923 and Johansson and Linde in 1925 were the first to observe these superlattice lines with x-ray diffraction.

13.2 LONG-RANGE ORDER

In a fully ordered alloy there are great distances within a crystal through which there is a perfect arrangement of A atoms on one set of lattice points and B atoms on another set. The ordering is consistent in steps through long distances. The degree of this long-range order may be defined by a fraction parameter S, which varies from zero at complete disorder to unity at complete order. This parameter can be defined as

$$S = (r_A - F_A)/(1 - F_A) \qquad (13.1)$$

where r_A is the fraction of A sites occupied by A atoms and F_A is the fraction of A atoms in the solid solution. As an illustration, consider a solid solution A_3B in which 75 A atoms and 25 B atoms are randomly arranged. Here, the total number of sites in the lattice is equal to the number of atoms present which is 100. Out of these 100 sites there are 75 "right" A sites for ordering to occur. Suppose, all these 75 "right" sites have A atoms, ordering is complete. From the equation 13.1 it can be observed that S becomes unity under such a condition because r_A is 1 and F_A is 0.75. If the solid solution is in a completely disordered state, then r_A becomes 0.75 because in a random distribution the probability of any "right" A site occupied by A atom is 0.75 as there are 75 A atoms for every 100 atoms in the solid solution. So, in a random solid solution, S takes a value of zero because the numerator of the equation 13.1 is zero (0.75 – 0.75). Thus, the long-range order parameter expresses the degree of order present in a solid solution.

13.3 SUPERLATTICES BASED ON F.C.C. STRUCTURE

Here, the discussion is centered on superlattices which are formed from high temperature disordered face centered cubic solid solutions.

13.3.1 The A_3B Type Superlattice

These superlattices are based on the compositions of solid solutions at or near 75 atomic %A and 25 atomic %B and the disordered crystal structure of these solid solutions is face centered cubic.

Historically copper-gold alloys containing 25 atomic % gold were among the first investigated. The phase diagram of copper-gold system is illustrated in Fig.13.1. The alloy of composition of 25 atomic % Au (Cu_3Au) exists in a state of disordered or random solid solution above 390° C (approx), which is the critical temperature (T_c) for this alloy. The unit cell of the disordered Cu_3Au is shown in Fig.13.2 (a). Here, if the structure is completely in a random state of solid solution, the probability of a given lattice site occupied by Au atom is 1/4 since for every 100 atoms of the solid solution there are 25 Au atoms. And there is no preferred site for either Au or Cu atoms in this random solid solution. Considering the structure as a whole, each site can be regarded as being occupied by a statistically

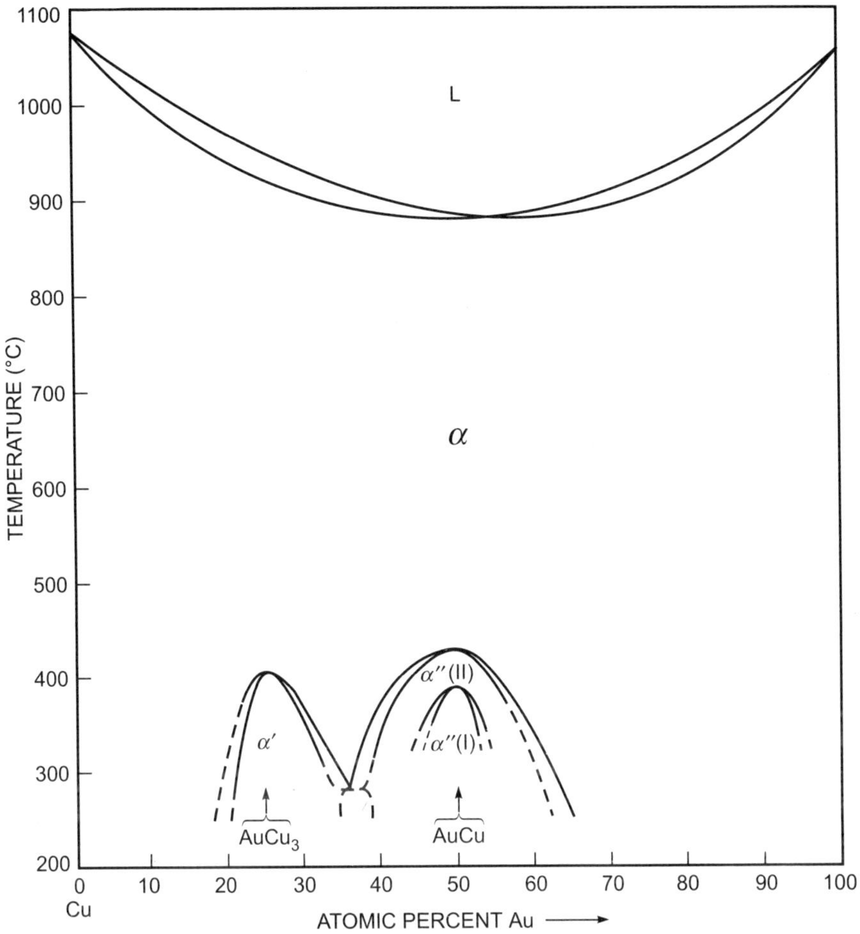

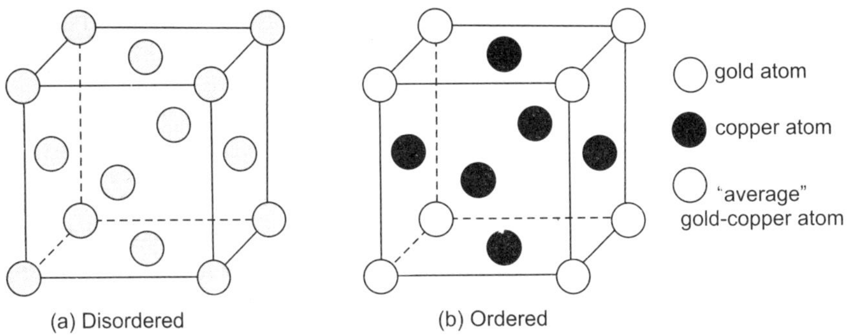

Fig. 13.1   Phase diagram of the gold-copper system. Two-phase fields not labeled for lack of room. (Compiled from *Metals Handbook*, American Society for Metals, 1948; J. B. Newkirk, *Trans*. A.I.M.E. 197, 823, 1953; F.N. Rhines, W. E. Bond, and R. A. Rummel, *Trans*. A.S.M. 47, 1955; R. A. Oriani. *Acta Metallurgica* 2, 608, 1954; and G. C. Kuczynski, unpublished results.)

Fig. 13.2   Unit cells of the disordered and ordered forms of $AuCu_3$.

average gold-copper atom as shown in Fig.13.2 (a). Below the critical temperature (T_c), the solid solution tends to get ordered with Au atoms preferring the cube corner sites and Cu atoms preferring face-center sites. The unit cell of completely ordered Cu_3Au structure is shown in Fig.13.2 (b).

It is the structure factor which causes the difference between the diffraction patterns of the ordered and the disordered alloy. In the completely disordered alloy structure, all the four unit cell sites can be thought of as being occupied by the average Au-Cu atom (i.e., 1/4Au + 3/4Cu) and the fractional indices positions of these atoms are (0, 0, 0), (1/2, 1/2, 0), (1/2, 0, 1/2) and (0, 1/2, 1/2). The atomic scattering factor of the average Au-Cu atom is given by

$$f_{av} = 1/4\ f_{Au} + 3/4\ f_{Cu}$$

Therefore, the structure factor is given by

$$F = \sum f \cdot e^{2\pi i(hu + kv + lw)}$$

By putting the fractional indices (uvw) of each of the four atoms

$$F = f_{av} \cdot [1 + e^{\pi i(h + k)} + e^{\pi i(k + l)} + e^{\pi i(l + h)}]$$

This equation reduces to (as seen in section 4.4.2 for f.c.c. case)

$$F = 4 \cdot f_{av} = 4 \cdot (1/4\ f_{Au} + 3/4\ f_{Cu})$$
$$= f_{Au} + 3f_{Cu}, \text{ for unmixed indices}$$

and
$$F = 0, \text{ for mixed indices.}$$

Thus, the disordered alloy produces a diffraction pattern similar to that of any f.c.c. metal.

On complete ordering at temperatures below T_c, the unit cell of the ordered Cu_3Au can be thought of as having totally four atoms per unit cell with one gold atom at fractional indices position (0, 0, 0) and three copper atoms at fractional indices positions (1/2, 1/2, 0), (1/2, 0, 1/2), and (0, 1/2, 1/2) as shown in Fig.13.2 (b). This also satisfies the atomic ratio of Au:Cu of 1:3 in the alloy. Therefore, the structure factor for the ordered Cu_3Au can be written as

$$F = f_{Au} + f_{Cu} \cdot [e^{\pi i(h + k)} + e^{\pi i(k + l)} + e^{\pi i(l + h)}]$$

Therefore

$$F = f_{Au} + 3f_{Cu}, \text{ for unmixed indices hkl}$$

and
$$F = f_{Au} - f_{Cu}, \text{ for mixed indices hkl.} \tag{13.2}$$

Thus, it can be seen that the diffraction pattern of the ordered Cu_3Au structure shows diffraction lines for all hkl values and thus the pattern resembles that of a simple cubic substance. In fact, the ordered structure can be imagined as a simple cubic lattice of gold interpenetrated by three simple cubic lattices of copper. So, there has been a change in Bravais lattice on ordering from a face centered cubic structure to a simple cubic structure. But, it should be noted that, for an ordered structure, the intensity of a line with unmixed indices is strong, since it is proportional to the square of $(f_{Au} + 3f_{Cu})$, whereas the intensity of a line with mixed indices is very weak since it is proportional to the square of $(f_{Au} - f_{Cu})$. Compared to this, in a simple cubic structure, the intensities of all the lines are proportional to the square of the atomic scattering factor of the elemental atom of the structure as seen in section 4.4.2. The lines with unmixed indices which are present in the diffraction patterns of both the ordered and the disordered structures are termed the fundamental lines and the additional lines of mixed indices which appear in the ordered structure are termed superlattice lines. Thus, it can be said that the presence of the superlattice lines in the diffraction pattern is an evidence of ordering. At low temperatures the long range order in Cu_3Au is almost perfect but as T_c is approached, some disorder sets in. The variation in long-range order parameter S with temperature is shown in Fig.13.3. Any departure from the long-range order in a superlattice causes the superlattice lines to become weaker and the structure factor for the superlattice lines can be rewritten as

$$F = S \cdot (f_{Au} - f_{Cu}) \tag{13.3}$$

The decrease in the order affects the superlattice lines very strongly, since the intensity of a superlattice line is proportional to the square of the changed structure factor. The basic difference between equation 13.2 and 13.3 is the parameter S which is present only in the latter. Since, S is a fraction, the square of it produces a value even lower than S and the intensities of the already weak superlattice lines further decrease.

One important observation is that since in this particular case of ordering of A_3B type structure, the ratio of the face centered site atoms to the corner site atom in the unit cell is 3:1, it follows that perfect order can be obtained in this case only

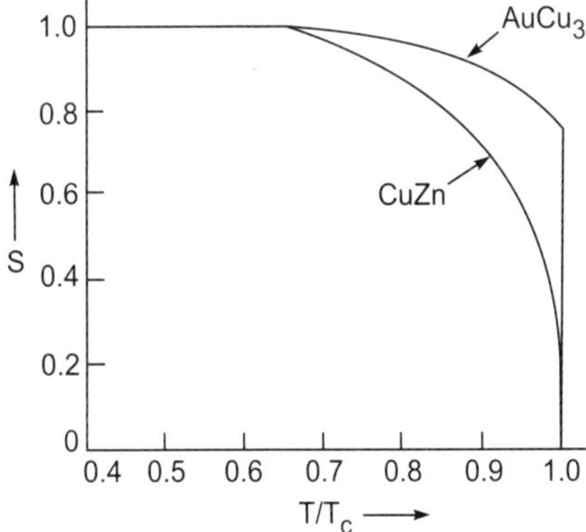

Fig. 13.3 Variation of the long-range order parameter *S* with temperature, for AuCu₃ and CuZu. (AuCu₃ data from D. T. Keating and B.E. Warren, *J. Appl. Phys.* 22, 286, 1951; CuZu data from D. Chipman and B.E. Warren, *J. Appl. Phy.* 21, 696, 1950.)

if the alloy has A and B atoms in the ratio of 3:1. But ordering can also take place in alloys containing slightly more or slightly less than 25 atomic %B. The Cu-Au phase diagram as shown in Fig.13.1 illustrates this fact where the ordered structure of Cu_3Au (α') is seen to be occupying a range of about 20 to 32% of Au at lower temperatures. Here it is found that whenever additional atoms of one kind are present over and above the required ratio for ordering, these additional atoms occupy the preferential positions of the other kind and vice versa.

Face centered cubic based Cu_3Au type of structures have been observed in a large number of alloy systems. Some examples of this type of structure are Au_3Cd, Co_3Al, Pt_3Sn, Al_3U, Zr_3Al, Co_3V, Ni_3Fe, Pd_3Fe, Ni_3Mn, Si_3U, Zn_3Ti, Tl_3U, etc.

13.3.2   The AB Type Superlattice

These types of superlattices occur at or near the composition of 50 atomic % of each kind of atoms A and B. Two types of AB superlattices exist in Cu-Au system. The phase diagram of Cu-Au (Fig.13.1) shows that below a critical temperature of approximately 410°C, a phase α'' (II) exists and in addition, at still lower temperatures below about 380°C another phase α'' (I) exists. Both of these phases are superlattices of the type AB. The α'' (I) is a tetragonal phase called CuAu I type superlattice and the orthorhombic phase α'' (II) is called the CuAu II type superlattice.

CuAu I type superlattice: This type of superlattice is shown in Fig.13.4. The unit cell of the disordered structure of this composition i.e., 50% A and 50% B shows the face centered cubic structure with all the sites occupied by a statistically average gold-copper atom. That means there is a 50% probability of finding a given kind of atom at a given site since 50% of the atoms are of one kind and remaining the other. But on ordering, the gold and copper atoms occupy alternate (002) planes as shown in Fig.13.4. It can

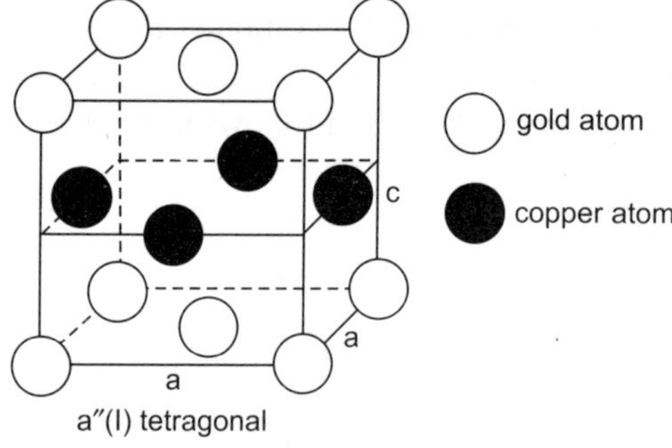

Fig. 13.4   CuAu I Super lattice

be noted that there are two Au and two Cu atoms per unit cell in this structure. Further, as the sizes of Au and Cu atoms are not the same, the shape of the unit cell is no longer cubic but changes over to tetragonal with a c/a ratio of 0.93. Here, the disorder to order transformation has not only changed the Bravais lattice but also has changed the crystal system from cubic to tetragonal. It is interesting to note that there is a change in the Bravais lattice but the crystal system remains the same in the case of A_3B superlattice discussed in the earlier section. In the A_3B

superlattice discussed earlier, it can be observed that, by ordering, the structure does not lose its original three-dimensional cubic symmetry, since the concentration of A and B atoms in any given plane does not differ from that in any other plane. But, in CuAu I type superlattice, the alternate (002) planes contain either Cu atoms or Au atoms completely. Thus, the lattice parameter c of the ordered CuAu I is slightly greater than the lattice parameter a because of the preference of a given kind of atom for a particular cube plane (002). This type of superlattice is present in a number of alloy systems: AgTi, CdPt, HgPt, AlTi, CoPt, HgTi, CuTi, HgZr, BiLi, FePd, InMg, BiNa, FePt, NiPt, CdPd, HgPd, PdZn, and PtZn.

CuPt type superlattice: In the Cu-Pt system, at a composition of 50 atomic % Pt, the f.c.c. disordered lattice transforms to the CuPt superlattice, shown in Fig. 13.5, at lower temperatures. This superlattice consists of alternating layers of Cu and Pt atoms on (111) planes because of which the lattice changes over from f.c.c. to rhombohedral. CuPt is the only known example of this type of superlattice. Since, each atom has six like and six unlike neighbours in both the ordered and the disordered state, there is no increase in the number of unlike neighbours on ordering. Copper atoms in excess of the 50-50 composition displace Pt atoms at random on the Pt (111) planes, but in alloys with Pt atoms in excess of the 50-50 ratio there is an interesting tendency for them to displace certain atoms in the Cu (111) planes in the manner illustrated by the sketch of the (111) plane in Fig.13.5 (b). Each Pt atom tends to be surrounded by Cu atoms, and a secondary type of ordering is thus produced. This arrangement is found to be complete in a Cu_3Pt_5 alloy.

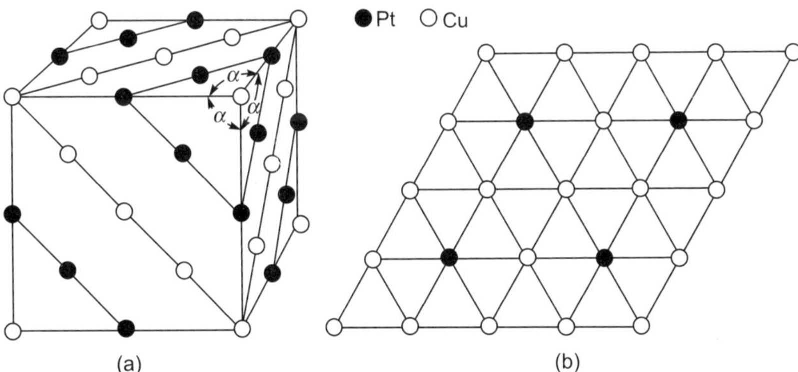

Fig. 13.5   The rhombohedral superlattice of CuPt, $L1_1$ type. (a) Distribution of atoms at the composition CuPt. Alternate (111) planes are occupied by Cu and Pt atoms. (b) Distribution of atoms on a single (111) plane, at the composition Cu_2Pt_3 (Ref: C.S. Barrett and T.B. Massalski, Structure of Metals).

13.3.3   Long-period Superlattices

There is a special group of superlattices which are known as long-period superlattices or anti-phase superlattices. A classical example for such a superlattice is available in the copper-gold system referred as CuAu II in section 13.3.2. This superlattice exists between approximately 385°C and 410°C and is orthorhombic. The unit cell of orthorhombic CuAu II, shown in Fig.13.6, consists of totally 10 unit cells of CuAu I stacked in the b direction and switching the contents of (001) planes from all gold to all copper atoms halfway along the new cell. This gives rise to an anti-phase or out-of-step boundary halfway along the cell (i.e., at intervals of five unit cells along the b axis) and at subsequent similar intervals along the b axis. The distance between two anti-phase boundaries is given by b · (M + δ) where M is the half period of the superlattice (in this case = 5) and δ is a slight expansion of the lattice in the direction of the superperiod (i.e., b axis). This structure may be called a one-dimensional long-period superlattice, with the superperiod M = 5 existing only in one direction. The orthorhombic unit cell of CuAu II has a b/a ratio of about 10.02 and a c/a ratio of about 0.92.

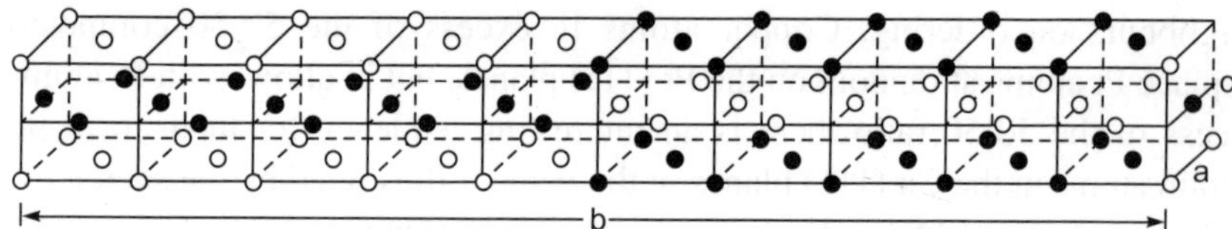

Fig. 13.6   CuAu II Superlattice.

In addition to the one-dimensional long-period superlattices of the CuAu II type, there exist two-dimensional long-period superlattices in A_3B type alloys with half periods M1 and M2 along two crystal axes, with M1 not necessarily equal to M2. Examples are found in Cu-Pd, Au-Zn, Au-Mn, etc. An illustration of one- and two-dimensional types of long-period superlattices for an alloy of A_3B composition is shown in Fig.13.7.

13.4   SUPERLATTICES BASED ON B.C.C. STRUCTURE

13.4.1   β Brass Type

The superlattice in β brass is typical of those based on high temperature disordered body centre cubic structures. The β brass (Cu-Zn alloy with 46-50 atomic % Zn) has a body centre cubic structure at high temperatures with the copper and zinc atoms statistically distributed in a random fashion. Below a critical temperature of about 465°C ordering sets in. The cell corners are then occupied only by copper atoms and the cell centres only by zinc atoms as illustrated in Fig.13.8. The ordered alloy, therefore has the CsCl structure and its Bravais lattice is simple cubic.

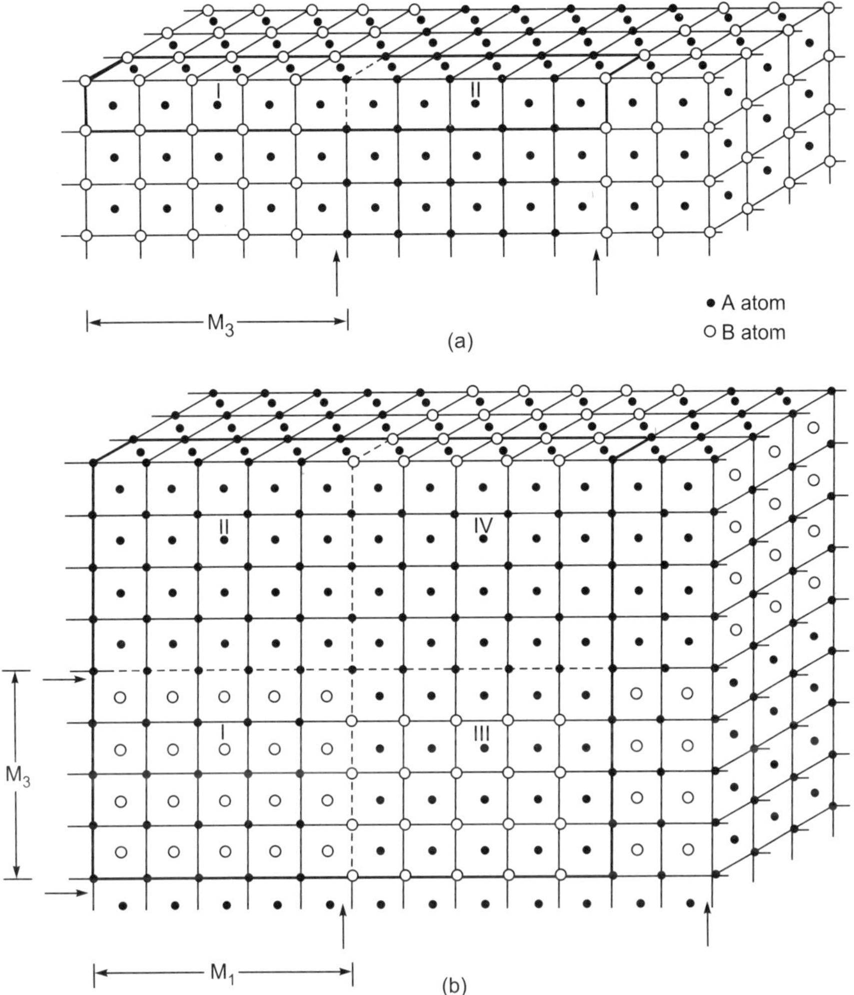

Fig. 13.7 One- and two-dimensional long-period superlattices in an A_3B alloy. The domain sizes are indicated by M_1 and M_3, and the domain boundaries are indicated by the arrows. (a) The unit cell of the one-dimensional superlattice is outlined by heavier lines and contains domains I and II of size of $M_3 = 5.0$. (b) The unit cell of the two-dimensional superlattice contains four domains, I, II, III, IV, of size $M_1 = 5.0$ by $M_3 = 4.0$. The A atom has different positions in the small cells of each of the four domains. (H. Sato and R.S. Toth.) (Ref: C.S. Barrett and T.B. Massalski, Structure of Metals).

By calculations similar to those made in the earlier section, the structure factor for β brass for the ideal composition of CuZn (i.e., 50 atomic % Cu and 50 atomic % Zn) can be shown to be

$$F = f_{Cu} + f_{Zn}, \text{ for } h + k + l \text{ even}$$
$$F = S \cdot (f_{Cu} - f_{Zn}), \text{ for } h + k + l \text{ odd} \qquad (13.4)$$

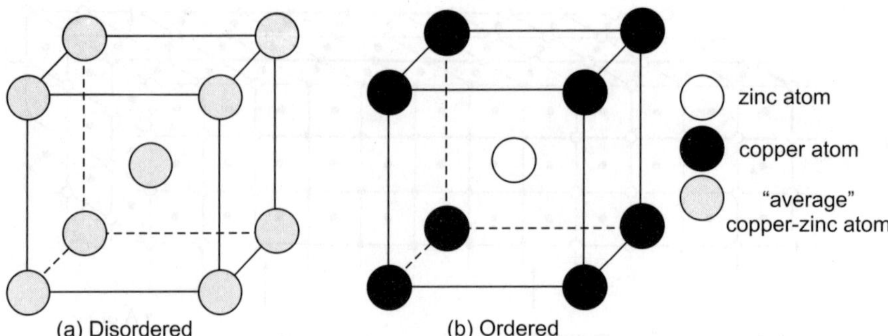

Fig. 13.8   Unit cells of the disordered and ordered forms of CuZn.

It can be noted that the diffraction pattern of the disordered b.c.c structure of CuZn consists of only reflections from planes with even values of (h + k + l) as for any b.c.c. structure. Therefore, these lines are fundamental lines. On ordering, additional lines appear in the diffraction pattern from planes with odd values of (h + k + l) which are superlattice lines. The intensity of the superlattice lines depend on the degree of order, represented by the long-range ordering parameter S.

Fig.13.3 indicates how the long-range ordering parameter S, in CuZn, varies with temperature. The long-range order parameter S decreases continuously to zero as the temperature approaches the critical temperature T_c whereas for Cu_3Au the parameter remains fairly close to unity right upto T_c and then abruptly drops to zero. Further, the disorder to order transformation in Cu_3Au is relatively so slow that the structure of this alloy at any temperature can be retained by quenching to room temperature. But, in CuZn the transformation is so fast that the disorder existing at any elevated temperature cannot be retained at room temperature and the room temperature structure can always be assumed to be ordered.

Typical examples of this type of superlattice are CuBe, CuPd, AgZn, FeCo, NiAl, AuCd, LiTl, NiZn, etc. It should be noted that some of these superlattice structures remain ordered right upto the melting point and do not undergo an order-disorder transformation.

13.4.2   FeAl and Fe_3Al Type

As aluminium atoms are added to the b.c.c. structure of α-iron, initially Al atoms replace iron atoms at random, but beyond 18% they concentrate at certain positions and leave others, giving rise to a number of ordered structures. Both FeAl and Fe_3Al superlattices are based upon the unit cell shown in Fig.13.9. From 18 to 25 atomic %, the Al atoms concentrate more and more in the positions labelled x in Fig.13.9, this process being completed at the composition Fe_3Al (i.e., at 25 atomic % Al). From 25 to 50 atomic % Al, increasing number of Al atoms go to y positions until both x and y positions are filled with Al atoms at the

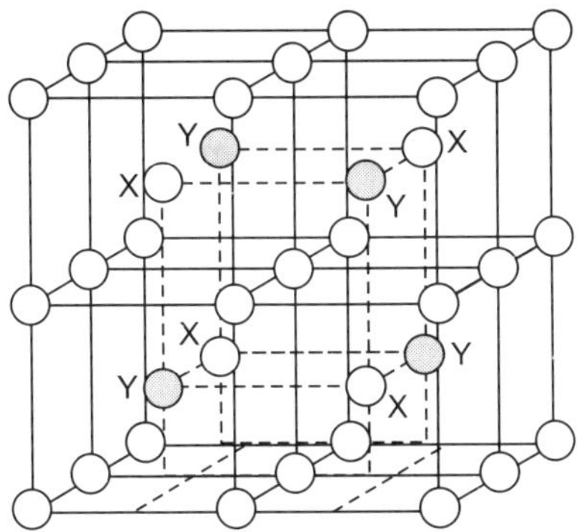

Fig. 13.9   The structure of Fe_3Al and FeAl. Al atoms are confined to the X positions in Fe_3Al and the X and Y positions in FeAl

composition FeAl (i.e., at 50 atomic % of Al). The structure is then similar to the ordered β brass structure. The fully ordered Fe_3Al can be obtained by cooling the 25 atomic % Al alloy slowly through 550°C to room temperature or annealing the alloy at around 300 to 350°C for a prolonged period.

A_2BC type of superlattice is closely related to Fe_3Al superlattice discussed above. The representative structure of A_2BC superlattice is found in Heusler alloy with the formula Cu_2MnAl. A related group of such structures was discovered by Heusler which was found to possess the remarkable property of being ferromagnetic when in ordered condition despite the fact that in most cases the constituent atoms are non-ferromagnetic. The ordered structure of Cu_2MnAl can be visualized from the Fig.13.9 in which y positions are occupied by Al atoms, x positions by Mn atoms and the rest of the positions by Cu atoms. Other ferromagnetic superlattices of this type are Cu_2MnSn, Cu_2MnIn and Cu_2MnGa.

Examples of Fe_3Al superlattices are found in the following alloys: Li_3Hg, Li_3Bi, Mg_3La, Cu_3Sb, Li_3Sb, Fe_3Si and Mg_3Pr. The A_2BC type of superlattices are found in Ni_2TiAl, Cu_2CoSn, Co_2MnSn, Cu_2FeSn, Cu_2MnSn Cu_2NiSn, Ni_2MgSb, Ni_2MgSn and Mg_2LiTl.

13.5   SUPERLATTICES BASED ON H.C.P. STRUCTURE

A large number of alloy phases have been discovered which possess the h.c.p. structure at high temperatures and which undergo ordering on cooling. Fig.13.10 shows one such superlattice unit cell with the general formula A_3B. The unit cell can be described in terms of four interpenetrating simple sublattices where each sublattice is h.c.p. with the 'a' spacing corresponding to twice that of the disordered structure and the 'c' spacing unchanged. B atoms occupy one or two of the four sublattices, resulting in the general formula A_3B or AB respectively. Some of the superlattices of this type are as follows: Al_3Th, Fe_3Sn, Cd_3Mg, Mg_3Cd, GeMn, Pt_3U, InNi, Co_3Mo, Co_3W, Ni_3Sn and SnTi.

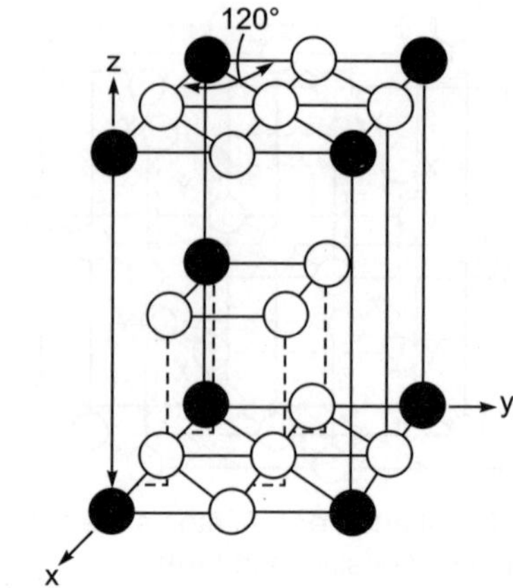

Fig. 13.10 Superlattice based on HCP structure

13.6 DETECTION OF SUPERLATTICE LINES

The intensities of the superlattice lines are generally much lower than those of fundamental lines, since the structure factor of superlattice lines is proportional to the difference in the atomic scattering factors of the constituent elements as seen in equations 13.2 and 13.4. Thus, in the case of fully ordered Cu_3Au, from equation 13.2

$$I_s/I_f = \frac{\text{Intensity of the superlative line}}{\text{Intensity of the fundamental line}}$$

$$\simeq |F_s|^2/|F_f|^2 = \frac{(f_{Au} - f_{Cu})^2}{(f_{Au} + 3f_{Cu})^2}$$

At $(\sin \theta)/\lambda$ value of zero, atomic scattering factor f can be equated to the atomic number Z. Since, the atomic numbers of gold and copper are 79 and 29 respectively, the above equation reduces to

$$I_s/I_f = \frac{(79-29)^2}{(79+3\times29)^2} \simeq 0.09.$$

Superlattice lines therefore have intensities about one-tenth of the intensities of the fundamental lines, but still they can be detected without any difficulty.

But in the fully ordered CuZn, there is some difficulty in detecting the superlattice lines. The atomic numbers of copper and zinc being 29 and 30 respectively, the intensity ratio

$$I_s/I_f = \frac{(f_{Au} - f_{Zn})^2}{(f_{Au} + f_{Zn})^2} \simeq \frac{(29-30)^2}{(29-30)^2}$$

$$\simeq 0.0003.$$

This ratio is so low that the superlattice lines of the ordered CuZn can be detected by x-ray diffraction only under special circumstances. This is true for any superlattice whose elements differ in atomic numbers by one or two units, since the superlattice line intensity is generally proportional to the square of the difference in atomic numbers of the elements. Under such conditions, it is possible to increase the intensity of superlattice lines by the proper choice of the incident wavelength. It was assumed in section 4.4.2 that the atomic scattering factor depends on the quantity $\sin\theta/\lambda$ and not directly on λ. But, it is seen that when the incident wavelength λ is nearly equal to the wavelength λ_K of the K-absorption edge of the scattering element, the atomic scattering factor of that element may be several units lower than it is when λ is very much shorter than λ_K. If f is the atomic scattering factor for $\lambda \ll \lambda_K$ (usually tabulated f value) and Δf is the change in f when the chosen λ is near λ_K, then the quantity $f' = f + \Delta f$ gives the value of the atomic scattering factor when λ is near λ_K. Fig. 13.11 shows the variation of Δf with λ/λ_K approximately. This type of curve may be used to estimate the correction Δf which is to be applied for a given wavelength and scattering element. When λ/λ_K is less than about 0.8, the correction factor Δf is practically nil and when λ/λ_K is more than about 1.4 the correction is almost constant and is independent of small variations in λ_K. But when λ is in the neighbourhood of λ_K, the correction factor Δf is quite substantial and varies sharply with the incident wavelength λ. This means that the correction factor can be quite different for two elements of nearly the same atomic number. This fact can be taken advantage of in increasing the intensities of the superlattice lines.

As an example, if ordered CuZn is examined with Mo Kα radiation, λ/λ_K is 0.52 for the copper atom and 0.55 for the zinc atom. The value of Δf is then about

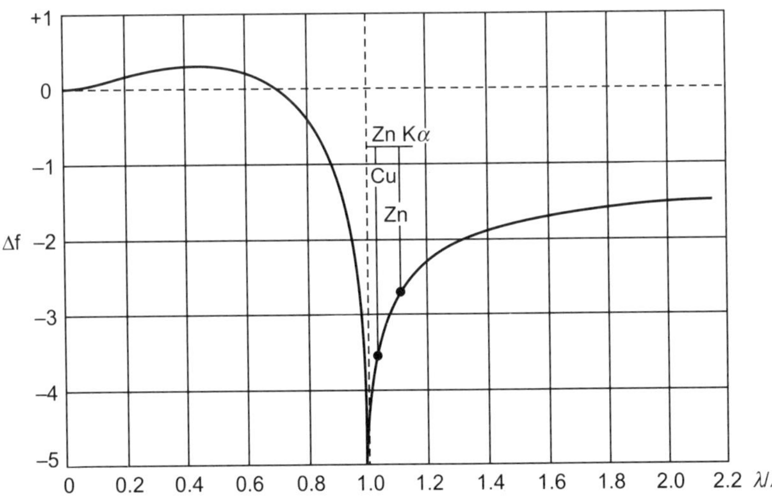

Fig. 13.11 Variation of Δf with λ/λ_K. (Data from R. W. James, *The Optical Principles of the Diffraction of X-Rays*, G. Bell and Sons, Ltd. London, 1948, p. 608.)

+ 0.3 for either atom and the intensity of the superlattice line would be proportional to $[(29 + 0.3) - (30 + 0.3)]^2 = 1$ at low values of 2θ. The superlattice line under these circumstances would be invisible. But, if Zn $K\alpha$ radiation is used, λ/λ_K becomes 1.04 and 1.11 for the copper and the zinc atoms respectively, and Fig.13.11 shows the corrections Δf are -3.6 and -2.7 respectively. Now, the superlattice line intensity is proportional to $[(29 - 3.6) - (30 - 2.7)]^2 = 3.6$, which is high enough for detection. It can be noted that to a good approximation, the change in atomic scattering factor Δf is independent of scattering angle and therefore a constant for all superlattice lines on the diffraction pattern. So, a corrected f′ curve can be constructed by adding algebraically the same value of Δf to all the ordinates of the usual f vs. $\sin\theta/\lambda$ curve as shown in Fig.13.12.

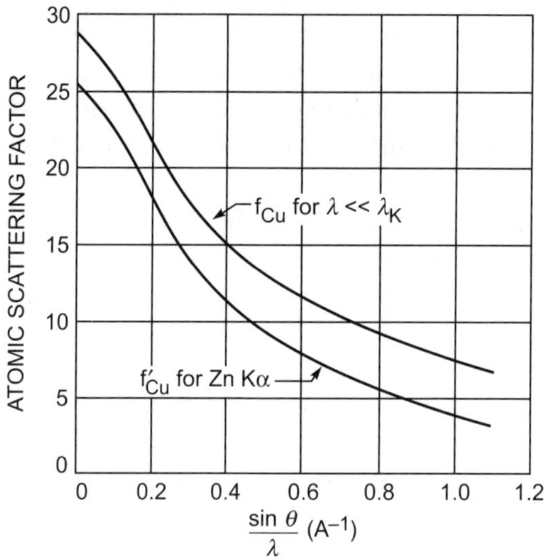

Fig. 13.12   Atomic scattering factors of copper for two different wavelengths.

Neutron diffraction is a better technique for the detection of the superlattice lines. The neutron scattering powers of any two elements, differing in atomic numbers by just one unit, can differ by large magnitudes, a situation very convenient for the detection of superlattice lines. It is better to go for neutron diffraction when the detection of superlattice lines by x-ray diffraction is impossible or difficult.

13.7   SHORT-RANGE ORDER AND CLUSTERING

Long-range order disappears and the atomic distribution becomes more or less random above the critical temperature T_c. The diffraction pattern also indicates the absence of superlattice lines above T_c. But, on careful scrutiny there seems to be a tendency for unlike atoms to be nearest neighbours. This kind of pairing of unlike atoms is known as short-range order. This state of affairs could be revealed by analyzing the diffuse scattering which forms the background of the diffraction pattern. Short-range order exists in all solid solutions which exhibit long-range order. The degree of short-range order decreases as the temperature is raised above T_c, that is to say, as the thermal agitation increases the atomic distribution becomes more random. Further, short-range order has also been found to exist in solid solutions which do not exhibit long-range ordering such as gold-silver and gold-nickel solid solutions.

Short-range order is defined in terms of the right number of "right pairs" of atoms. A right pair is a pair of unlike atoms i.e., an AB pair. As the temperature increases above T_c, the number of AB pairs (right pairs) decrease and the number of AA and BB pairs (wrong pairs) increases until a disordered state is reached in which half the pairs are right and half are wrong. The local or short-range order 'σ' may be defined as the probability of finding an unlike atom beside a given atom minus the probability of finding a like atom there. Considering a certain A atom, the probability that a nearest neighbour is a B atom is $(1 + \sigma)/2$, while the probability that it is an A atom is $(1 - \sigma)/2$. The ratio of these two when equilibrium is reached at a temperature T is the Boltzmann factor:

$$[(1 - \sigma)/2]/[(1 + \sigma)/2] = e^{-V/KT} \qquad (13.5)$$

where V is the change in energy of the crystal when one pair is changed from an AB to an AA pair. This energy V must be positive if a superlattice is to form; if it is negative, there will be a tendency for like atoms to cluster together and precipitate from solid solution.

Clustering is another kind of departure from disorder in solid solution, namely, a tendency of like atoms to be close neighbours. Here, the right pair is AA or BB pair and not AB pair. Generally, it can be said that there is no such thing as perfectly random solid solution since in any solid solution composed of unlike atoms (e.g., A and B), there is either an attraction or a repulsion between them which results in short-range order (increased AB pairs compared to AA or BB pairs) or clustering (increased AA or BB pairs compared to AB pairs) respectively. Clustering has been observed in aluminium-zinc and aluminium-silver solid solutions.

Von Laue showed that if two kinds of atoms A and B are distributed completely at random in a solid solution, then the intensity of the diffuse scattering produced is given by

$$I_D = k \cdot (f_A - f_B) \qquad (13.6)$$

where k is a constant for any one composition, and f_A and f_B are atomic scattering factors of A and B respectively. Both f_A and f_B decrease as $\sin \theta/\lambda$ increases, and so does their difference. Therefore, I_D is maximum at $2\theta = 0$ and decreases as 2θ increases. This diffuse scattering is weak in intensity and difficult to measure experimentally since other forms of diffuse scattering such as Compton modified scattering may also superimpose on this. The effects of short-range order and clustering on the diffraction pattern is shown in Fig.13.13. In this figure, the intensity of the diffuse scattering is plotted, not against 2θ, but against a function of $\sin\theta$ (here, $4\pi\sin\theta/\lambda$). The fundamental diffraction lines are not shown in this figure because their intensities are too high compared to that of diffuse scattering, but the positions of 111 and 200 lines are shown on the x-axis.

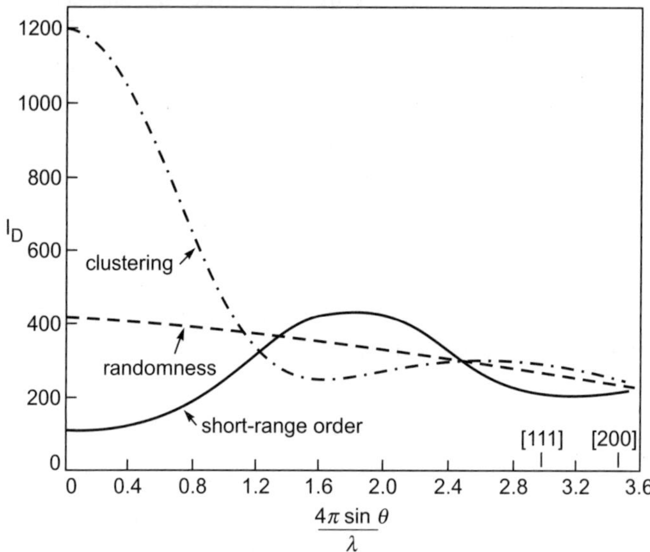

Fig. 13.13 Calculated intensity I_D of diffuse scattering in powder patterns of solid solutions (here, the face-centered cubic alloy Ni_4Au) which exhibit complete randomness, short-range order and clustering. The short-range order curve is calculated on the basis of one additional unlike neighbor over the random configuration, and the clustering curve on the basis of one less unlike neighbor. (B. E. Warren and B. L. Averbach, *Modern Research Techniques in Physical Metallurgy*, American Society for Metals, Cleveland, 1953, p. 95).

If the atomic distribution is perfectly random, the scattered intensity decreases gradually as 2θ or $\sin\theta$ increases from zero, according to equation 13.6. If short-range order exists, the scattering at small angles becomes less intense and low as well as broad maxima occur in the scattering curve. These maxima usually occur at the same angular positions as the sharp superlattice lines formed by long-range ordering. Clustering causes strong scattering at low angles. These effects are however very weak and are masked by other forms of diffuse scattering. It is therefore necessary to use strictly monochromatic radiation and to make allowances for the other forms of diffuse scattering so as to reveal the effects of short-range order or clustering.

14

CHAPTER

PHASE DIAGRAM DETERMINATION

14.1 INTRODUCTION

A phase is a physically and chemically homogeneous part of a system separated from other parts of the system by a boundary. A system consisting of two or more elements may exist in multiphase equilibrium under a given condition. It becomes necessary to determine the number and quantity of phases present in a system under a given temperature and composition conditions so as to study and ascertain the properties of the system. This requirement can be met by the phase diagram of the relevant system. A phase diagram or an equilibrium diagram is a pictorial representation of the phases present in a given system under different equilibrium temperature and composition conditions. A binary (two-component) phase diagram is a plot of temperature versus composition divided into areas where a single phase or two phases are stable. There are a number of methods available to determine phase diagrams of which the commonly used ones are metallography and thermal analysis. The x-ray diffraction method supplements these techniques by unambiguously establishing the solvus line positions in phase diagrams. The thermal analysis method is not capable of determining the solvus line, since the method is based on the heat effects involved in phase transformations and these heat effects in solid state reactions (as at the solvus line) are negligible.

14.2 GENERAL PRINCIPLES

Each phase in an alloy produces its own diffraction pattern irrespective of the presence or absence of other phases. The interpretation of powder patterns of alloys basically depends on this fact. A single-phase alloy produces a single pattern based on this phase whereas a two-phase alloy produces two superimposed patterns one corresponding to each phase. The lattice parameters of a solid solution change as the solute content increases. If the size of the solute atom is larger than that of solvent atom, the lattice parameters of the solid solution phase increase as the solute content increases. If the solute atom size is smaller, the lattice parameters decrease with the solute content. From Bragg's law it can be seen that the interplanar spacing d (which is related to the lattice parameters of a given crystal structure by a specific equation) is inversely related to the θ value for a given λ value. Thus, the diffraction patterns of the single-phase solid solution alloys with increasing solute content show no difference (since, there is no change in crystal structure) except for the shift in the positions of the diffraction lines which are dependent on the lattice parameters. If the solute content is increased beyond the solubility limit, a second phase precipitates and the diffraction pattern shows lines corresponding to the second phase in addition to those of original solid solution phase.

When an unknown phase diagram is being determined, it is better to make a preliminary survey of the system by preparing a series of alloys at definite composition intervals, like 5 or 10 atomic percent, from one pure component at one end to the other at the other end. The powder patterns of these alloys are then obtained by equilibriating each alloy at the temperature of interest. The following simple facts must be borne in mind while identifying the phase boundaries:

   (i) A constant temperature line drawn across a binary phase diagram must pass through single-phase and two-phase regions alternately.

  (ii) In a single-phase region, a change in composition produces a change in lattice parameter and consequently a shift in the positions of the diffraction lines of that phase.

(iii) In a two-phase region, a change in composition produces a change in the relative amounts of the two phases but no change in their compositions. The compositions of the two phases are fixed at the ends of the tie-line (and consequently the lattice parameters of the two phases are fixed) at the temperature of interest. Thus, the diffraction pattern of a two-phase alloy shows the diffraction lines of both the phases superimposed over each other. Further, as the composition of the alloy changes within a two-phase field, the intensities of the diffraction lines of one of the phases increase relative to the other but the positions of the lines do not change (since, the lattice parameters are fixed).

14.3   DETERMINATION OF SOLVUS CURVES—BINARY SYSTEMS

There are two methods to determine the solvus curves by x-ray diffraction. One is based on the lever rule and is called the disappearing phase method and the other is based on the variation of the lattice parameter and is known as the parametric method.

14.3.1 Disappearing Phase Method

The Fig.14.1 shows a two-phase field in a binary system AB in which a tie-line ab is drawn at a temperature of interest T_1. Let the point "c" indicate the composition

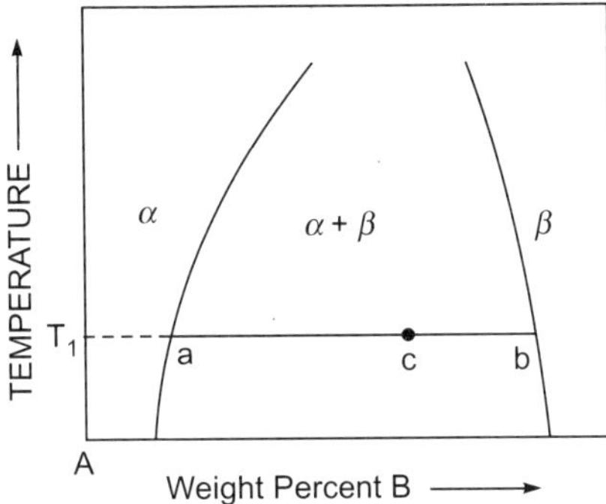

Fig. 14.1   Lever-law construction for finding the relative amounts of two phases in a two-phase field.

of the alloy under investigation. It is intended to identify the point "a" i.e., a point on the solvus line separating the two-phase field from the single-phase (α) field. By the lever rule, the ratio of the relative amounts of α and β phases are given by:

$$\text{amount of } \alpha / \text{amount of } \beta = \text{length cb/length ac} \tag{14.1}$$

As the intensities of the diffraction lines are related to the amounts of the phases present, the above equation can be conveniently used to establish the solvus line. If a diffraction pattern is obtained for the alloy of composition c, it would show the diffraction pattern of α superimposed over the diffraction pattern of β. As the alloy composition point "c" is moved towards the point "a", the diffraction pattern would show decreasing intensities for β lines (decreasing lengths of ac) and increasing intensities for α lines (increasing lengths of cb). The diffraction pattern for the alloy of composition exactly matching with "a" would consist of diffraction lines of "a" alone as the alloy is composed entirely of α phase. Thus, the diffraction lines of β phase would have zero intensity (disappearing phase) for the alloy of composition "a" and would keep on increasing as the alloy composition changes towards "c". This variation, though not linear, could be used to locate the solvus curve point "a" at a given temperature T_1. A set of alloys in the suspected two-phase region can be brought to equilibrium at temperature T_1 and quenched. Then the room temperature diffraction patterns of these alloys are studied. The ratio of the intensity I_β of a prominent line of the β phase to the intensity I_α of a prominent line of α phase can be plotted as a function of composition (weight % B) as done in Fig. 14.2. The composition at which the ratio I_β/I_α extrapolates to zero (i.e., where β phase disappears) can be taken as the point "a". The ratio I_β/I_α is taken, rather than I_β alone, in order to avoid the effect of the variation of the incident beam intensity on the individual diffraction patterns. A similar procedure can be adopted for getting a few more points on the solvus curve by quenching the alloys from a few other temperatures and studying the diffraction patterns thereafter.

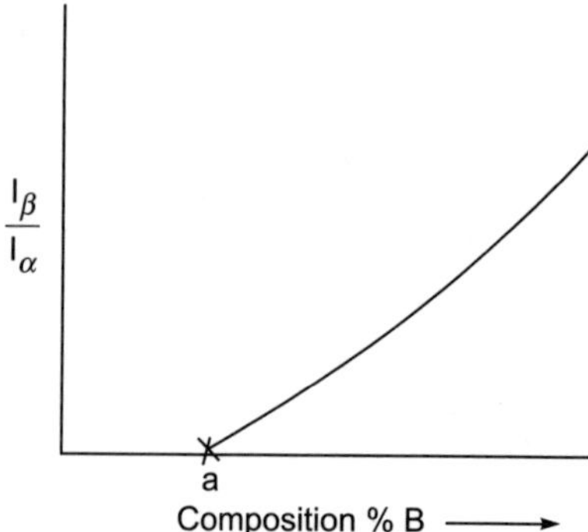

Fig. 14.2 I_β/I_α versus composition

It is necessary to select the alloys properly so as to obtain a high accuracy in the determination of solvus curves. Alloys with compositions as close as possible to the solvus curve are to be taken for the study in order to get high accuracy in extrapolation. The accuracy of the disappearing phase method is governed by the sensitivity of the x-ray diffraction method in detecting very small amounts of the second phase in a mixture. This sensitivity in turn depends on the ability to recognize very low intensity lines on the diffraction patterns. The intensities of diffraction lines depend on basically the atomic scattering factor which in turn depends on the atomic number of scattering element. In the binary system AB shown in Fig.14.1 if the elements A and B have very close atomic numbers, the phases α (A-rich solid solution) and β (B-rich solid solution) have nearly the same scattering powers and the diffraction patterns of α and β will have nearly the same intensities. Under such favourable conditions it is possible to detect the presence of even 1% of the second phase. But, if the atomic number of B is considerably less than that of A then a relatively large percentage of β (B-rich phase) may go unnoticed due to the extremely low intensities of β lines in the presence of high intensity α (A-rich phase) lines. In extreme cases where the atomic numbers differ by 70 to 80 units, even 50% of the second phase may go undetected and the method becomes worthless. The accuracy of the method also depends on the width of the two-phase region. If the two-phase region is only a few percent wide, the relative amounts of α and β will vary rapidly with slight changes in the total composition of the alloy and this rapid change in amounts of the phases will enable the phase boundary to be fixed accurately.

14.3.2   Parametric Method

Unlike the disappearing phase method which concentrates on the two-phase region, this method focuses on the single-phase region. The parametric method is based on the fact that the lattice parameter of a phase changes with composition up to the saturation limit and then remains constant in the two-phase region as discussed earlier.

Fig.14.3 shows a solvus curve, the location of which is to be determined accurately. A number of alloys 1 to 8 of varying compositions are brought to equilibrium at a temperature T_{max}, where the α field is thought to have the maximum width, and quenched to room temperature. The lattice parameter of α phase is determined by x-ray diffraction technique for each alloy and plotted against the alloy composition. The resulting graph will now have two branches, one horizontal (st) and the other inclined to the composition axis (uv) as shown in

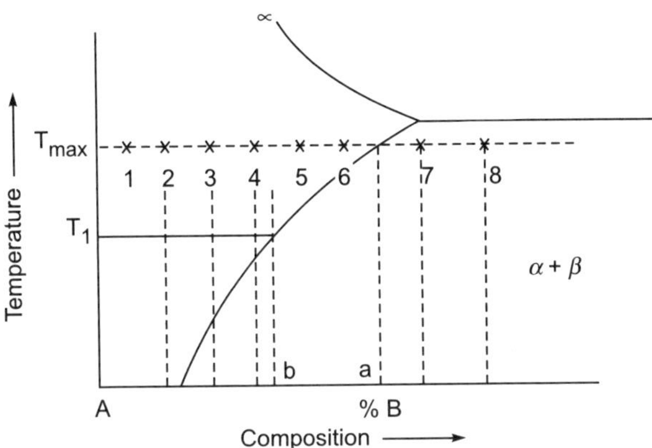

Fig. 14.3 Parametric method

Fig.14.4. The horizontal branch st indicates that in alloys 7 and 8 the phase α is saturated and that the lattice parameter does not vary with the composition. Thus, it can be concluded that the alloys 7 and 8 are in the two-phase region $\alpha + \beta$ where the lattice parameter of α is fixed. In alloys 1 to 6 the lattice parameter of α varies with the composition. The solvus point "a" at temperature T_{max} can be fixed by the point of intersection of the horizontal and inclined branches as indicated in Fig.14.4. The other points on the solvus curve at other temperatures can be located in a similar manner. The procedure for getting the solvus curve points at temperatures lower than T_{max} can be simpler. The lattice parameter versus composition curve obtained at T_{max} can be used as a reference curve for the purpose. If the solvus point at a temperature T_1 is required to be determined, only one two-phase alloy is needed for the study. The alloy 8 (or 7) can be brought to equilibrium at the temperature T_1 and quenched and then its lattice parameter measured by x-ray diffraction. Let the measured parameter be a_1. From the established reference curve of Fig. 14.4, it can be found that the alloy phase α with parameter a_1 has a composition b percent B. This point b is nothing but the point on the solvus curve at the temperature T_1. Similarly, other solvus curve points at other temperatures can be determined by using the same two-phase alloy 8 (or 7). Thus, the lattice parameter versus composition curve once established at the highest temperature T_{max} (at which solvus curve exists) it can be used as a master curve for the determination of the entire solvus curve.

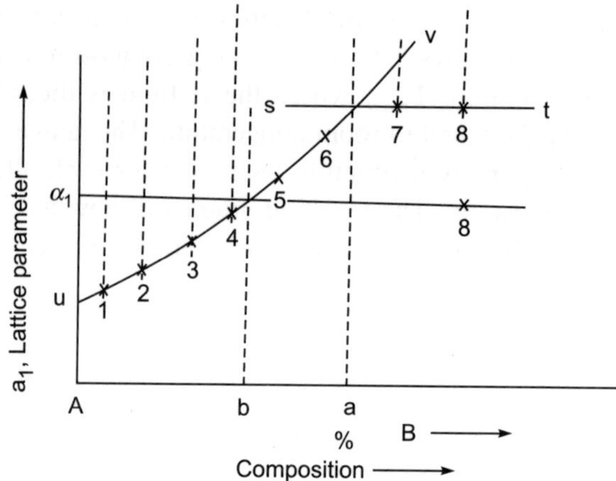

Fig. 14.4 Lattice parameter versus composition

Though the sensitivity of this method basically depends on the accuracy in measuring the lattice parameter, for a given accuracy the sensitivity of the method depends on the slope of the parameter versus composition curve. If a large variation in the composition produces a small variation in the lattice parameter (i.e., almost a flat curve), then the composition, as determined from the parameter, and consequently the location of the solvus will be subject to considerable error. But, if the curve is almost vertical (having a very high slope) then a high accuracy can be obtained in fixing the solvus curve.

In general, the parametric method is more accurate than the disappearing phase method in the determination of the solvus curve. The determination of the presence or absence of the second phase (β in the case considered) is very difficult in most of the alloys when the second phase is present in extremely small quantities. And it is this fact which makes the disappearing phase method less accurate compared to the parametric one. Further, this same fact is also responsible in identifying a solvus point at a higher composition than the actual. This kind of error never occurs in the parametric method as it is based on the lattice parameter of the most abundant phase (α in this case) in the alloys and the presence (in small or large quantities) or the absence of the second phase never affects the lattice parameter of the abundant phase. The parametric method is applicable even when the lattice parameter is difficult to determine due to the complicated crystal structure of the abundant phase. In such cases either the plane spacing 'd' or the 2θ value of some high angle diffraction line can be plotted against composition resulting in a curve almost resembling the parameter composition curve and can be used in a similar fashion to obtain a point on the solvus curve.

14.4   DETERMINATION OF SOLVUS CURVES: TERNARY SYSTEMS

In place of solvus lines in binary diagrams one encounters solvus surfaces in ternary diagrams which are in fact three-dimensional diagrams because of an additional composition variable. The disappearing phase method as well as the parametric method can both be used in the study of ternary diagrams for the determination of phase boundaries of solid phases.

Generally, a ternary system can be represented by plotting the composition in an equilateral triangle with the three corners representing the pure components A, B and C which make up the system ABC as shown in Fig.14.5 and plotting the temperature perpendicular to the composition plane ABC. Any isothermal (constant temperature) section of the ternary diagram thus shows an equilateral triangle on which phases that are in equilibrium at that temperature can be depicted in two dimensions. It is generally convenient to study ternary diagrams by equilibriating the alloys at certain constant temperatures to determine the phase equilibria and by plotting the isothermal sections. The Fig.14.5 shows a possible low temperature isothermal section of a ternary system ABC consisting of three known binary eutectic systems AB, BC and CA. Suppose, the precise positions of the phase boundaries of this isothermal section is to be determined. A number of ternary alloys with different compositions can be equilibriated at the temperature of the desired isothermal section and their diffraction patterns can be studied. This will reveal the number of phases present at the chosen temperature at equilibrium in each of these alloys. This survey would roughly fix the phase boundaries on the isothermal section. Let such a preliminary work reveal three single-phase regions α, β and γ, three two-phase regions $\alpha + \beta$, $\beta + \gamma$ and $\gamma + \alpha$ and one three-phase region $\alpha + \beta + \gamma$ as shown in Fig.14.5.

Three-phase regions in isothermal sections of ternary systems have the shape of a triangle (called the tie triangle) with the three corners indicating the compositions of three phases in equilibrium at the temperature of the isothermal section. So, any alloy of composition selected within this three-phase triangle

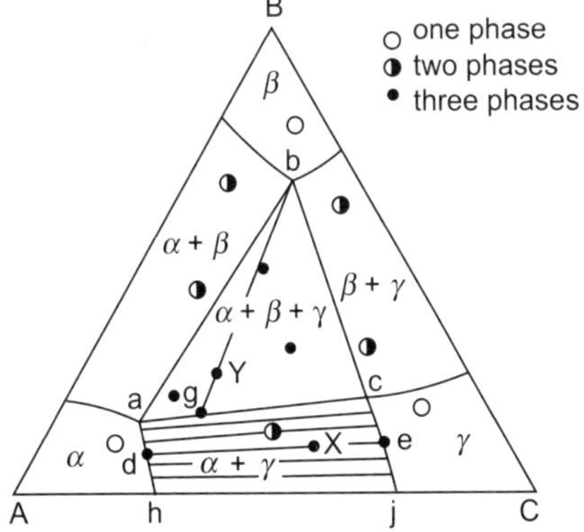

Fig. 14.5   Isothermal section of hypothetical ternary diagram

would show three phases in equilibrium with each other (at the temperature of the isothermal section) with compositions fixed by the three corners of the tie-triangle. But, the amounts of the individual phases in the alloy will depend on the location of the alloy composition. In Fig.14.5, the three-phase region $\alpha + \beta + \gamma$ is located inside the tie-triangle abc. The corner points a, b and c of the triangle indicate the compositions of α, β and γ respectively which are in equilibrium with each other at the temperature of the section. Alloys of compositions taken along any line bg drawn from the corner b to the side ac would show decreasing amounts of β as the composition changes from b towards g. Exactly at the point g (the boundary between the three-phase region $\alpha + \beta + \gamma$ and the two-phase region $\alpha + \gamma$) the ternary alloy shows only two phases α and γ in equilibrium with each other. In other words, as the composition of the alloy changes from b to g along the line bg the amount of β phase keeps on decreasing and at g the β phase disappears. This fact can be utilized to locate the boundary line ac by the disappearing phase method. It is not necessary to select the alloys strictly on the basis of the location of the line bg, since the tie-triangle position itself is not known precisely in the beginning. On the basis of the preliminary survey conducted, it is possible to select a straight line in the isothermal section such that this line has the possibility of encountering the boundary line ac which is good enough for the purpose of the disappearing phase method. All the boundaries ac, ab and bc can be located in a similar way.

The two-phase region in an isothermal section of a ternary system has two solvus curves (intersections of solvus surfaces of the two involved phases by the isothermal section plane) as its boundaries. The two-phase region $\alpha + \gamma$ is shown in Fig. 14.5 has solvus boundaries ah and cj. The other boundaries of this two-phase region are ac, separating the three-phase region from this region, and hj on the binary system AC. A number of tie-lines are drawn in the $\alpha + \gamma$ region to connect the compositions of α and γ, lying on the respective solvus, which are in equilibrium. The alloys lying along any tie-line de would consist of varying amounts of α and γ, but the compositions of α and γ are fixed on the points d and e respectively. The solvus boundary ah can be located by the disappearing phase method by the disappearance of the phase γ in the diffraction patterns of the alloys lying on the line de. The selected alloys for the experiments need not be exactly lying on the line de as the location of the boundary is based on the disappearance of the phase γ which would anyway occur as soon as the boundary of the two-phase region is reached.

The parametric method can determine the phase boundaries more accurately than the disappearing phase method and is more useful. Suppose, it is necessary to determine the point b, the boundary point between α and $\alpha + \gamma$ regions shown in Fig.14.6 a. Then a series of alloys along the line abc (where, bc is a part of a tie-line in the $\alpha + \gamma$ region) may be examined for the parameter variation of α phase with composition. The curve would then resemble Fig.14.6 (b), similar to the case

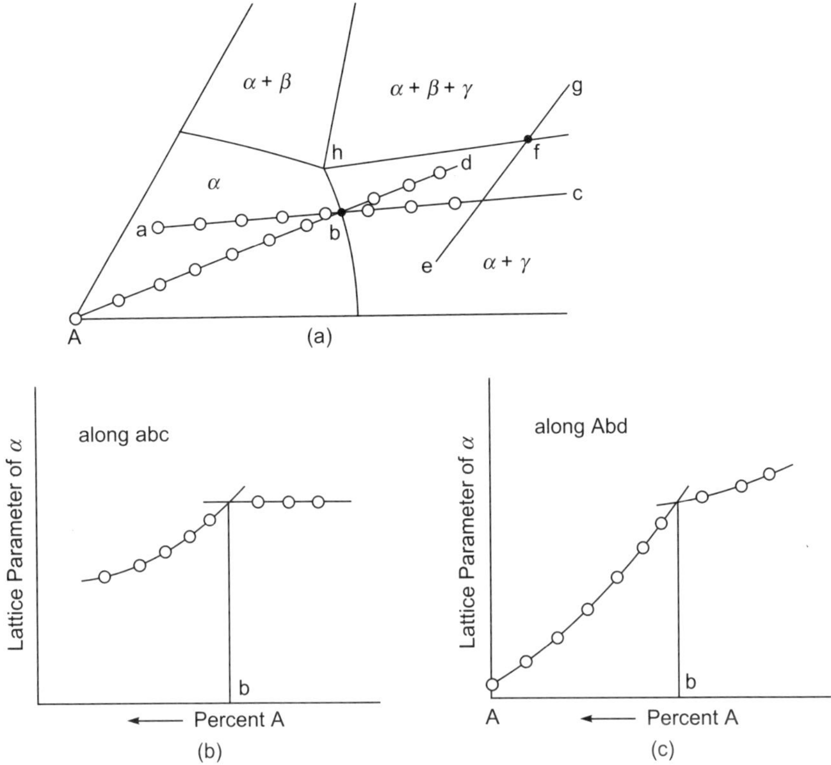

Fig. 14.6 Parametric method of locating phase boundaries in ternary diagrams.

in binary phase diagrams. It may not be always possible to know the positions of the tie-lines prior to the investigation in order to precisely select the alloys for the study. But, even in cases where the alloys are selected along an arbitrary line Abd, the point b can be located by the method as shown in Fig.14.6 (c). Though, the parameter of the phase α in the two-phase region $\alpha + \gamma$ does not remain constant, the variation of the parameter with composition is different from that in the single-phase region α which enables one to locate the boundary point b.

The three-phase boundary point f can be located by the parametric method by selecting alloys along a line efg (Fig.14.6 a) for study. The alloys lying in the three-phase region, would show no change in the lattice parameter of the phase α (since, the composition of α in the three-phase region is fixed), but the alloys in the two-phase region $\alpha + \gamma$ do show a change in the parameter of a because in a two-phase region the composition of a phase changes along any line other than the tie-line. So, the boundary point between the two-phase and the three-phase regions is located by plotting the lattice parameter of α versus composition (which will resemble Fig.14.6 b with point b corresponding to point f here) along the line efg.

15

CHAPTER

CHEMICAL ANALYSIS

15.1 GENERAL

The diffraction method of x-ray analysis is based on the fact that a given crystalline phase produces its characteristic diffraction pattern irrespective of (i) its quantity in the substance concerned (quantity may affect the intensities of the lines and not the pattern) and (ii) the presence or absence of other crystalline phases. It is to be noted here that the diffraction pattern is characteristic of the crystal structure and as such indicates the presence of the concerned crystalline phase, be it an element or a compound. An ordinary chemical analysis is capable of identifying only the individual elements or radicals present in a substance and does not indicate the form in which the elements exists i.e., in pure elemental form or compound form. For example, if a substance contains a compound $A_{2x}B_y$, a chemical analysis would identify elements A and B and does not tell us whether the elements exist in free form or as a compound. But, an x-ray diffraction analysis would directly indicate the presence of the compound $A_{2x}B_y$ by its characteristic diffraction pattern. As an additional example, a chemical analysis of a steel would identify elements Fe, C, Mn, Si, P, and S, but would not tell us anything regarding the phases present in the sample like ferrite, austenite, martensite and cementite. But, it is possible to identify the phases like ferrite, austenite, cementite and martensite by x-ray diffraction. Thus, diffraction analysis is useful to identify the phases as it exists in the unknown sample.

The other advantages of diffraction analysis are economy in sample size, high speed of detection and its non-destructive nature. So, it is widely used for the analysis of materials such as alloys, ores, clays, refractories, corrosion products, industrial dusts, etc. The diffraction analysis can be done qualitatively (identification of the phases present) or quantitatively (determination of weight percentages of phases present) depending on the requirement.

15.2 QUALITATIVE ANALYSIS

15.2.1 Principles

A given substance has a characteristic powder pattern in a fashion comparable to the characteristic fingerprint of a human being. Thus, if a collection of known powder diffraction patterns are available, it is possible to identify the unknown pattern by comparing it with the available patterns. The collection of the diffraction patterns has to be exhaustive, so that most of the substances are covered. Further, there must be a proper system of classification of the existing patterns in the collection, so that locating a given pattern and matching it with the unknown one becomes a quick and efficient process. Hanawalt devised such a system in 1936 in which a given pattern is filed on the basis of 'd' values of its reflecting planes and the relative intensity values of diffraction lines. He devised a method wherein a pattern could be located by noting the three strongest lines on the pattern.

15.2.2 The Hanawalt Method

Hanawalt and his associates built up a collection of powder patterns of about 1000 different substances and classified them, which was later extended by the American Society for Testing Materials on an international scale with the assistance of a number of other scientific societies. In 1941, the Joint Committee on Chemical Analysis by X-ray Diffraction Methods was founded. Under the guidance of this Joint Committee, a collection of diffraction data on about 1300 substances was published in the form of a set of $3 \times 5"$ data cards. A second set of 500 patterns followed in 1945. These compilations are more commonly known as Powder Diffraction Files (PDFs). In 1969, the organization was constituted as a Pennsylvania non-profit Corporation under the title of The Joint Committee on Powder Diffraction Standards (JCPDS). The name of the organization was changed again in 1978 to "JCPDS-International Centre for Diffraction Data" (JCPDS-ICDD) and is located at Swarthmore, Pennsylvania, USA. In order to meet the individual needs of diffractionists worldwide, PDF is available on a wide variety of media, e.g., books, microfiche, magnetic tapes, floppies and CD-ROM (Compact Disc-Read Only Memory). At present, PDF has 44 sets including about 59,800 active patterns.

Hanawalt characterized each substance by the 'd' values of its three strongest lines, namely, d_1, d_2 and d_3 for the strongest, second-strongest and the third-strongest line, respectively. The values of d_1, d_2 and d_3, together with relative intensities are usually sufficient to characterize the pattern of an unknown and enable the corresponding pattern in the file to be located. In each section of the PDF, the pattern data are arranged in groups characterized by a certain range of d_1 spacings. Within each group of such substances, the patterns are arranged in the order of decreasing d_2 values. When several substances in the same group have identical d_2 values, the order of decreasing d_3 values is followed. The groups themselves are arranged in the decreasing order of their d_1 ranges.

Though, it is possible to directly search the required card from the file, it is easier and less time-consuming to go in for the index books or files which accompany the regular card files. These are of two types:

1. Alphabetical index: This is a listing of chemical formulae of substances arranged in an alphabetical fashion along with the 'd' values of the three strongest lines in the pattern and their relative intensities. It also gives the serial number of the card in the file which can be referred later for detailed data. All entries are cross-indexed; e.g., both 'aluminium silicate' and 'silicate, aluminium' are listed. This index can be used only when the chemical elements (one or more) present in the sample are known.

2. Numerical index: This lists the spacings and intensities of eight strongest lines, the chemical formula, name and the card serial number of the substance involved. Each substance is listed three times, once with the three strongest lines listed in the usual order d_1, d_2 and d_3, second time in the order d_2, d_1 and d_3 and finally in the order d_3, d_1 and d_2. All entries are divided into groups according to the d_1 values listed. The arrangement within each group is in the decreasing order of d_2 values listed. The purpose of these additional listings is to enable the investigator to match an unknown with an entry in the index even when complicating factors have altered the relative intensities of the three strongest lines of the unknown substance. These complicating factors are usually due to the presence of more than one phase in the sample which leads to additional lines and even superimposed lines.

The Hanawalt method of qualitative analysis begins with the preparation of the powder diffraction pattern of the unknown sample. This may be done with a Debye-Scherrer camera or, more commonly nowadays, with a diffractometer using any convenient characteristic radiation. The characteristic radiation may be chosen so as to minimize fluorescent radiation and to obtain adequate lines on the pattern. Most of the earlier data of PDF were obtained using a Debye-Scherrer camera and molybdenum $K\alpha$ radiation. Though 'd' values do not change with the wavelength of the radiation chosen or with the type of equipment (D-S camera or diffractometer) chosen, the relative intensities of the lines depend on both of these factors. But, it is possible to convert the intensity values from one experimental condition to the other by using proper conversion procedures. Presently, most of the laboratories are equipped with diffractometers and the recent PDF data are based on this equipment. During specimen preparation, care should be taken to see that preferred orientation is minimized. The pattern obtained is analyzed to determine the 'd' values of every line and the corresponding relative intensities. A rough eye estimation of the intensities can be done which is sufficient for the purpose of qualitative analysis if D-S camera is used and numerical values can be assigned for the line intensities on a scale of 1 to 10. If a diffractometer is used to obtain the pattern, automatic recording of the intensity will provide sufficient accuracy. After the tabulation of the 'd' and relative intensity values, identify the three strongest lines in the pattern and name them as d_1, d_2 and d_3 based on decreasing intensity values from d_1 to d_3. The procedure for identification of the unknown sample is as follows:

(i) Locate the proper d_1 group in the numerical index.

(ii) Read down the second column of decreasing 'd' values to find the closest match to d_2.

(iii) Once the closest match has been found for d_1, d_2 and d_3, compare the other strong 5 'd' values tabulated and the relative intensities of all the eight lines in the index.

(iv) When the 'd' values and the intensity values agree within reasonable limits, locate the proper data card in the file by the number available in the index. Compare all the lines of the unknown sample with those of the card file. If full agreement for all the lines of the unknown pattern is obtained, the identification is complete.

In recent days, the above procedure of identifying the unknown pattern by comparing and matching the observed one with those available in the PDF is being done with a personal computer.

Fig.15.1 shows a typical data card of PDF. At the top left, 'd' values for the three strongest lines along with the largest 'd' value for this structure are given together with their relative intensity (I/I_1) values, in percentage of the strongest line in the pattern, below each. The left half of the card also contains the serial

8 – 262 MINOR CORRECTION

d	2.70	2.37	2.49	2.70	(A$_L$N) 4$\underline{H}$						
I/I_1	100	70	60	100	ALUMINUM NITRIDE						

Rad. CuKa λ 1.542 Filter Ni Dia	d Å	I/I_1	hkl	d Å	I/I_1	hkl
Cut off 1.I₁ DIFFRACTOMETER	2.70	100	100			
Ref. PARETZKIN, POLYTECHNIC INST. OF BROOKLYN. Brooklyn, NEW YORK	2.49	60	002			
	2.372	70	101			
Sys. HEXAGONAL S.G. P6₃MC (186)	1.829	20	102			
a₀ 3.114 **b₀** **c₀** 4.986 A C 1.602	1.557	30	110			
α β γ Z 2 Dx 3. 255	1.414	20	103			
Ref. V. STACKELBERG, AND SPIESZ, Z. PHYS. CHEM. 175A 127 (1933)	1.348	6	200			
	1.320	18	112			
& a n ω β $\varepsilon \gamma$ Sign	1.3θ1	8	201			
2V D mp Color	1.186	4	202			
Ref.	1.047	6	202			
	1.019	4	210			
PREPARED BY HEATING POWDERED AL FOR 2 HOURS AT 1000°C. IN A NITROGEN ATMOSPHERE, CRUSHING THE PRODUCT AND RE-PEATING THE HEAT TREATMENT.	0.9984	6	211			
	.9345	6	105			
	.8684	6	213			

Fig. 15.1   Data card of Aluminium Nitride (PDF of JCPDS)

number of the card, the radiation used, D-S camera diameter, method of measuring intensity, crystallographic, chemical and optical data and a reference to the original work. The right hand portion of the card lists the 'd' values of the diffraction lines and the corresponding intensities in terms of the percentages of the maximum.

15.2.3   Difficulties Encountered

In practice, various difficulties are encountered while using the Hanawalt method of qualitative analysis which are either due to the errors in the pattern or to the errors in the card file.

It should be remembered that in a D-S camera, the absorption decreases line intensities more at small angles than at large angles and this difference in absorption does not occur in a diffractometer. This factor should be kept in mind when comparing the unknown pattern with a data card in PDF. It is possible that the PDF data card might have been prepared by looking into a D-S pattern and the pattern of the unknown by a diffractometer or vice versa. In such cases, care should be taken to compensate for the absorption differences of the individual equipment.

More serious difficulties arise due to the errors in the card file since they are not anticipated by the analyst and may lead to mistaken identifications. Whenever, any doubt arises in identification, a standard pattern of a pure substance, identified, has to be prepared and the 'd' values and the intensities of the diffraction lines have to be checked to verify the validity of the identification.

The Hanawalt method fails when the substance taken for identification is not listed in PDF or when the substance, though listed, is not present in sufficient quantity (where, the unknown is a mixture of a number of components) to give visible diffraction lines.

Identification of surface deposits (scales) or coatings can be done conveniently by the reflection method of x-ray diffraction using a diffractometer. A reflection method like Schultz is particularly suitable because of the shallow penetration of x-rays into most metals and alloys as discussed in earlier chapters. In these cases, most of the recorded pattern is produced by an extremely thin surface layer favourable to the detection of small amounts of surface deposits.

15.3   QUANTITATIVE ANALYSIS

15.3.1   General Principles

If the sample contains only a single phase solid solution, the lattice parameter itself can be a factor which can be studied for the analysis. The lattice parameter of a solid solution of component B in component A depends only on the percentage of B in the solid solution as long as the solution remains unsaturated. So, if a curve of composition versus lattice parameter of the required solid solution is known (like the one in Fig.14.4), it is a simple matter to establish the solute content of the solid solution by the determination of its lattice parameter. The accuracy of this method depends on the slope of composition versus parameter curve. With a steep slope, the accuracy can be as good as $\pm 1\%$ or better. It is to be noted that this method is applicable only to binary alloy solid solutions and not to ternary or quaternary systems since in these latter systems it is possible to have the same lattice parameters with different combinations of components.

Multiphase quantitative analysis is based on the fact that the intensity of diffraction of a particular phase in a multiphase mixture depends on the concentration of that phase in the mixture. The relation between the intensity and concentration of any given phase is not generally linear because the diffracted intensity varies with the absorption coefficient of the mixture which itself varies with the concentration.

15.3.2 Direct Comparison Method

This method is of great interest to metallurgists since it can be applied to massive, polycrystalline samples. The method has been commonly used to determine the amount of retained austenite in hardened steels.

Generally, the diffractometer is used for quantitative analysis since this instrument has the advantage of quicker intensity measurements and the independence of absorption factor with the Bragg angle θ compared to the D-S camera. The expression for the intensity diffracted by a single phase powder specimen in a diffractometer is:

$$I = \left(\frac{I_o \cdot e^4}{m^2 \cdot e^4}\right) \cdot \left(\frac{\lambda^3 \cdot A}{32\pi r}\right) \cdot \left(\frac{1}{V^2}\right)\left[|F|^2 \cdot p \cdot \left(\frac{1 + \cos^2 2\theta}{\sin^2\theta \cdot \cos\theta}\right)\right]\left(\frac{e^{-2M}}{2\mu}\right) \quad (15.1)$$

where I is the integrated intensity per unit length of diffraction line, I_o is the intensity of the incident beam, e, m are charge and mass of the electron respectively, c is the velocity of light, λ, A are wavelength and cross-sectional area respectively of the incident beam, r is the radius of the diffractometer circle, V is the volume of the unit cell, θ is the Bragg angle, F, p, e^{-2M} and $1/2\mu$ are structure, multiplicity, temperature and absorption factors respectively. This equation is applicable to powder specimen in the form of a flat plate, of effectively infinite thickness, making equal angles with the incident and diffracted beams. The equation 15.1 can be rewritten as

$$I = \frac{K \cdot R}{2\mu} \quad (15.2)$$

where $\quad K = \left(\frac{I_o \cdot e^4}{m^2 \cdot C^4}\right)\left(\frac{\lambda^3 \cdot A}{32\pi r}\right)$

= a constant independent of the diffracting specimen taken and

$$R = \left(\frac{1}{V^2}\right)\left[|F|^2 \cdot p \cdot \left(\frac{1 + \cos^2 2\theta}{\sin^2 \theta \cdot \cos\theta}\right)\right] \cdot e^{-2M}$$

= a constant which depends on the diffracting specimen and the Bragg angle θ.

Let us assume that a specimen contains two phases A and B and it is intended to determine the amounts of each phase in the specimen. For a particular diffraction line of phase A, the equation 15.2 can be modified as

$$I_A = \frac{K \cdot R_A \cdot C_A}{2\mu_m} \tag{15.3}$$

where R_A is a constant dependent on the phase A, C_A is the volume fraction of the phase A in the specimen and μ_m is the absorption coefficient of the specimen which is a mixture of phases A and B. Similarly, for a particular diffraction line of phase B,

$$I_B = \frac{K \cdot R_B \cdot C_B}{2\mu_m} \tag{15.4}$$

Dividing equation 15.3 by 15.4 results in

$$\frac{I_A}{I_B} = \frac{R_A \cdot C_A}{R_B \cdot C_B} \tag{15.5}$$

Thus, the ratio C_A/C_B can be determined by the measurement of I_A/I_B from the diffraction pattern of the specimen along with the calculated values of R_A and R_B from respective diffraction lines considered. Since, it is known that the specimen contains only two phases A and B,

$$C_A + C_B = 1.$$

With the ratio C_A/C_B being known along with the above equation, it is now possible to determine the volume fractions C_A and C_B. The above method can be used to make an absolute measurement of retained austenite in a quenched steel by measuring the integrated intensity of austenite and martensite lines. It is possible to obtain several independent values of austenite content by comparing several independent pairs of austenite-martensite lines in the diffraction pattern of the quenched steel. Any serious disagreement between these independent values thus determined indicates errors in observations and/or calculations.

This method can also be used when the sample for analysis contains more than two phases. For example, consider a sample having three phases A, B and C. After conducting a diffractometer experiment, it is possible to determine the volume fraction ratios, C_A/C_B, C_B/C_C and C_C/C_A from the corresponding intensity ratios of the respective lines with the help of equations of the type 15.5. Further, it is also known that $C_A + C_B + C_C = 1$, since only three phases are present in the sample. So, the individual volume fractions C_A, C_B and C_C can be determined.

Proper care must be taken while choosing the diffraction lines for intensity measurements. Overlapping lines of different phases must not be chosen and the lines to be measured should be distinct, quite strong in intensity and well separated from the lines of other phases. The value of 'R' should be calculated as accurately as possible. That is, the unit cell volume 'V' must be determined taking into account the actual lattice parameter calculated from the experimental diffraction pattern obtained for quantitative analysis; the structure factor must be calculated on the basis of atomic scattering factor duly corrected for anomalous scattering effect discussed under section 13.6; the Lorentz-Polarization factor

$$\left(\frac{1 + \cos^2 2\theta}{\sin^2 2\theta \cdot \cos\theta} \right),$$ as written in this fashion, is applicable only to unpolarized

incident radiation and if polarized radiation like the crystal monochromated one is

used, this factor has to be modified to $\left(\dfrac{1 + \cos^2 2\alpha \cdot \cos^2 2\theta}{\sin^2 2\theta \cdot \cos\theta} \right)$ where 2α is the

diffraction angle in the monochromator; the value of the temperature factor 'e^{-2M}' can be selected from plots similar to Fig.15.2 for the appropriate value of sin θ/λ.

The sensitivity of the x-ray diffraction method in determining small amounts of phases is limited mainly by the intensity of the background radiation present. The lower the background radiation, the easier it is to detect and measure weak lines of the desired phase. The sensitivity can therefore be improved by making use of the crystal monochromated radiation which permits the detection of as low as 0.1 volume percent (V%) of the desired phase. With ordinary filtered radiation the minimum detectable amount is 5 to 10 V%.

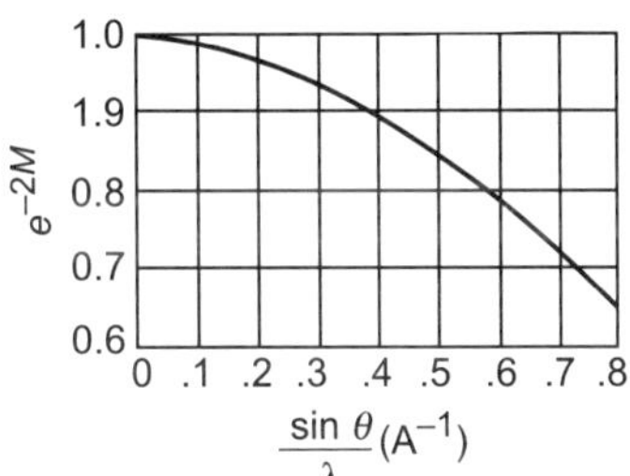

Fig. 15.2 Temperature factor e^{-2M} of iron at 20°C as a function of (sin θ)/λ (Ref.: B.D. Cullity, Elements of X-ray Diffraction)

15.3.3 Internal Standard Method

A diffraction line from the phase being determined is compared with a line from a standard substance mixed with the sample in known proportions in this method. Since, the method inherently needs mixing of an internal standard substance with the sample, it is necessary that the sample is in the form of powder.

Let a sample contain a mixture of phases A, B, C, D, ... etc. It is intended to determine the amount of the phase A in such samples in which the relative amounts of other phases present (B, C, D, etc.) may vary. A known amount of a standard substance 'S' is mixed with a known amount of the sample to form a composite sample. Let C_A and C_A' be the volume fractions of phase A in the original and the composite samples respectively. Further, let C_S be the volume fraction of the standard sample 'S' in the composite sample. A diffraction pattern of the composite sample is prepared using a diffractometer. The relative intensity of a particular diffraction line from phase A is given by the equation 15.1 which can be simplified as

$$I_A = \frac{K_1\,C_A'}{\mu_m} \qquad (15.6)$$

where K_1 is connected to K and R_A (which are already defined earlier in equation 15.3) by the relation

$$K_1 = K \cdot R_A/2.$$

and similarly, the intensity of a particular line from the standard 'S' can be written as

$$I_S = \frac{K_2\,C_S}{\mu_m} \qquad (15.7)$$

where $\qquad K_2 = K \cdot R_S/2.$

Dividing equation 15.6 by 15.7,

$$\frac{I_A}{I_S} = \frac{K_1\,C_A'}{K_2\,C_S} \qquad (15.8)$$

volume fractions may be converted into weight fractions (W_A', W_B', W_C', W_D'....and W_S) by making use of density values (ρ_A, ρ_B, ρ_C, ρ_D ......and ρ_S), i.e.,

$$C_A' = \frac{W_A'/\rho_A}{W_A'/\rho_A + W_B'/\rho_B + W_C'/\rho_C + W_D'/\rho_D + ... W_S/\rho_S}$$

A similar expression may also be written for CS, i.e.,

$$C_S = \frac{W_S'/\rho_S}{W_A'/\rho_A + W_B'/\rho_B + W_C'/\rho_C + W_D'/\rho_D + ... W_S/\rho_S}$$

Dividing C_A' by C_S,

$$\frac{C_A'}{C_S} = \frac{W_A' \cdot \rho_S}{W_S \cdot \rho_A}$$

Substituting this value of C_A'/C_S into equation 15.8 gives

$$\frac{I_A}{I_S} = \frac{K_1 \cdot W_A' \cdot \rho_S}{K_2 \cdot W_S \cdot \rho_A}$$

But, in this equation, it can be observed that K_1, K_2, ρ_S, ρ_A are all constants and W_S is also a constant by maintaining a known weight of a standard substance being mixed with a known weight of the sample. Thus, all these constants can be represented by a single constant, say, K_3, and the above relation can be rewritten as

$$\frac{I_A}{I_S} = K_3 \cdot W_A{}'$$

(15.9)

The weight fractions of A in the original and the composite samples are related by

$$W_A{}' = W_A \cdot (1 - W_S)$$

(15.10)

By combining equations 15.9 and 15.10

$$\frac{I_A}{I_S} = K_4 \cdot W_A$$

(15.11)

where K_4 is another constant. So, the intensity ratio of a line from phase A and a line from standard S is a linear function of the weight fraction of A, W_A, in the original sample. In practice the following procedure is adopted in making use of the above equation 15.11. Several composite samples can be prepared by mixing known constant weights of samples having different concentrations of A individually with known constant weights of an internal standard substance S. The diffraction patterns of all these composite samples are analyzed to get a calibration curve (linear as per the equation 15.11) for the intensity ratio I_A/I_S versus weight fraction W_A. Once this calibration curve is determined, the weight fraction of A in the unknown sample is obtained by measuring the ratio I_A/I_S in the diffraction pattern of the composite sample prepared by mixing the same proportion of the standard sample with the unknown as was used in calibration. This method is used commonly for the measurement of quartz content of industrial dusts which is essential for monitoring industrial health programmes.

15.3.4 Difficulties in Quantitative Analysis

There are certain complicating factors which can give considerable errors in quantitative phase analysis. They are discussed below:

1. Texture: The intensity equation 15.1 is derived for the case of a sample with randomly oriented crystals or grains. If preferred orientation exists in a sample (which usually does, if it is a solid one), the referred intensity equation is not valid. So, the analyst may preferably use powder samples and ascertain that preferred orientation does not exist. Whenever, a sheet sample is to be quantitatively analyzed, the analyst must bear in mind the effect of texture on diffraction line intensities discussed in chapter 11.

2. Microabsorption: If a sample containing two phases α and β is used for quantitative analysis, the incident beam and the diffracted beam both must pass through the mixture of phases α and β and a single value of linear absorption coefficient of the mixture, μ_m, could generally be used in the intensity equation 15.1. But, it should be noted that if an α grain or crystal in the sample is diffracting then a small part of the path length of the incident as well as the diffracting beams would lie completely in the single diffracting grain of α. If μ_α is largely different

from μ_β or if the grain size of α is largely different from that of β then using an average value of mm in the intensity equation 15.1 may lead to erroneous result compared to the actual intensity diffracted. To minimize this difficulty, it is necessary to use very finely ground powder sample, so that the path length in any given phase is very small to lessen the microabsorption differences from phase to phase.

3. Extinction: All real crystals have a specific imperfection known as mosaic structure as shown in Fig.15.3 and the degree of this imperfection may vary from crystal to crystal. The intensity equation 15.1 is derived for real crystals having fine mosaic structure (individual mosaic blocks of the order of 10^{-4} to 10^{-5} cm in thickness) with a high

Fig. 15.3   The mosaic structure of a real crystal (Ref: B.D. Cullity, Elements of x-ray Diffraction).

degree of disorientation (mosaic blocks essentially non-parallel). Such crystals have a high diffracting power. A crystal made up of large mosaic blocks all or most of which are parallel to each other is nearly perfect and has much less reflecting power. This decrease in intensity of the diffracted beam as the crystal becomes more nearly perfect is called 'extinction'. Extinction is absent for an ideally imperfect crystal and the equation 15.1 indicates the intensity under such a situation. So, a certain amount extinction occurs in crystals during diffraction depending on the extent of perfection. Here again grinding the sample into fine powder will help in minimizing the extinction by reducing the mosaic block size, disorienting the blocks and straining them non-uniformly.

Microabsorption and extinction can seriously decrease the accuracy of the direct comparison method because this method is an absolute method wherein the values of the constant 'R' are to be calculated accurately for the individual phases. But, fortunately in the case of hardened steel, both these effects are negligible and a fairly accurate determination of retained austenite is possible. The two phases present in hardened steel are martensite and austenite and both of these phases have practically identical linear absorption coefficients and particle sizes which means that microabsorption effect is absent. Further, extinction is also absent in this case since martensite transformation during quenching sets up severe non-uniform strains in both the phases so as to make these phases highly imperfect. Except in such fortunate circumstances, the direct comparison method must be used with caution.

But, the presence of microabsorption and extinction does not cause errors in the internal standard method provided these effects are constant from sample to sample including the calibration samples. Microabsorption and extinction affect only the values of constants K_1 and K_2 and therefore the constant K_4 in equation 15.11 is modified which alters the slope of the calibration curve, but this does not affect the use of it.

REFERENCE BOOKS

1. L.V. Azaroff and M.J. Buerger, *The Powder Method in X-ray Crystallography*, McGraw-Hill, New York, 1958.
2. L.V. Azaroff, *Elements of X-ray Crystallography*, McGraw-Hill, New York, 1968.
3. C.S. Barrett and T.B. Massalski, *Structure of Metals*, third edition, S.Chand & Company, New Delhi, 1968.
4. M. J. Buerger, *X-ray Crystallography*, Wiley, New York, 1942.
5. B.D. Cullity, *Elements of X-ray Diffraction*, second edition, Addison Wesley Publishing Company Inc., California, 1978.
6. H.P. Klug and L.E. Alexander, *X-ray Diffraction Procedures*, second edition, Wiley, New York, 1974.
7. C. Suryanarayana and N.C. Grant, *X-ray Diffraction—A Practical Approach*, Springer, 1998.
8. A. Taylor, *X-ray Metallography*, Wiley, New York, 1961.
9. B.E. Warren, *X-ray Diffraction*, Addison Wesley, Reading Mass. , 1969.

INDEX